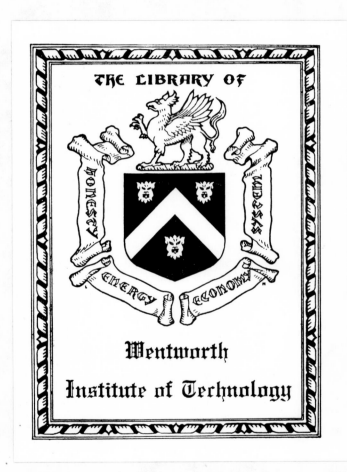

FABRICATION OF COMPOSITE MATERIALS

Source Book

A collection of outstanding articles from the technical literature

Other Source Books in This Series:

FABRICATION OF COMPOSITE MATERIALS

Source Book

A collection of outstanding articles from the technical literature

Compiled by
Consulting Editor

MEL M. SCHWARTZ

Chief of Metals Engineering
Sikorsky Aircraft
United Technologies

American Society for Metals
Metals Park, Ohio 44073

Library of Congress Catalog Card No.: 84-73363
ISBN: 0-87170-198-7
SAN 204-7586

PRINTED IN THE UNITED STATES OF AMERICA

Contributors to This Source Book

R. T. BEALL
Lockheed-Georgia Co.

E. M. BREINAN
United Aircraft Research Laboratories

W. D. BRENTNALL
TRW, Inc.

C. C. CHAMIS
NASA

J. B. CHEUNG
Flow Industries, Inc.

J. L. CHRISTIAN
Convair Division
General Dynamics Corp.

D. W. CLIFFORD
McDonnell Aircraft Co.

J. CUSUMANO
Grumman Aerospace Corp.

JANET DEVINE
Sonobond Corp.

ROD DOERR
Hughes Helicopters, Inc.

JAY H. DORAN
General Dynamics Corp.

DONALD R. DREGER
Machine Design

J. D. FOREST
Convair Division
General Dynamics Corp.

R. A. GARRETT
McDonnell Douglas Astronautics Co.

P. GRASS
FHG

ED GREENE
Hughes Helicopters, Inc.

FISKE HANLEY
General Dynamics Corp.

A. HEINTZE
Du Pont Co.

JOHN HUBER
Grumman Aerospace Corp.

G. H. HURLBURT
Flow Industries, Inc.

R. W. JECH
NASA

J. R. KENNEDY
Grumman Aerospace Corp.

W. KÖNIG
FHG

S. KOSTURAK
Sikorsky Aircraft
United Technologies

K. G. KREIDER
United Aircraft Research Laboratories

R. F. LARK
NASA

B. LYON
Hughes Helicopters, Inc.

C. R. MAIKISH
Convair Aerospace Division
General Dynamics Corp.

K. T. MARSHALL
Grumman Aerospace Corp.

G. D. MENKE
TRW, Inc.

G. E. METZGER
Air Force Materials Laboratory
Wright-Patterson, AFB

CARL MICILLO
Grumman Aerospace Corp.

M. F. MILLER
Convair Aerospace Division
General Dynamics Corp.

L. H. MINER
E. I. du Pont de Nemours & Co.

J. T. NIEMANN
Argonne National Laboratory

F. OKCU
MBB-UT

J. H. PAYER
Battelle Columbus Laboratories

DONALD W. PETRASEK
NASA

A. R. ROBERTSON
Convair Aerospace Division
General Dynamics Corp.

Cl. SCHMITZ-JUSTEN
FHG

E. H. SCHULTE
McDonnell Aircraft Co.

MEL M. SCHWARTZ
Sikorsky Aircraft Division
United Technologies Corp.

ROBERT A. SIGNORELLI
NASA

P. G. SULLIVAN
NETCO

T. L. SULLIVAN
NASA

S. TAHA
Hughes Helicopters, Inc.

I. J. TOTH
TRW, Inc.

ROBERT H. WEHRENBERG II
Materials Engineering

M. D. WEISINGER
Convair Division
General Dynamics Corp.

NOTE: Affiliations given were applicable at date of contribution.

PREFACE

Basic composite principles have been used for centuries, ever since man recognized that certain combinations of materials behaved advantageously under specific conditions. For example, early wheels were built using an iron band to contain the wooden rim, and primitive houses were made of clay bricks reinforced with straw. Some of the ancients were quite sophisticated in their use of composites for weapons as well. Mongol bows were constructed of a composite of bulls' tendons, wood and silk bonded together with animal adhesive. The so-called Damascus gun barrels were fabricated from iron and steel laminates, as were Japanese ceremonial swords.

Continuing into modern times, linoleum, plasterboard, asphalt and concrete are all composites that have been widely used for many years, although few people have looked upon them as composite materials. And in the short space of two decades, glass-reinforced plastics, plastics-metal laminates and clad metals have found common use as engineering materials.

Despite their presence and use for many years, composite materials have occupied an ambiguous and ill-defined position in materials technology. More often than not they have been considered for use only as a last resort, when monolithic materials have not adequately applied. And even when used, they have usually been treated by engineers as though they were monolithic materials. Thus, composites have been relegated to lesser degrees of development. As a result, each group of workers has tended to construct a specialized framework and terminology supporting its specific, and usually narrow, viewpoint. Consequently, no established discipline, no systematized body of knowledge, and few consistent general theories and models have developed.

Industrial innovation, improved energy planning, uncertain availability and skyrocketing cost of materials are several reasons for a new attitude and greater interest in composites. One major consideration is the increasing severity of performance requirements in many areas, which is taxing to the limit conventional monolithic materials. Another reason is the new point of view the user has gained over the past decade. Traditionally, the engineer's approach has been to fit the design or product to the properties of a material. Now the primary concern is finding and applying a material or materials with the right properties to meet the design, economic and service conditions. As a logical consequence of this new orientation, engineers are turning more and more to composites, for composites contain the promise of providing in a single material system the combination of properties sought for any application.

The composites concept offers even more than this. It provides the point of view and the means to develop and design materials that in themselves are the finished component, structure or product. This new and fresh way of thinking about materials is a concept of far-reaching significance for the materials development and application fields. It seems clear that in the future more and more of our engineering materials, as well as our structures and products, will be composites.

Currently, the design engineer's decision-making process is endless. If he thinks he has made the right decision one day, the events of the next day, or the next week, may prove him wrong. These events may take the form of product misuse, government regulations on material, or inconsistent manufacture of the material. In short, material selection and processing is a highly complex process. What applies to steels in the decision-making process is equally applicable to aluminum, copper, nickel, magnesium and titanium.

The selection process for plastic materials and their fabrication is no less difficult. Plastic families exist by the scores, and new plastics seem to be developed on a weekly schedule. Adding to the design engineer's woes are other engineering materials — carbon, fibers of all sorts, glass and particulates, to name a few. What's more, the lists of composite materials differ considerably from each other in form — honeycomb materials, fiber-reinforced plastics, laminates of all kinds, clad materials, graphite compounds and others — making the selection process even more difficult.

Looking at the future from the composites manufacturing and productivity viewpoint, basically, we are not going to see much new technology applied that is not available today. The philosophy must be to do more with less, and therefore, to be more innovative. The problem is not a lack of technology and tools, but a shortage of people with the skills and know-how to apply these assets more productively. Therefore, the challenge of the next decade is to train the engineers, analysts, programmers, manufacturing specialists and shop technicians who will be needed to select the correct material for the design requirements; to tool and plan the fabrication sequence; and to machine, trim, drill, paint, etc., to complete the finished product. It is already encouraging to see what is starting to happen at some of our universities in the development of materials science composite programs and manufacturing systems engineering curricula.

If there is any lesson to be learned from history, it is that the truly great forward and quantum leaps come not as the culmination of a series of small, practical advances, but as the result of a brilliant and useful generalization. While analytical minds are evolving the theory of composites, there is much more that engineers can do today to take advantage of what is already known and, for many practical purposes, quite well understood. Once the theory of composites is more adequately developed, disseminated and understood, we will begin to fully realize their promise.

This Source Book is but one step in the development of a coherent body of knowledge in composite fabrication methods, one that increases understanding and leads to useful, predictable information. The book describes and answers the practical questions that arise as engineers and technicians find more and more uses for composites, and find that, unlike metals, these materials have been developed so rapidly that in many cases the manufacturing operations are developmental, and not clearly repeatable.

The various sections of this book will cover data on machining, drilling, trimming, painting, coating, reaming and other processes integral to the fabrication of composites. Since the bounds of what can be considered composite materials are almost endless, this Source Book will devote itself primarily to polymer-reinforced composites (PMC) and metal-reinforced matrix composites (MMC).

*   *   *   *   *   *

The American Society for Metals extends most grateful acknowledgment to the many authors whose work is presented in this Source Book, and to their publishers.

MEL M. SCHWARTZ
Chief of Metals Engineering
Sikorsky Aircraft
United Technologies

CONTENTS

SECTION V: DRILLING

SECTION VI: MACHINING

SECTION I
General Overview

Part II: Secondary fabrication. Forming, joining, machining and processing composites is quite complex. The physical properties of aluminum matrix composites are reviewed.

Making a product from composites

by I. J. Toth, W. D. Brentnall, and G. D. Menke

Anisotropic properties limit secondary fabrication processes—Machining via erosion.

The strongly anisotropic properties of composite materials complicate their secondary fabrication, relative to processes used for less exotic materials. As briefly described in this section, however, processing limitations are being determined and fabrication techniques are being developed and optimized.

Forming

Composite tape is rapidly becoming the customary starting material for forming into the final product shape. For most shapes the tapes are formed into the desired configuration and then joined together by brazing, diffusion bonding, or other processes. Several approaches were investigated by Kreider[35] in the fabrication of Borsic-Al cylinders. Weisinger has fabricated complex shapes by high-pressure bonding,[43] as illustrated in Fig. 5. Fig. 6 illustrates the Pratt & Whitney Aircraft process for fabricating the Borsic-Al TF 30 blade. The blade, which includes a splayed root, leading edge protection, and cross-ply stiffening, has successfully passed the preliminary test requirements.

The anisotropic properties of composites and their stiff, low ductility reinforcement severely limit their formability. Forming which requires elongation in the filament direction is not feasible; however, bending in any direction can be accomplished. Elevated temperature forming produces the best results, as the decreased matrix strength minimizes stress buildup and lowers force requirements. Weisinger,[43] Happe and Yeast,[44] and Schaefer et al[45] have studied the formability of B-Al composites in detail. Weisinger[43] reports success in forming cross-ply material, using forming temperatures in the 850°F (450°C) range.

Twist forming of composites requires elevated temperatures also, as the filaments must move relative to each other, requiring a low shear strength matrix. Fig. 7 illustrates twist formed multilayer B-Al composites. Forming of the ±45° cross-ply panel required a higher temperature, as a greater degree of interlaminar shear was required. Both panels were formed without measurable filament damage.[79]

Advanced concepts in composite forming development include determination of time/temperature/pressure/geometry requirements for changing the dimensions and shape of the part simultaneously; that is, learning the parameters for successful "creep forging".

Joining

Use of composites in most structural applications requires that many pieces be joined to form an assembly. In unreinforced metals, the techniques commonly used are resistance welding, brazing, and fusion welding. Joining of composites requires increased sophistication due to the presence of easily damaged filaments and due to the requirements for higher joint strength, in order that the composite strength can be fully utilized.

Joint efficiencies pose the biggest problem when joints are in the filament direction, where composite strength is normally over 160 Ksi (11250 kg/cm²) and can exceed 300 Ksi (21000 kg/cm²) in laboratory specimens. Breinan and Kreider[46] experimented with brazed scarf joints, achieving 93 Ksi (6540 kg/cm²) strength for a 2° scarf and only 47 Ksi (3300 kg/cm²) for a 5° scarf, the latter strength being controlled by shear strength of the braze. Successful joints required large areas of overlapped reinforce-

Fig. 5—Aluminum-boron angle, T-stringer, and tube made by high-pressure braze bonding of tapes.

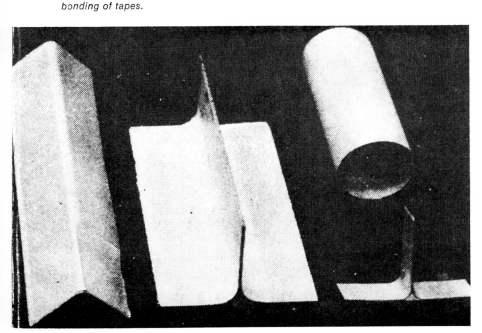

Reprinted with permission from Journal of Metals, Vol. 24, October 1972, 37-42, © 1972 Metallurgical Society/AIME

a) Tape cutting, and lay-up

b) Wedge forming die and compacted wedge

c) Airfoil bonding die

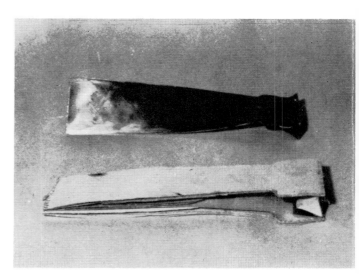

d) Splayed root airfoil ply bundle and bonded airfoil

Fig. 6—Borsic-aluminum airfoil fabrication sequence.

e) Instrumented die installed in vacuum retort

f) Vacuum retort mounted in press

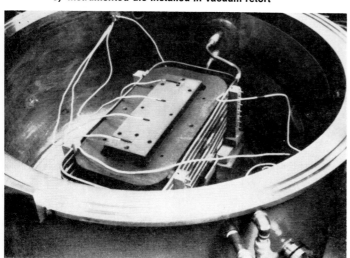

ment, reducing system weight efficiency. Happe and Yeast[44] evaluated TIG, EB, and spot welds, observing good weldments but low strengths. Hersh and Duffy[47] also evaluated spot welding, observing poor cross-tension strength and encountering difficulty in determining weld quality by NDT techniques. As far as riveting is concerned, the mechanical filament damage in making the required holes, the decreased composite effectiveness, and stress concentration effects, severely limit use of this approach. Diffusion bonding currently is the best joining technique, although expensive and geometry limited. Olster[48] successfully diffusion bonded scarf joints in an Al-SS system; due to filament damage, the technique is not as readily applied to B-Al. For most applications, furnace brazing or adhesive bonding, using overlapping reinforcement, will remain the most practical composite joining technique.

Machining

The B-Al composites pose special problems in machining because of the high hardness and uneven shear fracture of the boron filaments. Weisinger[43] compared damage resulting from nine different processes, and ranked electro-discharge machining (EDM) as least damaging; Happe and Yeast[44] had best results using ultrasonic machining, a process not evaluated by Weisinger. Both of these processes remove material by erosion the filaments are not stressed during machining, and therefore do not exhibit splitting. Boron filaments with appreciably higher transverse strengths are now available and the filament splitting tendency is reduced. All machining processes should probably be re-evaluated in the light of this very significant improvement in filament characteristics.

Thermo-Mechanical Processing

The use of heat treatments or cold working to improve properties, when applied to composite materials, solve some problems but create others. The effect of transverse rolling on B-Al tensile strength was investigated by Getten.[49] Fig. 8 illustrates some of the data generated. Two points should be noted: (a) the effect is small, for practical reductions, and (b) fila-

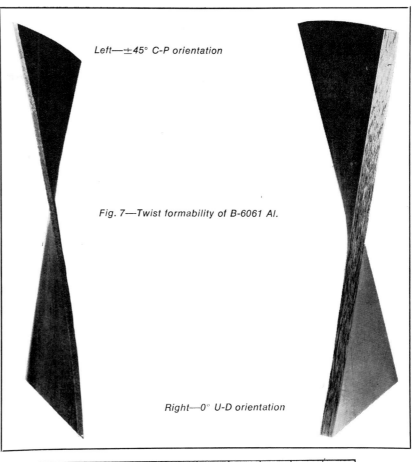

Left—±45° C-P orientation

Fig. 7—Twist formability of B-6061 Al.

Right—0° U-D orientation

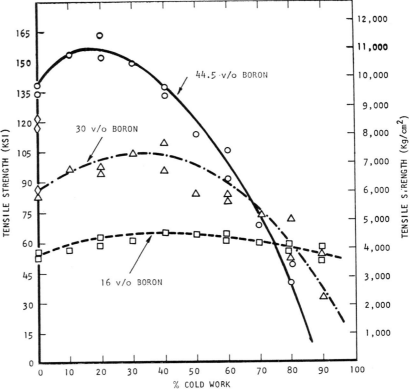

Fig. 8—The effect of transverse rolling on the tensile strength of B-Al composites.

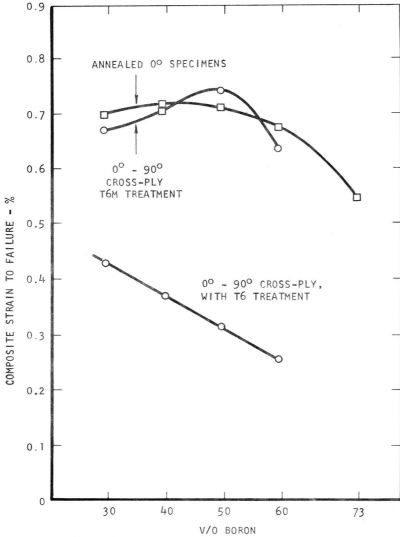

Fig. 9—*Effect of modified heat treatment on failure strain.*

ment breakage resulted at larger reductions. It should be mentioned that a significant amount of damage was present in the as-fabricated composite.

Heat treatment is the accepted technique used for strengthening Al-matrix composites, primarily to improve the transverse properties. Naturally, heat treatment is not to be considered where the final application involves elevated temperatures, as the advantage will be rapidly lost due to overaging. Also, selection of the heat treatment conditions must be made with care. As shown by Forest and Christian,[50] solution annealing of B-Al must be performed below 1000°F (540°C) as excessive filament degradation otherwise occurs.

Sumner[9] reports significant gains in longitudinal tensile strength prop-

erties for B-7178-T6 Al, reaching strengths of 209 Ksi (14700 kg/cm²) for 45 v/0 reinforcement; this is approximately the theoretical strength, based on use of fiber with 400 Ksi average strength.

Use of a modified T6 treatment has significantly improved the properties of cross-plied material.[1] In Fig. 9, the failure strain of T6 treated 0°-90° cross-ply B-6061 Al is compared to that of the modified treatment, T6M: The T6M treated specimens achieved full utilization of the strength of the 0° fibers. The T6M treatment adds a liquid nitrogen (LN$_2$) quench after the water quench, changing the residual stress state of the triaxially restrained matrix prior to aging. The exact operating mechanism is not known; however, Hancock reports[64] that the LN$_2$ subzero

cooling alters the metallurgical structure in the matrix. The subcooling technique, used by itself, may prove valuable for improving properties in massive cross-plied parts, where T6 treatment is not feasible due to quenching stresses.

PHYSICAL PROPERTIES

There are little published data on properties such as thermal expansion, thermal conductivity, damping, etc., for aluminum matrix composites and those that are available reflect the stage of development of the respective composite systems. Thus, practically all of the experimentally determined data relate to the most advanced system, boron-aluminum. Theoretical analyses have been made, however, which in some cases have subsequently been compared to experimental data. Rosen[51] recently reviewed theoretical analyses of the thermo-mechanical properties of unidirectional fibrous composites and concluded "The published literature contains analyses of the thermo-elastic properties of fibrous composites which yield results that are theoretically sound, and in reasonable agreement with available experimental data".

Density

Since composites are mechanical mixtures of two or more components, rather than precipitate or dispersoid type alloys, the density is given by the rule of mixtures, providing a void-free material is produced.

The composite density ρ_c is given by

$$\rho_c = \rho_f V_f + \rho_m (1 - V_f)$$

where V is the volume fraction and subscripts m and f refer to matrix and fiber. Fig. 10 illustrates the density of various aluminum composites as a function of volume reinforcement. The shaded band shows the range of volume percent reinforcement normally of interest because of the engineering design requirements with regard to specific stiffness and specific strength.

Thermal Expansion

Thermal expansion characteristics of unidirectional Borsic-1100 Al and Borsic-2024 Al composites were determined by Kreider et al[52] as a function of fiber volume fraction and at various orientations with respect to the fiber axis. These results are reproduced in Fig. 11. The data were analyzed with reference to models due to Turner,[53] Kerner,[54] Thomas,[55] Levin[56] and Shapery.[57] Using elastic theory, Levin's predicted values for axial and transverse coefficients $\overset{*}{a}$ and $\overset{*}{t}$ are given by

$$\alpha^*_a = \bar{a} + \left(\frac{\alpha_1 - \alpha_2}{(1/K_1) - (1/K_2)} \right)$$
$$\left[\frac{3(1 - 2\nu^*_a)}{E^*_a} - \left(\overline{\frac{1}{K}} \right) \right],$$

$$\alpha^*_t = \bar{a} + \left(\frac{\alpha_1 - \alpha_2}{(1/K_1) - (1/K_2)} \right)$$
$$\left[\frac{3}{2K^*_t} - \frac{3\nu^*_a(1 - 2\nu^*_a)}{E^*_a} - \left(\overline{\frac{1}{K}} \right) \right].$$

where K is the bulk modulus and ν is Poisson's ratio for the respective phases, Kreider observed that in aluminum matrix composites the matrix can undergo plastic deformation and developed the plasticity approximation

$$\alpha_c = \alpha_f + \frac{A_m \sigma_m Y}{A_f E_f}$$

where A_m and A_f are respective areas of matrix and filament and $\sigma_m Y$ is the matrix yield stress. Good agreement was observed between calculated and experimental values for Borsic-1100 Al composites as shown in Fig. 11. Shapery's prediction for thermal expansion in the transverse direction

$$\alpha_t = (1 + \nu_m)\alpha_m V_m$$
$$+ (1 + \nu_f)\alpha_f V_f - (\alpha_c)_1 \nu_c$$

showed good agreement with experimental values. It is interesting to observe the initial (predicted) increase of $\bar{a}$ at low volume fractions.

Kreider concluded that care should be exercised in applying theoretical equations which predict composite thermal expansion on the basis of component properties and assume elastic behavior.

Specific Heat and Thermal Conductivity

Rosen[51] observed that the effective specific heat of a composite at constant pressure or constant volume is not the volume-weighted average of the respective component specific heats. This is because a temperature change occurring at constant volume of the bulk composite takes place with volume changes of each separate phase.

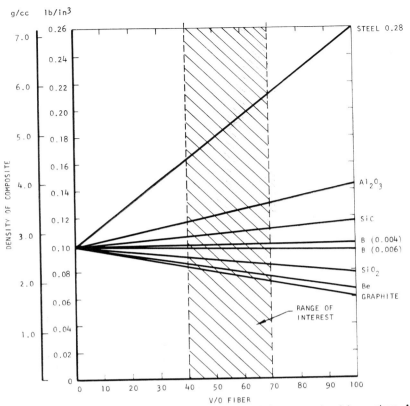

Fig. 10—Composite density as a function of fiber reinforcement level for various Al matrix composite systems.

For a two-phase composite the effective specific heat at constant pressure C^*_p is given by

$$\frac{C^* - \bar{C}}{T_a} = \frac{9(\alpha_f - \alpha_m)^2}{[(1/K_f) - (1/K_m)]^2}$$
$$\left[\left(\overline{\frac{1}{K}} \right) - S^*_{iijj} \right]$$

where the sum of the effective compliances

$$S^*_{iijj} = \frac{(1 - 2\nu^*_a)^2}{E^*_a} + \frac{1}{k^*_t}$$

The effective thermal conductivities of unidirectional composites has also been expressed in terms of the individual phase properties.[58] The axial and transverse conductivities (μ^*_a and μ^*_t) were defined simply as

$$\mu^*_a = \bar{\mu} = V_f \mu_f + V_m \mu_m$$
$$\mu^*_t = \mu_m \frac{V_m \mu_m + (1 + V_f)\mu_f}{(1 + V_f)\mu_m + V_m \mu_f}$$

No experimental data on aluminum matrix fiber reinforced composites appears to have been published.

Damping Characteristics

Most of the experimental studies on material damping in composites have been carried out on glass fiber/epoxy combinations. Pottinger[59] investigated damping in unidirectional B-6061 Al at various orientations to filaments, over the frequency range 1 KHz to 100 KHz. No details regarding material quality and volume percent reinforcement were reported. It was concluded that the material damping basically follows the behavior of the matrix for boron-aluminum composites, although there was evidence of a secondary maximum in damping presumed to be associated with the fiber at a frequency of about 100 KHz. □

REFERENCES

(References listed cover Part II. A complete bibliography will be listed at the end of this series.)

[43] Weisinger, M.D.; "Forming and Machining Aluminum-Boron Composites", 1970 Western Metal and Tool Conference and Exposition, 9-12 Mar. 1970, Los Angeles, Calif., ASM Paper W70-5.2.

[44] Happe, R.A. and Yeast, A.J.; "Evaluation of Boron-Aluminum Composite Material for Space Structures", Materials and Processes for the 70s, Vol. 15, SAMPE, Apr. 29—May 1, 1969.

[45] Schaefer, W.H., Christian, J.L., et al.; "Evaluation of the Structural Behavior of Filament Reinforced Metal Matrix Composites", AFML-TR-69-36, Vol. I—III, Jan. 1969.

[46] Breinan, E.M. and Kreider, K.G.; "Braze Bonding and Joining of Aluminum Boron Composites", Metals Eng. Qtly. (ASM), pp. 5-15, Nov. 1969.

[47] Hersh, M.S. and Duffy, E.R.; "Development of Fabrication Methods for Aluminum-Boron Composite Aircraft Structures", in DMIC Memorandum 243, May 1969.

48 Olster, E.F. and Jones, R.C.; "Diffusion Bonded Scarf Joints in a Metal Matrix Composite", in *Composite Materials: Testing and Design*, ASTM STP 460, pp. 393-404, 1969.
49 Getten, J.R. and Ebert, L.J.; The Cold Rolling Characteristics of Aluminum-Boron Fiber Composites", ASM Trans., Vol. 62, p. 869, 1969.
50 Forest, J.D. and Christian, J.L.; "Development and Application of High Matrix Strength Aluminum-Boron", *Metals Eng. Quarterly*, Feb. 1970.
51 Rosen, B.W.; "Thermochemical Properties of Fibrous Composites", Proc. Roy. Soc., Lond. Series A, 319, pp. 79-94, 1970.
52 Kreider, K. G., et al; "Thermal Expansion of Boron Fiber-Aluminum Composites", Met. Trans., 1, p. 3431, Dec. 1970.
53 Turner, P.S.; Nat. Bur. Std. Res. Paper RP 1745, Vol. 37, Oct. 1946.
54 Kerner, E.H.; *Proc. Phys. Soc.*, Vol. 69, 808, 1956.
55 Thomas, J.P.; Convair (Fort Worth) Report No. FGT-2713.

56 Levin, V.M.; *Mekhanika Tverdogo Tela*, Vol. 88, 1967.
57 Shapery, R.A.; *J. Composite Mat.*, Vol. 1, 2, 1968.
58 Hashin, Z.; "Mechanics of Composite Materials", Pergammon Press, N. Y.
59 Pottinger, M.G.; "Material Damping of Glass Fiber-Epoxy and Boron Fiber-Aluminum Composites", ARL 70-0237, Oct. 1970.
60 Moore, C.I.; "The Strength of Composite Materials Reinforced with Brittle Fiber", DMIC Memorandum 243, May 1969.
61 Jones, B.H.; "Strength and Fracture Characteristics of Filamentary Composites", ASME Paper 70-DE-31.
62 Pepper, R.T., Upp, J.W., Rossi, R.C. and Kendall, E.G.; "The Tensile Properties of a Graphite-Fiber-Reinforced Al-Si Alloy", *Met. Trans.*, Vol. 2, pp. 117-120, Jan. 1971.
63 Menke, G.D. and Toth, I.J.; "The Time Dependent Mechanical Behavior of Composite Materials", AFML-TR-70-174, June 1970.

64 Toth, I.J.; "Tensile and Fatigue Behavior of B-Al Composites", presented at the 1970 Spring Meeting of the Metallurgical Society of AIME at Las Vegas, Nevada, May 1970.
79 Menke, G.D., Brentnall, W.D. and Toth, I.J.; "Creep Formability of Metallic Composites", presented at the 1971 Metal Show, Detroit, Oct. 21, 1971.

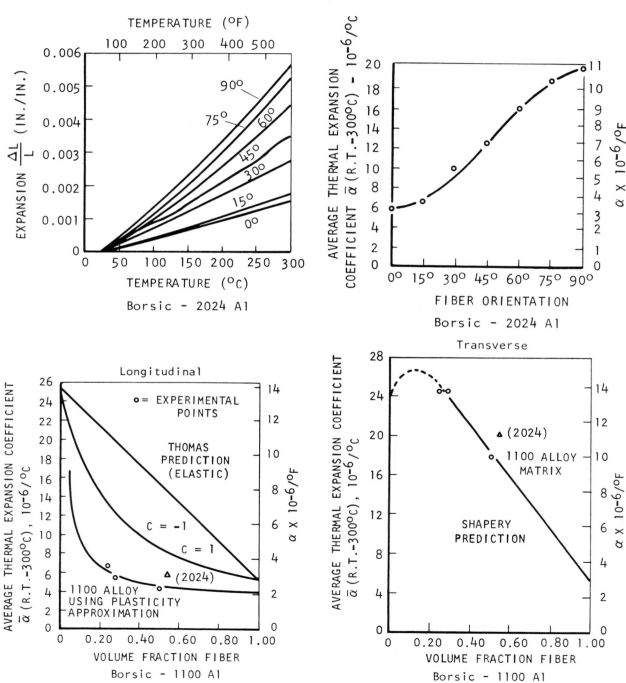

Fig. 11—Thermal expansion behavior of borsic-Al.

Along with high strength and outstanding rigidity, laminated carbon and glass-fiber composites have a serious shortcoming: They are weakened appreciably when cut or machined. This strength degradation is particularly troublesome when composites must be fastened together and still maintain structural integrity.

Design guidelines for

JOINING ADVANCED COMPOSITES

DONALD R. DREGER Staff Editor

ADVANCED composites, by virtue of their high-modulus reinforcing fibers, have engendered an entirely new concept of design: a materials technology that combines high strength, stiffness, and controlled anisotropy with light weight. This capability, which is so vital in aircraft, naval vessels, and military vehicles, is also becoming important in industrial and consumer products—especially automobiles and other mobile equipment.

Advanced composites also introduce design problems different from those common to metal and molded-plastic construction, particularly where components must be fastened together. Because most types of composite joints—be they adhe-

Why 'Advanced' Composites?

The term "advanced composite" came into being about ten years ago to designate certain composite materials having properties considerably superior to those of earlier composites. The term is ambiguous, however, because it does not identify specific material combinations, nor does it indicate their arrangement or configuration in the composite. The term can include materials having reinforcing fibers that are either continuous or chopped, or both, and they may be oriented or randomly distributed in the matrix. And a composite can include metal, wood, foam, or other material layers, in addition to the fiber/resin components.

Nevertheless, an advanced composite has come to denote, to most engineers, a resin-matrix material reinforced with high-strength, high-modulus fibers of glass, carbon, aramid, or even boron, and usually laid up in layers to form an engineered component. More specifically, the term has come

to apply principally to epoxy-resin-matrix materials reinforced with oriented, continuous fibers of carbon or of a combination of carbon and glass fibers, laid up in multilayer fashion to form extremely rigid, strong structures.

Early uses of these materials in sports and recreational equipment provided excellent test products. Sports enthusiasts were willing to pay several times the cost of conventional wood or metal construction for exotic "graphite" products whose manufacturers suggested that vastly improved performance would result from using their tennis racquets, archery bows, or skis.

The aircraft industry was also quick to see the advantages of using epoxy/carbon composites which offered light weight, high strength and elastic modulus, and—most important—excellent fatigue performance. Most of the engineering and manufacturing experience with these composites

has been gained in developing components for military aircraft. While the manufacture of advanced-composite components is still highly labor intensive and expensive, plans are being implemented by at least one aircraft manufacturer to build an automated factory for producing composite structures of epoxy/carbon-fiber construction.

Some of the technology of composites has begun to move into the high-production areas—automobiles, specifically. Automakers are cautiously testing components made from these high modulus materials. Some applications have been announced; many more are still under wraps. More uses will emerge as more is learned about the special design requirements involved in the construction of advanced composites and as more automation is applied confidently to manufacturing and joining these structures.

When Holes are Necessary. . .

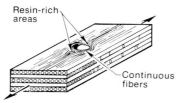

Avoid molded-in holes. In molding, continuous fiber is forced around the plug that forms the hole, leaving resin-rich areas, which are relatively weak.

Drilled holes are better. The cross fibers, at 45° or 90° to the load, maintain transverse integrity, keeping cracks from forming.

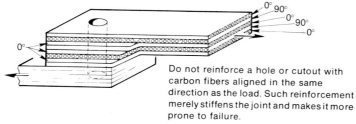

Do not reinforce a hole or cutout with carbon fibers aligned in the same direction as the load. Such reinforcement merely stiffens the joint and makes it more prone to failure.

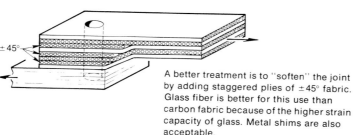

A better treatment is to "soften" the joint by adding staggered plies of ±45° fabric. Glass fiber is better for this use than carbon fabric because of the higher strain capacity of glass. Metal shims are also acceptable.

Bonded joints in advanced-composite structures are preferred where possible. But when strength requirements exceed those available from adhesives alone, bolts are used, often in conjunction with an adhesive. Because of their low strain capacity and inhomogeneity, composites require special design treatment when they are to be joined by mechanical fasteners.

Slight dimensional inaccuracies in bolt patterns in materials that deform plastically—aluminum or steel, for example—seldom cause problems. The material around the holes can yield to distribute the load to adjacent bolts. This is not true of the advanced composites, however. These materials have no yield point, and if one bolt is tighter in its hole than others, the bearing stress at that hole remains higher.

The brittleness of an epoxy/carbon-fiber structural composite could cause ultimate failure of the material around a hole and the sudden distribution of its entire load to the other holes. One way to avoid such problems is to soften the laminate in the bolt area through the use of staggered fabric plies. Another way is to coat the bolts with uncured epoxy resin prior to attachment. The resin fills the clearances between bolts and holes and, when cured, helps distribute the load evenly.

sively bonded or mechanically fastened—involve some cutting or machining of the strength-providing fibers, joint configurations require careful planning to minimize the possibility of failure.

Adhesives or Bolts?

Because structural composites are made with thermosetting resins, they cannot be joined by welding methods as can thermoplastics and metals. The choice for composites lies between mechanical methods and adhesive bonding. Each technique produces a joint with significant differences both in production and in function; each has its advantages and limitations.

Mechanical joints are generally not as adversely affected by thermal cycling or humidity as are bonded joints. They permit disassembly without destroying the substrate, and they are readily inspected for joint quality. Also, mechanical joints require little or no surface preparation of the substrate, and they do not require white-glove and clean-room production conditions.

On the less favorable side, mechanical methods add weight and bulk to a joint and, since they require machining of holes in the substrate members to accommodate bolts or screws, the members are weakened. In addition, the stress concentrations produced by mechanical fasteners can cause joint failure.

Advantages of adhesive-bonded joints include light weight, distribution of the load over a larger area than in mechanical joints, and elimination of the need for drilled holes that weaken the structural members.

Drawbacks to bonded joints are the difficulty of inspection for bond integrity, the possibility of degradation in service from temperature and humidity cycling, and their relative permanence, which does not permit disassembly without destruction of the joined members. In addition, bonded joints require a rigorous cleaning and preparing of adherend surfaces.

Bonded-Joint Design

Performance and safety requirements of various applications for advanced composites differ widely, and the design approaches differ accordingly. For example, the importance of a failure in a carbon-fiber composite tennis-racquet frame in no way approaches that of a failure in a composite helicopter rotor blade. Thus, the consequences of component failure determine the type of design data needed and the nature of testing that should be done before the design is finalized.

The comparisons of joining-method characteristics listed pertain to practically all materials, composite or not. The special difference that sets apart orthotropic materials, such as

Line up the Fibers with the Load

Multilayer composites must not be stressed in a manner likely to cause delamination. Proper orientation of the surface fibers in a bonded joint—so the fibers are in the direction of the load—must be included in the design specifications. Otherwise, machining of bond-line areas can result in a fiber orientation that weakens the joint drastically. The sketches show this principle in use on components having 0° and 90° fiber orientation.

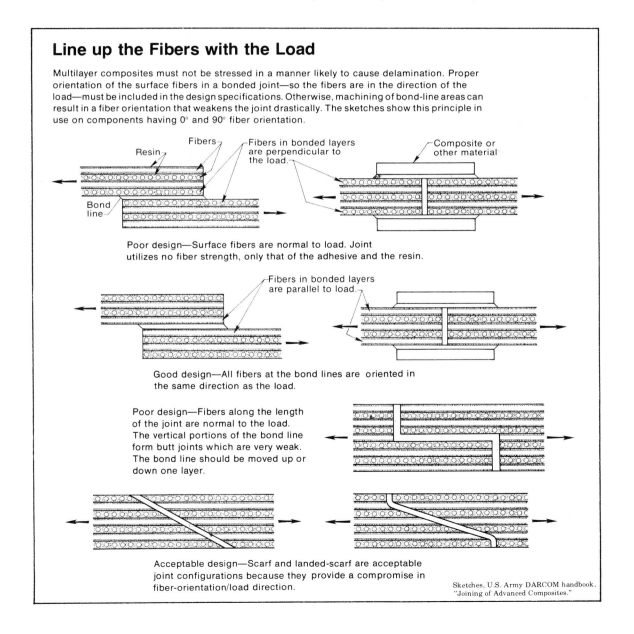

Poor design—Surface fibers are normal to load. Joint utilizes no fiber strength, only that of the adhesive and the resin.

Good design—All fibers at the bond lines are oriented in the same direction as the load.

Poor design—Fibers along the length of the joint are normal to the load. The vertical portions of the bond line form butt joints which are very weak. The bond line should be moved up or down one layer.

Acceptable design—Scarf and landed-scarf are acceptable joint configurations because they provide a compromise in fiber-orientation/load direction.

Sketches, U.S. Army DARCOM handbook, "Joining of Advanced Composites."

the advanced composites, from isotropic materials is that damage (from drilling or machining) weakens the composites to an extent that makes them susceptible to interlaminar shear, delamination, and peeling.

For this reason, bonded joints are usually preferred over mechanically fastened joints for assembling advanced-composite components to each other or to metal parts. More specifically, configurations such as lap joints and strap joints are usually best because they require no machining of the adherends. These joints do, of course, impose a slight weight penalty and are unacceptable if

a smooth, uninterrupted surface is absolutely necessary.

Design techniques for bonded composite joints are essentially extensions of those for isotropic materials, adjusted for the manner in which isotropic characteristics alter failure modes. Combining joint stress analyses with data on material properties and failure criteria can provide a fairly accurate prediction of the strength and deflection characteristics of the joint. Values of both shear·and tensile properties of the composite and the cured adhesive film are required. These property data should reflect the pertinent environmental conditions

and any special lay-up and cure characteristics of the composite as well as of the adhesive.

With so many variables involved it is unlikely that such material-characterization data will be readily available in handbook form in the near future. Eventually, all new data may some day be reported in a standardized format and placed in a central data bank, accessible for use by engineers conducting computer-aided design projects.

Surface Preparation

As with any bonding operation, careful preparation of adherend surfaces is essential to

Glossary of Bonded-Joint Configurations

Design alternatives for bonded joints are more numerous than for bolted joints. Illustrated here are various forms of lap, strap, and scarf joints, plus some less common configurations.

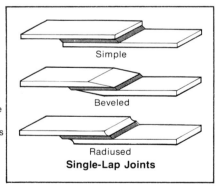

Single-Lap Joints

Simple

Beveled

Radiused

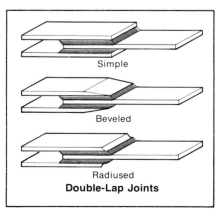

Double-Lap Joints

Simple

Beveled

Radiused

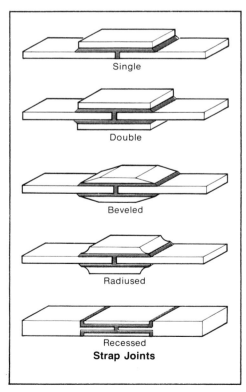

Strap Joints

Single

Double

Beveled

Radiused

Recessed

achieving dependable bonds in advanced composites. Recommended preparation generally consists of a solvent wipe to remove loose surface dirt and oil, followed by an abrading operation. The first step removes loosely adhered boundary layers, and the second increases the surface energy as well as the bond area of the faying surface. Abrading must be done carefully, however, to avoid damaging surface fibers. One recommended sequence for surface preparation is:

1. Scrub the entire panel surface with a stiff brush, using a 2% solution of detergent at 104 to 122° F.

2. Rinse thoroughly in tap water, then immerse in distilled water.

3. Dry in a recirculating-air oven at 150° F for 30 min.

4. Abrade bonding area plus a ½-in. margin with 240-grit sandpaper.

5. Vacuum-clean abraded area.

6. Swab abraded area with acetone-wetted gauze.

7. Rinse entire panel in distilled water.

8. Dry in a recirculating-air oven at 212° F for 60 min.

9. Bond the components immediately.

Another surface-preparation technique in use is the peel-ply method. In this technique, a portion of woven nylon cloth is incorporated into the outer layer of the composite during lay-up. When the component is prepared for bonding, the nylon peel ply is simply torn or peeled away, exposing a clean, roughened surface. The degree of roughness can be varied to some extent by the choice of weave of the peel ply.　　Ⓜ🄳

Technical assistance for this article was furnished by Dr. Howard S. Kliger, Research Supervisor, Celanese Research Co., Summit, N. J., and by the DoD Plastics Technical Evaluation Center (PLASTEC), Dover, N.J.

Comprehensive design information on composite joints is contained in "Joining of Advanced Composites," an Engineering Design Handbook published by the U.S. Army Materiel Development and Readiness Command, Alexandria, Va., and available from the National Technical Information Service, 5285 Port Royal Rd., Springfield, Va. 22161, $16.50. Call number is AD A072362.

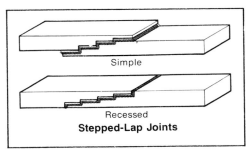

Stepped-Lap Joints

Simple

Recessed

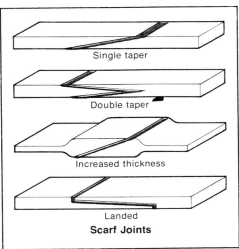

Single taper

Double taper

Increased thickness

Landed

Scarf Joints

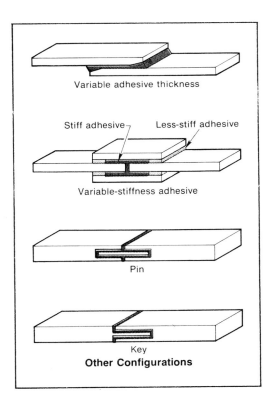

Variable adhesive thickness

Stiff adhesive — Less-stiff adhesive

Variable-stiffness adhesive

Pin

Key

Other Configurations

Sketches, U.S. Army DARCOM handbook, "Joining of Advanced Composites."

TUNGSTEN FIBER REINFORCED SUPERALLOYS - A STATUS REVIEW

by Donald W. Petrasek and Robert A. Signorelli

National Aeronautics and Space Administration

Lewis Research Center

Cleveland, Ohio 44135

ABSTRACT

Improved performance of heat engines is largely dependent upon maximum cycle temperatures. Tungsten fiber reinforced superalloys (TFRS) are the first of a family of high temperature composites that offer the potential for significantly raising hot component operating temperatures and thus leading to improved heat engine performance. This status review of TFRS research emphasizes the promising property data developed to date, the status of TFRS composite airfoil fabrication technology, and the areas requiring more attention to assure their applicability to hot section components of aircraft gas-turbine engines.

INTRODUCTION

The key to the development of more efficient heat engines lies in the use of higher operating temperatures which in turn relies on the development of higher temperature materials. Heat engine efficiency is proportional to maximum cycle operating temperature. Increases in these operating temperatures have been accomplished through improvements in the properties of nickel and cobalt base superalloys as well as by the incorporation of sophisticated schemes for air cooling turbine components. The gap between the highest temperature at which superalloys can be used for long time service, 870° to 950° C (1600° to 1750° F) and the gas inlet temperatures for which engines are being designed, 1300°-1600° C (2370°-2910° F), is progressively widening. Thus, there is an increasing need to employ more air cooling in

Reprinted courtesy of NASA Lewis Research Center, Technical Memorandum 82590, January 1981, 68 pages

the turbine section. However, as increasing amounts of compressor discharge air are diverted to cool hot-section components, the efficiency of a gas turbine engine drops. This coupled with the fact that nickel and cobalt base superalloys have been developed nearly to their technological limits, indicates that new types of materials are necessary to further raise engine efficiency. The challenge for materials engineering is to develop a new generation of high temperature materials which are qualified for long life application at material operating temperatures higher than 1000° C (1830° F).

The need for improved materials at elevated temperatures has stimulated research in many areas including efforts to develop fiber reinforced superalloy matrix composites. A number of fibers have been studied for such use including submicron diameter ceramic whiskers, continuous length ceramic filaments, boron filaments, carbon filaments, and refractory metal alloy wires. Attainment of high temperature strength with superalloy matrix composites using ceramic whiskers, continous length ceramic filaments, boron filaments or carbon filaments as the reinforcing fiber have been unsuccessful to date. Boron and carbon react and are virtually dissolved when incorporated into superalloy materials. Attempts to develop coatings to prevent such reactions have been unsuccessful. Ceramic whiskers and continuous length ceramic filaments also have suffered from degraded fiber strength because of fiber-matrix reaction. The fiber-matrix reaction causes fiber surface flaws which act as local stress concentrations. These stress concentrations become nucleation sites for premature failure. The thermal expansion mismatch between ceramic fibers and typical candidate matrix alloys can induce severe stresses in composite materials. The combination of fiber surface roughening from reaction of the fiber with the matrix combined with thermally induced stress can cause catastrophic fracture of com-

Source: NASA Technical Memorandum 82590, January 1981, 68 pages

posite test specimens even during fabrication. Because of the above factors, relatively low strength values have been achieved with such non-metallic fibers compared with those theoretically possible. However, due to the high use temperature potential of such composites, research efforts should and will continue in an attempt to overcome these problems.

The theoretical specific strength potential of refractory alloy fiber reinforced superalloys is less than that of ceramic fiber reinforced superalloys. However, the more ductile metal fiber systems are more tolerant of fiber-matrix reactions and thermal expansion mismatches. Also, the superalloy matrices can protect high strength refractory metal fibers from environmental attack. In laboratory tests, refractory fiber reinforced superalloy composites have demonstrated stress-rupture strengths significantly above those of the strongest superalloys. Tungsten fiber reinforced superalloy composites, in particular, are potentially useful as turbine blade materials because they have many desirable properties such as good stress-rupture and creep resistance, oxidation resistance, ductility, impact damage resistance, and microstructural stability. The potential of tungsten fiber reinforced superalloys (TFRS) has been recognized and has stimulated research to develop this material for use in heat engines. This status review of TFRS research emphasizes the promising property data developed to date, the status of TFRS composite airfoil fabrication technology, and the areas requiring more attention to assure their applicability to hot section components of aircraft gas-turbine engines. The approach will be to review first refractory metal fiber and matrix alloy development. This will be followed by a discussion of fabrication techniques for TFRS and property results of importance to their use at high temperatures. Specific criteria for employing TFRS as a turbine blade material are then presented. From these criteria emerged a first generation TFRS Material, tungsten alloy fiber/FeCrAlY, which is currently under evaluation. Property, design, fabrication and fabrication cost data for this system are discussed in the final section.

Fiber Development

Refractory metal wires have received a great deal of attention as fiber reinforcement materials in spite of their poor oxidation resistance and high density. When used to reinforce a ductile and oxidation resistant matrix, they are protected from oxidation and their specific strength is much higher than that of superalloys at elevated temperatures. The majority of the studies conducted on refractory wire/superalloy composites have used tungsten or molybdenum wire, available as lamp filament or thermocouple wire, as the reinforcement material. These refractory alloys were not designed for use in composites nor for optimum mechanical properties in the temperature range of interest for thermal engine application, $1000°-1200°$ C ($1830°$ to $2190°$ F). Lamp-filament wire such as 218CS tungsten was most extensively used in early studies. The stress-rupture properties of 218CS tungsten wire were superior to those of rod and bulk forms of tungsten and showed promise for use as reinforcement of superalloys. The need for stronger wire was recognized, and high strength tungsten, tantalum, molybdenum, and niobium alloys for which rod and/or sheet-fabrication procedures had already been developed were included in a wire fabrication and test program, (Refs. 1 to 4). The chemical compositions of these alloys are given in Table I. The above approach precluded development of new alloys specifically designed for strength at the intended composite use temperatures. The stress-rupture and tensile properties determined for the wires developed are summarized in

Table II and are compared with commercially available wire (218CS, W-1 percent ThO_2 and W-3Re).

Excellent progress was made in providing wires with increased strength compared to the strongest wires which were previously available. The ultimate tensile strengths obtained for the wires at 1093° and 1204° C (2000° and 2200° F) are plotted in Fig. 1. Tungsten alloy wires were fabricated having tensile strengths 2-1/2 times that obtained for 218CS tungsten wire. The strongest wire fabricated, W-Re-Hf-C, had a tensile strength of 2165 MN/m^2 (314 ksi) at 1093° C (2000° F) which is more than 6 times as strong as the strongest nickel or cobalt base superalloy. The ultimate tensile strength values obtained for the tungsten alloy wires were much higher than those obtained for molybdenum, tantalum or niobium alloy wire. When density is taken into account, the tungsten alloy wires show a decrease in advantage compared to tantalum, niobium or molybdenum wire, Fig. 2. Still the high strength tungsten alloy wires as well as molybdenum wires offered the most promise.

The elevated stress-rupture strength of such wire is more significant than the tensile strength since the intended use of the material is for long time applications. The 100 hour rupture strength at 1093° and 1204° C (2000° and 2200° F) is plotted for the various wire materials and compared to superalloys in Fig. 3. The rupture strength of tungsten alloy fibers was increased by a factor of 3 at 1093° C (2000° F) from about 434 MN/m^2 (63 ksi) for 218CS tungsten to 1413 MN/m^2 (205 ksi) for W-Re-Hf-C wire. The tungsten alloy wire was superior in stress-rupture strength to the other refractory wire materials with the exception of a tantalum alloy, ASTAR 811C, which was stronger than most of the tungsten alloy materials at 1093° C (2000° F). The strongest tungsten alloy wire, W-Re-Hf-C, was over

16 times as strong as superalloys at 1093° C (2000° F). The ratio of the 100 hour rupture strength to density for refractory metal wires and super-alloys is plotted in Fig. 4. Again the stronger tungsten wire materials are superior to the other refractory metal wires. When density is taken into account the strongest tungsten wire material, W-Re-Hf-C, is more than 7 times as strong as the strongest superalloys at 1093° C (2000° F) and more than 13 times as strong at 1204° C (2200° F).

The processing schedules used to fabricate the newer high strength wires were not optimized to provide maximum strength at 1093° and 1204° C (2000° and 2200° F). Much more work is needed to maximize their proper-ties. Considerable opportunity exists to develop wire processing schedules tailored for fiber-matrix composite use. The eventual application of TFRS composites will justify the added effort to further improve wire properties.

Matrix-Alloy Development

The matrix is the exposed component of fiber reinforced composites and therefore must be able to withstand high temperatures and an environment which can result in catastrophic oxidation and hot corrosion. The primary function of the matrix is to bind the fibers into a useful body and to pro-tect the fibers from oxidation and hot corrosion. The matrix must be rela-tively ductile compared to the fibers to facilitate load transfer from the matrix to the fiber. It also must be capable of evenly redistributing local stress concentrations and resisting abrasion and impact damage from foreign objects. The matrix and reinforcing fiber must be able to co-exist without mutually induced disintegration that can result from chemical interactions that can degrade both the fiber and matrix properties. The most important factor in the initial selection of matrix composition is the ability of the matrix to form a good bond with the fiber without excessive reaction occurring which could degrade the fibers properties.

For high temperature use, nickel-base, cobalt-base and iron-base superalloys are preferred as the matrices for refractory metal fiber composites because they have demonstrated strength and ductility at elevated temperatures as well as good oxidation and hot corrosion resistance.

A large proportion of the research effort conducted on refractory fiber composites has been on fiber-matrix compatibility with an effort being made to develop structurally stable composites by choosing a matrix composition which does not severely degrade the properties of the reinforcing fiber. One of the first systematic examinations to determine the effect of alloying reactions on the strength and microstructure of refractory metal fiber composites was reported in Ref. 5 where copper-based binary alloys were used as a matrix for tungsten-fiber composites. The effects of alloying element additions to copper on the strength and microstructure of tungsten fiber composites were compared with mutually insoluble pure copper matrix composites exposed under the same conditions. The alloying elements studied were aluminum, chromium, cobalt, niobium, nickel, zirconium and titanium. Data obtained for solute elements in this system can be related to the expected behavior of these same elements in superalloys. These effects served as the basis for modifying superalloy matrix composition to control fiber-matrix reaction. Three types of fiber-matrix reaction were found to occur: (1) diffusion-penetration reaction accompanied by a recrystallization of a peripheral zone of the tungsten fiber (2) precipitation of a second phase with no accompanying recrystallization (3) a solid solution reaction with no accompanying recrystallization in the fiber. Peripheral recrystallization was caused by diffusion of cobalt, aluminum, or nickel into the tungsten wire. Compound formation occurred with titanium and zirconium. Chromium and niobium in copper formed a solid solution with tungsten with no accompanying recrystallization of the tungsten fiber. The greatest damage to composite properties occurred with the penetration-recrystallization reaction while the two-phase and solid solution reactions caused relatively little damage. Recrystallization of tungsten fibers in a Cu-10 percent Ni matrix is shown in Fig. 5.

A number of studies have been conducted on nickel induced recrystallization of tungsten fibers, Refs. 6 to 8. Recrystallization could be induced at low temperature by the presence of solid nickel on the surface of the tungsten wire. Once initiated, nickel-induced recrystallization required a continued source of nickel for propagation of the recrystallization front. Work reported in Ref. 6 found that palladium, aluminum, manganese, platinum and iron also greatly lowered the recrystallization temperature of tungsten. Based on such findings, superalloy matrix compositions were developed that caused limited reaction with the fiber and minimal fiber-property loss on work reported in Ref. 9. These superalloys contained high weight percentages of refractory metals to reduce diffusion penetration of nickel into tungsten. Additions of Ti and Al to the matrix were also made to form intermetallic compounds which would further reduce the diffusion of nickel into tungsten. A typical alloy that was developed was Ni-25W-15Cr-2Al-2Ti. The fiber stress to cause rupture in 100 hours at 1090° C (2000° F) was reduced only 10 percent in the composite compared to the equivalent fiber rupture strength tested in a vacuum outside a composite.

The problem of obtaining structure-stable composite materials from the nickel-tungsten and nickel-molybdenum systems were further examined in Ref. 10. These experiments showed that in reinforced metal composite materials in which the matrix and fiber form restricted solid solutions in the absence of intermetallic compounds, minimal fiber dissolution can be achieved by alloying the matrix with the fiber material up to a concentration that is close to the solubility limit. However, when the matrix and the fiber react to form intermetallic compounds, matrix saturation is not effective in controlling fiber attack by dissolution.

The effect of the composition of nickel and cobalt alloys on the structural stability of composite materials reinforced with tungsten fibers was determined in Ref. 11. The rate of interaction between the fiber and the matrix was determined from the extent of recrystallization of the fiber; the formation of intermediate phases at the interface; the solution of the fiber in the matrix and the formation of diffusional porosity. In those composites having a nickel matrix, a zone of accelerated recrystallization of the fiber was observed which increased with annealing time. The recrystallization rate of the fiber was highest with a matrix of pure nickel and somewhat lower with a nickel-chromium matrix. Alloying of nickel-chromium with iron, titanium, aluminum, and especially tungsten and niobium slowed down the recrystallization process considerably. The stability of the fibrous structure of the tungsten fiber was highest in a matrix of cobalt or cobalt alloys. The accelerated recrystallization of tungsten fibers in a matrix of nickel-cobalt was due to the presence of nickel. Unfortunately intensive solution of tungsten fibers occurred in a matrix of cobalt alloys and especially in pure cobalt.

Reaction of tungsten fibers with cobalt-base matrices was also determined in Ref. 12. Fiber-reinforced binary and multicomponent cobalt-base alloys were found to have a strong propensity to intermetallic compound formation at the matrix fiber interface. Four cobalt-base alloys were identified that did not react with tungsten fibers after exposure for 1 hour at 1300° C (2370° F).

Work reported in Ref. 13 on the reaction of tungsten fibers with iron-base matrices concluded that certain such alloys can be successfully employed as matrices for composite materials reinforced with fibers of tungsten and its alloys. The matrix composition can be chosen so that it forms no intermetallic compounds with the tungsten fibers and the fibrous structure of the drawn tungsten wire is not affected by heating to 1300° C (2370° F). In most of the binary alloys investigated no recrystallization of the tungsten fibers occurred during annealing for 1 hour at 1300° C (2370° F). Dissolution of the fibers in the matrices was detected only rarely. However, in the majority of the matrices studied, layers of intermetallic compounds formed at the matrix-fiber interface. In the case of binary iron matrices alloyed with 5 and 10 percent aluminum no reactions were observed. After annealing for 1 hour at 1300° C (2370° F) no intermetallic compounds were formed in the binary iron matrices alloyed with 5-10 percent aluminum, 10 percent silicon, or 25 percent titanium, molybdenum, or nickel. The presence of nickel in multicomponent iron-base alloy matrix reinforced tungsten fiber composites accelerates the recrystallization of the fibers while increasing the amount of chromium and zirconium inhibits recrystallization. Complex alloying therefore offers a means of suppressing fiber recrystallization and the formation of intermetallic compounds at the interface between tungsten fibers and the iron-base matrix. Studies conducted on the reaction between tungsten fibers and a matrix of Fe-24Cr-5Al-1Y, Ref. 14, showed that only 10 percent of a 0.38-mm (0.015 in.) diameter fiber was reacted after exposure for 1000 hours at 1090° C (2000° F).

The reaction of tungsten fibers with binary alloys of iron, nickel and cobalt alloyed with 5, 10, or 25 weight percent of one of the following elements; aluminum, copper, silicon, zirconium, niobium, molybdenum, or tungsten were further investigated in Ref. 15. The composite samples were annealed at 1200°-1400° C (2190°-2550° F) for 1 hour. The interaction between the fiber and matrix was investigated by metallographic analysis and by change in hardness. The results of this study for samples annealed at 1200° and 1300° C (2190° and 2370° F) are summarized in Table III. Shown in Table III are the number of compositions investigated for each matrix system and the relative number of cases, based on a percentage, that resulted in the following reactions with the tungsten fiber, recrystallization, formation of an intermetallic compound, a diffusional penetration into the fiber and no detectable recrystallization or reaction with the fiber. At 1200° and 1300° C (2190° and 2370° F) contact between the tungsten fiber and the nickel binary alloys almost always induced recrystallization of the fiber. Nickel binary alloys containing 25 percent aluminum and 25 percent copper did not cause recrystallization of the fibers. As can be seen in Table III, the interaction of the tungsten fibers with binary alloys of cobalt generally did not lead to recrystallization of the fibers. At all annealing temperatures, however, a layer of intermetallic compounds was formed at the fiber-matrix interface. This process was suppressed at 1200° and 1300° C (2190° and 2370° F) only by alloying of the cobalt matrix with aluminum. Alloying of cobalt with nickel induced recrystallization of the fibers in addition to the formation of intermetallic compounds. The tungsten fiber was not recrystallized in most of the iron binary alloys studied. There was no interaction of any kind at 1200° C (2190° F) with alloys of iron and aluminum, silicon, titanium (10 and 25 percent), zirconium, niobium, chromium (10 and 25 percent), molybdenum, and tungsten. An intermetallic layer was formed in most of the iron alloys at the fiber-matrix interface at 1300° C (2370° F). At 1300° C (2370° F) no interaction occurred only in the case of the matrix with 5-25 percent aluminum or 25 percent titanium.

The results of the preceding investigations show that selective alloying additions to nickel-base, cobalt-base and iron-base matrix materials offer a means of suppressing fiber recrystallization and the formation of intermetallic compounds at the fiber-matrix interface. A number of matrix compositions have been identified, particularly for iron base alloys, in which no detectable reaction occurs with tungsten fibers after short time exposures at temperatures up to 1200° C (2190° F).

The selection of matrix composition based on compatibility requires a compromise between two opposing requirements. A strongly bonded interface must be formed between the matrix and fiber to allow efficient stress transfer and maintain continuity during heating-cooling cycles, while at the same time a destructive reaction must be prevented while at the operating temperature. The first requirement implies the initiation of a chemical reaction, while the second requirement involves the prevention of chemical reaction. A continuing reaction at the operating temperature would be acceptable if the rate of reaction is sufficiently slow to give an adequate service life. Studies conducted on the rate of reaction between tungsten fibers and nickel-base, cobalt-base and iron-base alloy matrix materials indicate that compositions can be obtained satisfying the above requirements.

Most of the matrix compositions investigated that resulted in minimum reaction with the fibers involved the formation of intermetallic compounds which served to reduce interdiffusion. Thus, intermetallics might be sought as a natural occurring diffusion barrier. Use of a suitable protective barrier between the fiber and matrix offers the possibility of a wider range of composition selection for composites for high temperature application. However, as pointed out in Ref. 16, the introduction of a second interface and a deposited coating, whose possible breakdown in service at high temperatures would cause a catastrophic decrease in strength, is not an attractive proposition to engine manufacturers and operators. It was pointed out in Ref. 17 that although diffusion barrier coatings on reinforcing wire are a potentially effective way to achieve control of fiber-matrix interaction, techniques attempted to date have not resulted in reproducible, successful barrier coatings for refractory alloy wire. Optimism continues, however, that such natural or deposited coatings are possible and will offer increases in both strength and use temperature. Trade-offs in compound composition and ductility offer a fruitful area for continued studies.

Sufficient evidence has been accumulated on matrix-fiber reactions to project that matrix alloy compositions with chemical compatibility and adequate bonding can be obtained with iron base, nickel base and cobalt base alloy systems. However, as will be discussed later in this paper, several additional requirements must be satisfied in selecting a matrix alloy, including ductility, thermal expansion compatibility, creep strength, strain hardening rate, oxidation and corrosion resistance and ease of fabrication. The basis for selecting matrix compositions to fill the host of requirements is in an early state of development. The approach taken in the NASA-Lewis program has been to select alloys with chemical compatibility to control fiber-matrix reactions and with ductility and low strain hardening to resist cracking from thermally induced cyclic stresses. Other requirements have been given consideration but were subordinated to those two needs. Variation of matrix composition within the composite geometry has also been used to better satisfy the variation in requirements within a component. For example, a higher creep strength material with less thermal strain and lower oxidation resistance at moderate temperatures can be used in the base portion of a blade because it is a better match to fill the requirements in this portion of the component. Matrix alloy selection is a complex activity and much more experience is needed before guidelines can be established. Matrix-fiber reaction is one requirement that can be met, but the challenge is to satisfy simultaneously the many additional needs.

Composite Fabrication

The fabrication of matrix and fibers into a composite with useful properties is one of the most difficult task in developing refractory-wire-reinforced superalloys. Fabrication methods for refractory-wire-superalloy composites must be considered to be in the laboratory phase of development. Production techniques for fabrication of large numbers of specimens for extensive property characterizations have not yet been developed.

Fabrication methods can be classified as either solid phase or liquid phase depending upon the condition of the matrix phase during its penetration into a fibrous bundle. Liquid phase methods consists of casting the molten matrix using investment casting techniques so that the matrix infiltrates the bundle of fibers in the form of parallel stacks or mats. The molten metal must wet the fibers, form a chemical bond and yet be controlled so as not to degrade the fibers by dissolution, reaction or recrystallization. This control is difficult at very high temperatures since liquid metal temperatures of 1500° C (2730° F) are involved and contact time at temperature must be less than 1 minute, Ref. 18. Larger fiber diameters have been used to increase the size of the unreacted wire core and matrix alloy composition have been selected to reduce solute diffusion into the fibers. Liquid phase methods are particularly suited to the preparation of fiber reinforced superalloys for gas turbine application because casting is the basic fabrication technique presently used for turbine blades and vanes. The technology developed for casting cooled blades via shell molds containing silica inserts could be applied directly to the fabrication of a reinforced blade, Ref. 16. Experience has shown, however, that the use of fibers having a diameter less than 0.075 cm (0.030 in.) in composites prepared by liquid metal infiltration leads to displacement of the wires. This is explained by the lack of rigidity of the unsupported span of the wire permitting surface tension to pull the fibers into a tight bundle near the mid-span, Ref. 16. Thus, control of alignment and spatial distribution of fibers at low volume fractions is difficult. A similar technique was attempted in Ref. 19 to cast tungsten fiber reinforced cobalt-base alloy test specimens. It was reported, however, that the matrix composition inside the fiber bundle became richer in tungsten than the matrix surrounding the fiber bundle and there appeared to be separation of the filament bundle from the matrix. An alternate means of using liquid-state technology is to coat the wire with the matrix by passing it as single strand or small bundles rapidly through a molten bath of matrix. The coated wires could then be diffusion bonded in closed dies to form the composite component.

Solid phase processing requires diffusion, which is time-temperature dependent. Solid phase processing temperatures are much lower than liquid phase processing temperatures; diffusion rates are much lower and reaction with the fiber can be less severe. The prerequisite for solid state processing is that the matrix be in either sheet, foil or powder form. Hot pressing or cold pressing followed by sintering is used for consolidation of the matrix and fiber into a composite component.

Use of matrix materials in the form of sheet or foil involves placing the reinforcing fibers between layers of the matrix sheet or foil which are then pressed together. They may be hot pressed or alternately cold pressed followed by diffusion bonding. An example of this type of processing is reported in Ref. 20. One of the most promising methods of manufacture of composite sheet materials is that of vacuum hot rolling, which gives high productivity and enables large sized sheets to be manufactured. In Ref. 21 a study was reported of the processing parameters for the manufacture of composite sheet material by vacuum hot rolling. A pack of several layers of tungsten fibers were placed in alternate layers between matrix alloy sheets; the pack was wrapped up in nickel foil, covered with thin mica sheets; and placed in a carbon steel sheath. The sheath was sealed by gas welding. The resultant pack was evacuated, heated and rolled. On the basis of the results obtained in this study the authors claim that it is possible to select rolling conditions ensuring maximum high-temperature strength in the resultant material, adequate adhesion between the fibers and matrix, freedom from porosity and strong welding between matrix sheets in the composite. A disadvantage of the use of thin sheet or foil is that most high strength superalloys are not available in this form, although economics rather than technical limitations cause the problem.

The powder metallurgy approach is one of the most versatile methods for producing refractory fiber-superalloy composites and has yielded some excellent results. Almost all alloy metals can be produced in powder form. However, the large surface area of the fine powders is easily contaminated and introduces impurities that must be removed. High capital cost equipment is necessary to apply pressure and temperature in an inert atmosphere. Most powder-fabrication techniques limit fiber content to 40–50 volume percent.

Despite these disadvantages, powder processing has been used to achieve control of matrix-fiber reactions and has resulted in excellent composite properties.

Slip casting of metal alloy powders around bundles of fibers followed by sintering and hot pressing was developed for the solid state fabrication of refractory fiber-superalloy composites, Ref. 9. The slipcast slurries consisted of a mixture of powders and an organic gel in water. Slurries were converted in a vibratory mold to a solid "green" composite which was sintered using a heating schedule designed to drive off the organic binder material and sinter the powders and fiber together without oxidizing either. The sintered composite was then isostatically hot pressed to full density. This method is capable of achieving good matrix consolidation and bonding between fiber and matrix without excursions into the liquid metal region which would greatly increase fiber matrix reactions. Although the slip-casting + sintering + hot pressing technique has demonstrated excellent success for uniaxially reinforced specimens, it is not regarded as an ideal method for component fabrication because most applications require some cross-ply fiber orientation, which is not easily accomplished with slip casting. Control of fiber distribution was also found to a problem using the slip casting technique.

In Ref. 22 a fabrication procedure was developed utilizing solid phase processing in which fiber distribution, alignment and fiber-matrix reaction could be accurately controlled. Matrix alloy powders were blended with a small quantity of organic binder (Teflon) and warm rolled into high density sheets. During rolling the Teflon formed an interlocking network of fibers which held the powder particles together. Fiber mats were made by winding the fibers on a drum, and then spraying them with a binder. The fiber array was cut from the drum and flattened to form a fiber mat. Precollimated fibers in mat form were sandwiched between layers of matrix powder sheet and the matrix was densified and extruded between fibers by hot pressing. Fiber-matrix and matrix-matrix metallurgical bonding was achieved while preserving uniform fiber distribution and eliminating any voids. This procedure resulted in the fabrication of a single layer of fibers contained in the matrix material which was termed a monotape. Monotapes can be cut into any shape desired with any orientation of fiber desired and subsequently stacked up and hot pressed into a desired component. This approach appears to be the most viable method currently available for the fabrication of tungsten fiber reinforced superalloy composite components for use in heat engines.

Composite Properties

The principal reason for most of the work on refractory fiber superalloy composites has been to produce a material capable of operation as highly stressed components such as turbine blades in advanced aircraft and industrial gas turbine engines at temperatures of 1100° to 1200° C (2010° to 2190° F) or higher. Such an increase in temperature above the current limit of about 950° C (1740° F) for superalloys would permit higher turbine inlet temperatures, and markedly decreased cooling, thus improving engine performance and efficiency. An increase in blade temperature of 50° C (90° F) over current limits would be considered a significant improvement, Ref. 18. A review of gas turbine blade material property requirements such as presented in Refs. 23-26 indicates creep resistance, stress-rupture strength, low cycle fatigue, thermal fatigue resistance, impact strength, and oxidation resistance as properties of primary concern for turbine blade application. The following section reviews the results obtained for refractory fiber/superalloy composites for the above property areas.

Stress-Rupture Strength

At temperatures of 1100° C (2010° F) and above, a superalloy matrix contributes very little to the rupture strength of the composite compared to the contribution of the refractory fibers. Fiber stress-rupture strength, volume fraction of fiber and the degree of fiber-matrix reaction all control the stress-rupture strength of the composite. Figure 6 is a plot comparing the 100 hour rupture strength at 1093° C (2000° F) for various fibers and composites containing 70 volume percent of these fibers, from studies conducted in Refs. 9, 27, and 28. The matrix composition (Ni-15Cr-25W-2Al-2Ti) was the same for all of the composites and as indicated in the plot, the stronger the fiber the greater the stress-rupture strength of the composite. The effect of fiber content on the stress-rupture strength of a composite is shown in Fig. 7, (Ref. 29). Stress-rupture strength increases linearly as the fiber content increases.

A comparison of the 100 hour rupture strength at 1093° C (2000° F) for some of the composite systems that have been investigated Refs. 9, 14, 19, 27, 28, 29, 30, 31 and 32 are given in Table IV and plotted in Fig. 8. Where possible, comparisons were made for composites containing 40 volume percent fiber. It should be noted that higher values would be obtained for these composite systems if the fiber content was increased. Also shown in the plot are the values for the 100 hour rupture strength for unreinforced alloys and for the strongest commercially available superalloys. The 100 hour stress-rupture strength of all of the alloys investigated was substantially increased by the addition of tungsten fibers. All of the 40 volume percent fiber composites had a 100 hour rupture strength greater than that for the strongest commercially available superalloys. The W–Hf–C fiber composite system is the strongest composite system obtained to date. A 40 volume percent W–Hf–C fiber content superalloy composite is over 3 and 1/2 times as strong in rupture for 100 hours at 1100° C (2010° F) as the strongest commercially available superalloys. The composite containing a larger amount of fiber reinforcement (56 volume percent W–1 percent ThO_2 wire) in FeCrAlY also had an impressive stress-rupture strength, over 2 and 1/2 times that for the strongest commercially available superalloys.

The density of these composite materials is greater than that of superalloys and this factor must be taken into consideration. The stresses in turbine blades, for example, are a result of centrifugal loading; therefore, the density of the material is important. A comparison of the specific strength properties of composites and superalloys is therefore significant. Figure 9 is a plot comparing the ratio of the 1100° C (2010° F) 100 hour rupture strength to density for composites and superalloys. Even when density is taken into account, the stronger composites are much superior to the strongest commercially available superalloys. The composite containing 40 volume percent W–Hf–C wire is almost 2 and 1/2 times as strong as the strongest superalloys.

The comparison of stress-rupture strength between composites and superalloys is even more favorable for the composite when long application times are involved. Figure 10 is a plot of stress to rupture versus time to rupture for three different fiber compositions, each having the same matrix material, compared to the strongest superalloys. All of the fiber composite systems are stronger relative to superalloys for rupture in 1000 hours than for rupture in 100 hours at 1093° C (2000° F). The stress (to cause rupture) to density ratio versus time to rupture is plotted in Fig. 11. The

Source: NASA Technical Memorandum 82590, January 1981, 68 pages

specific stress-rupture strength advantage for the composite also increases with time to rupture. The 40 volume percent tungsten fiber composite, for example, has about the same specific (density corrected) strength for rupture in 100 hours compared to superalloys but is almost twice as strong as superalloys for rupture in 1000 hours. For currently required blade lives of 5,000 - 10,000 hours this advantage becomes even greater.

A comparison of the range of values for the 100 hour rupture strength for tungsten fiber reinforced superalloy composites tested at 1093° C (2000° F) with the range for the stronger cast superalloys as a function of temperature is shown in Fig. 12. The strongest TFRS composite has the same rupture strength at 1093° C (2000° F) as does the strongest superalloy at 915° C (1680° F). This represents a material use temperature advantage for the composite of 145° C (320° F) compared to the strongest superalloy. Figure 13 shows the density corrected values for rupture in 100 hours as a function of temperature. When density is taken into consideration the composite has a material use temperature advantage of 110° C (200° F) over the strongest superalloys.

Creep Resistance

The creep-rupture properties of Nimocast 713C reinforced with tungsten or tungsten-5 percent rhenium wire were evaluated and compared with the data determined for vacuum-cast Nimocast 713C in Ref. 31. Typical composite creep curves are shown in Fig. 14 together with a comparative curve for the unreinforced matrix. The creep curves for both materials exhibit the three characteristic stages of creep associated with conventional materials. Essentially, reinforcement reduces the second stage minimum creep rate markedly for a given applied stress due the presence of the more creep resistant fibers. The reduction in minimum creep rates observed on reinforcing Nimocast 713C suggests that the stronger, more creep resistant component, the fiber, controls the creep behavior. The lack of evidence of creep deformation in the matrix of the composite, except at the matrix-fiber interface adjacent to the fracture surface, also suggested that the behavior is controlled by the reinforcement. Similar results were obtained with tungsten-1 percent ThO_2 reinforced Hastelloy X composites, Ref. 33, tungsten-1 percent ThO_2 reinforced FeCrAlY composites, Ref. 14, and with tungsten-nickel composites, Ref. 34.

Fatigue

High-temperature materials in gas turbines are subject to cyclic stresses and strains. These can lead to the development of cracks and failures which conventionally are discussed in three separate groupings, depending on the magnitude and cause of the stresses:

High cycle fatigue

Low cycle fatigue

Thermal fatigue

a. High Cycle Fatigue

High Cycle fatigue failures can occur due to cyclic mechanical loads that lead to small scale yielding (for example, applied stresses which cause only localized plastic deformation while the rest of the structure is loaded in the elastic range). High cycle fatigue is often simulated in the laboratory by testing specimens under cyclic loads, the results being plotted as an S-N curve in which the applied cyclic stress amplitude (S) is plotted as a function of the number of cycles to failure (N). Typically, the number of cycles to failure is greater than 10^5. Rarely are tests carried beyond 5×10^8 cycles, while turbine blades in service may experience 10^{12} or more cycles, Ref. 35.

High cycle fatigue tests have been conducted on W-1 percent ThO$_2$/Hastelloy X composite specimens (Ref. 33). Fatigue tests were performed at 10 to 15 Hz using direct stress, tension-tension, axially loaded specimens. The specimens were cycled from a minimum stress to a selected maximum stress and back to the minimum stress. The load ratio, R, ranged from 0.1 to 0.7 for unreinforced Hastelloy X and 0.1 to 0.2 for the composite. The composite load ratios represented a more severe test. The stress ratio, A, ranged from 0.16-0.82 for unreinforced Hastelloy X and 0.52-0.91 for the composite. Test temperatures were room temperature, 816°, 900° and 980° C (1500°, 1650°, and 1800° F). Composite specimens were unidirectionally reinforced and contained 23, 30, and 37 volume percent fibers. The stress to cause failure in 1x10^6 cycles versus temperature is plotted in Fig. 15. Unreinforced Hastelloy X data are plotted for comparison. The composites were stronger at all temperatures, ranging from 1.2 times as strong at room temperature to 4 times as strong at 980° C (1800° F). The ratio of fatigue strength to ultimate tensile strength for the same materials is plotted in Fig. 16. For all test temperatures, the ratio for the composite was higher than that for the Hastelloy X, indicating that high cycle fatigue resistance of the composite is controlled by the fiber.

The high cycle fatigue strength for W-1 percent ThO$_2$/FeCrAlY composites was determined at 760° and 1038° C (1400° and 1900° F) in Ref. 36. Fatigue tests were performed at 30 Hz using direct tension-tension, axially loaded specimens. The specimens were cycled from a minimum stress to a maximum stress and back to the minimum stress. The load ratio, R, was 0.3 and the stress ratio, A, was 0.5. Composite specimens containing 20, 35, and 40 volume percent fibers were tested. Figure 17 is a plot of maximum stress versus the number of cycles to failure for specimens tested at 760° C

(1400° F). The maximum stress versus number of cycles to failure for specimens tested at 1038° C (1900° F) is plotted in Fig. 18. The results again indicate that fatigue is controlled by the fiber. The composite containing 40 volume percent fiber had a 1×10^6 cycle fatigue strength to ultimate tensile strength ratio of 0.9, at 760° C (1400° F). Similar ratios for some superalloys range from 0.52 to 0.67 at the same test temperature, Ref. 37. Thus the composites' response to high cycle fatigue appears to be superior to superalloys. Figure 19 is a plot of the 1×10^6 cycle fatigue strength to ultimate tensile strength ratio for some superalloys and the range of values obtained for TFRS composites showing the advantage for the composite. Push/pull and reverse bend fatigue strength data were determined for a W/superalloy composite in Ref. 29. The fatigue strength measured in push/pull tests at 20°, 300°, and 500° C (70°, 570°, and 930° F) was substantially increased by the introduction of 40 volume percent tungsten wires. With cantilever specimens tested in reverse bending, a significant increase in fatigue strength also resulted from the incorporation of tungsten wires.

b. Low Cycle Fatigue

Low cycle fatigue is characterized by high cyclic loads that lead to failure after a relatively small number of load cycles, usually less than 10^4 to 10^5. Fracture under low cycle fatigue conditions is generally associated with largescale yielding.

Limited work has been reported on the low cycle fatigue behavior of refractory fiber/superalloy composites. The low cycle fatigue behavior for tungsten fiber reinforced nickel was determined at room temperature in Ref. 38. Specimens containing 11 to 25 volume percent, 500 micrometer (0.020 in.) diameter, tungsten fibers or 20 to 28 volume percent, 100 micro-

Source: NASA Technical Memorandum 82590, January 1981, 68 pages

meter (0.004 in.) diameter, tungsten fibers were fabricated by a liquid metal infiltration process and tested in fatigue. Specimens containing 8 or 10 volume percent of 300 micrometer (0.012 in.) diameter tungsten fibers were fabricated by a powder and subsequent forging process and also tested in fatigue. Fatigue tests were performed at about 150 Hz using direct stress, tension-tension, axially loaded specimens. Figure 20 is a plot of the ratio of the maximum stress for fatigue failure to ultimate tensile strength for the range of cycles investigated. The observed fatigue ratios shown for the composite specimens were much higher in comparison to some superalloys referenced by the author. The fatigue ratio reported for Nimocast 713C for 10^8 cycles was 0.24 and for Incoloy 901 and Udimet 700 for 10^7 cycles the ratio was 0.14 and 0.17, respectively. As shown in Fig. 20 the fatigue ratio obtained for the composites was greater than 0.65 at 10^6 cycles.

Low cycle fatigue tests were conducted at 760° and 980° C (1400° and 1800° F) on 20 and 35 volume percent W- 1 percent ThO_2 fiber reinforced FeCrAlY composites, Ref. 36. Fatigue tests were performed at 0.65 Hz and a stress cycle of from 5.5 MN/m^2 (0.8 ksi) to a maximum stress. The load ratio, R, was approximately 0.01 and the stress ratio, A, was 1. The low cycle fatigue results are plotted in Fig. 21. The results indicate that the fiber controls low cycle fatigue strength as was the case for high cycle fatigue behavior. The 35 volume percent fiber content specimens had much higher values of fatigue strength versus cycles to failure than did the 20 volume percent fiber specimens. The ratio of fatigue strength to ultimate tensile strength versus cycles to failure is plotted in Fig. 22. Very high values were obtained at both 760° and 980° C (1400° and 1800° F) indicating that the composite has a high resistance to low cycle fatigue in this temperature range.

c. Thermal Fatigue

Thermal fatigue failures are caused by the repeated application of stress that is thermal in origin. Rapid changes in the temperature of the environment can cause transient temperature gradients in turbine engine components. Such temperature gradients give rise to thermal stresses and strains. Thermal fatigue failure is the cracking of materials caused by repeated rapid temperature changes.

Superimposed on stresses generated by temperature gradients, in the case of the composite, are internal stresses caused by the difference in expansion coefficients between the fibers and matrix. The mean coefficient of thermal expansion from room temperature to 1100° C (2010° F) for super-alloys ranges from 15.8 to $19.3x10^6/°C$ (8.8 to $10.7x10^{-6}/°F$) and is approximately $5x10^{-6}/°C$ ($2.7x10^{-6}°F$) for tungsten. Because of the large difference in expansion coefficients between the fiber and matrix and the resulting strains, thermal fatigue is believed to be the most serious limitation on composite usefulness.

A number of investigators have developed analytical methods to calculate the dependence of composite deformation on cyclic, geometric and constituent deformation parameters, Refs. 39 to 41. The results of these calculations illustrate the possible effects of several variables on deformation damage parameters. Because of the difference in expansion coefficients, the matrix is strained in tension upon cooling and in compression upon heating while the fiber is strained in compression upon cooling and tension upon heating.

Figure 23 is a plot showing the stress on the matrix as a result of heating and cooling. During cooling (A to B) the matrix stress increases continuously in tension because the matrix contracts faster than does the

Source: NASA Technical Memorandum 82590, January 1981, 68 pages

fiber. After reaching the lowest temperature, point B, and upon reheating the matrix is strained elastically in compression up to point C and the stress falls linerally. At point D the temperature and stress are high enough to allow matrix creep, and yielding of the matrix occurs. Because the temperature continually increases above point D, creep becomes much easier and the stress on the matrix continuously decreases until the original maximum temperature is achieved, point E. If, upon reaching point E, the temperature is held constant the matrix stress would fall further due to matrix creep with time. The magnitude of the stress on the matrix is dependent upon the volume fraction fiber present in the composite. The larger the volume fraction fiber present in the composite the higher the stress on the matrix and the lower the temperature at which the matrix would yield upon heating. Yielding of the matrix could also occur in tension during cooling. Work reported in Ref. 39 indicates that the hysteresis loop of matrix stress versus temperature caused by plastic deformation of the matrix stabilizes after a few cycles so that a steady-state plastic compression-tension fatigue results when no external stress is present. Total cyclic plastic strain increases by a law of the form, total strain = strain per cycle x number of cycles. The ability of the matrix to accommodate plastic strain thus controls the number of cycles to failure for the composite if plastic deformation of the matrix governs the failure mode of the composite in fatigue.

A similar type plot could be constructed for the stress on the fiber as a function of temperature. The stress on the fiber during cooling would be compressive and upon heating the stress would be tension. At low volume fraction fiber contents the compressive stress on the fiber during cooling could be large enough to cause plastic flow of the fiber as indicated in

Ref. 40. The stress on the matrix and fiber is also dependent upon the
maximum cycling temperature, heating-cooling rate and creep rate. Strain
induced damage increases markedly with increases in the maximum cycling temperature. Decreasing the heating-cooling rate or increasing the creep rate
has the same effect. A decrease in heating-cooling rate increases the range
of plastic strain per cycle, simultaneously decreasing the maximum matrix
stress. Three types of cycling damage have been noted to date: plastic
flow of the fiber in compression, matrix fracturing and fiber-matrix interface debonding.

A number of studies have been conducted on the response of tungsten
fiber/superalloy composites to thermal cycling. Table V compares the data
obtained for several composite systems. Cylindrical specimens of 40 percent
W/Nimocast 258 were cycled between room temperature and 1100° C (2010° F) in
a fluidized bed to obtain rapid heating and cooling, Ref. 29. Metallor-
graphic examination after 400 cycles revealed no apparent damage at the
fiber-matrix interface. Presumably no matrix or fiber cracks were observed
or they would have been reported as such. Cylindrical specimens of 13 percent W/Nimocast 713C were cycled in a fluidized bed in the temperature
ranges shown in Table V, Ref. 16. Cracking occurred after relatively few
cycles with the exception of the specimens cycled from 20° to 600° C (70° to
1110° F). The cracking was not extensive for the 550° to 1050° C (1020° to
1920° F) cycle tests, but was extensive in the 20° to 1050° C (70° to
1920° F) cycle tests. The bond between the fiber and matrix was reported to
be severely degraded by thermal cycling to 1050° C (1920° F). Thermal cycle
tests were conducted on specimens of reinforced sheet material having a
matrix of EI435 (Nichrome) and volume fiber contents of 14, 24 or 35 percent, Ref. 42. The specimens were heated in an electric resistance furnace

for 2.5 minutes up to a temperature of 1100° C (2010° F) followed by a water quench to room temperature. The number of cycles for debonding between the fiber and matrix to occur was determined as a function of fiber content. As shown in Table V the number of cycles for debonding to occur decreased with increasing fiber content. The tests also indicated that one of the inherent requirements of the fiber, from the viewpoint of obtaining the best thermal fatigue response, is a strict uniformity of spacing of the fibers throughout the body of the material. Thermal fatigue failure occurred first where fibers were in contact with one another. Concurrently, where fibers were congested, cracks formed in the matrix along surfaces paralleling the fiber axis. Tests were also conducted on reinforced EI435 sheet material in Ref. 43. The fiber contents investigated were 15 or 32 percent. The specimens were heated by passage of an electric current. Specimens were heated and cooled in 30 seconds in the temperature ranges shown in Table V. Irreversible deformation occurred after cycling for all of the 15 volume fiber content specimens but not for the 32 volume fiber content specimens. During the initial stages of cycling, warpage and bending were observed. A length decrease was observed during the entire test. With an increase of the number of cycles the length and rate of the dimensional change diminished. After 1000 cycles from 600° to 1100° C (1110° to 2010° F) the length of the specimen decreased 20 percent. Heat treatment had a considerable effect on dimensional instability of the composite. As a result of annealing the specimen their propensity for deformation during thermocycling decreased. After annealing at 1100° C (2000° F) for 4 hours specimens cycled from 570° to 1000° C (1050° to 1830° F) for 1000 cycles did not change their shape. Annealing reduced the yield strength of the matrix. The level of stresses arising in the fibers as a consequence of the difference in expansion coefficients was determined by the resistance to plastic deformation of the matrix. With a decrease in the yield strength of the matrix the level of stress on the fiber decreased and the fibers did not plastically deform. Cracks at the interface were observed for specimens containing 15 volume percent fiber while cracks in the matrix between the fibers were observed for specimens containing 32 volume percent fibers. Several different nickel base composite systems were thermally cycled in Ref. 44. Specimens containing 35 or 50 volume percent fiber were heated by passage of an electric current. The specimens were heated to 1093° C (2000° F) in 1 minute and cooled to room temperature in 4 minutes. All of the 35 volume fiber content specimens were warped after 100 cycles, while the 50 volume percent fiber content specimens were not. Two of the nickel base composite materials containing 35 volume percent fiber experienced a decrease in specimen length after 100 cycles. The most ductile matrix materials NiCrAlY showed the least amount of damage after 100 cycles. The 50 volume percent fiber content specimen containing a NiCrAlY matrix was cycled from 427° to 1093° C (800° to 2000° F) for 1000 cycles and experienced internal microcracking in the matrix between the fibers. Specimens containing 30 volume percent W-1 percent ThO_2 fibers in a matrix of FeCrAlY were exposed to 1000 cycles from room temperature to 1204° C (2200° F), Ref. 45. The specimens were heated up to 1204° C (2200° F) in 1 minute and cooled to room temperature in 4 minutes. As shown in Fig. 24, surface roughening occurred, but there was no matrix or fiber cracking after the 1000 cycle exposure.

As indicated in Table V, a composite system has been identified, W-1 ThO_2/FeCrAlY, that can be thermally cycled though a large number of cycles without any apparent damage. With the exception of the 40 percent

W/Nimocast 258 composite system which withstood 400 cycles without any apparent damage, all of the other systems investigated indicated that some type of damage occurred. These systems would be limited to applications where the component would be exposed to very few thermal cycles. Only a limited number of systems have been investigated to date and a need exists to identify other thermal fatigue resistant systems. The results obtained indicate that a ductile matrix which can relieve thermally induced strains by plastic deformation is required for composite thermal fatigue resistance.

Impact Strength

Composite materials must be capable of resisting impact failure from foreign objects or from failed components that may pass through the engine if they are to be considered for use as a blade or vane.

Factors affecting the impact strength of tungsten fiber metal matrix composites were investigated in Ref. 46. Miniature Izod and standard Charpy impact strength data were obtained for a tungsten fiber reinforced nickel base alloy (Ni-25W-15Cr-2Al-2Ti). It has been found that composite properties as measured by the miniature Izod impact test correlate closely with composite properties as measured by various ballistic impact tests and it was concluded that the miniature Izod test is a reasonable screening test for candidate turbine blade and vane materials, Ref. 47. The Izod impact strength of unnotched and notched specimens as a function of fiber content is plotted in Fig. 25 for two test temperatures, (75° and 1000° F). Impact strength decreased with increasing fiber content at the lower temperature, but increased with increasing fiber content at the higher temperature. Figure 26 is a plot of impact strength as a function of temperatures. There is a sharp increase in impact strength for the 60 volume percent unnotched specimen at 260° C (500° F), the ductile-brittle transition temperature

(DBTT) for the fiber. In general unnotched composites had higher impact strength compared to the matrix at temperatures above the DBTT of the fiber and lower impact strength that the matrix below the DBTT of the fiber. The matrix's contribution to impact strength for the composite is most significant at low temperatures while the fiber controls higher temperature impact strength, above 260° C (500° F). The effect of fiber content on notch sensitivity was also determined. The ratio of the composite's notched impact strength per unit area to its unnotched impact strength per unit area is plotted as a function of fiber content in Fig. 27. The notch sensitivity of the composite decreased with increasing fiber content both above and below the DBTT of the fiber. Heat treatment or hot rolling improved the room temperature impact strength of the composite. Heat treatment increased the impact strength of the notched unreinforced matrix by almost four times and nearly doubled the impact strength of a 45 volume percent fiber content composite. Round rolling increased the impact strength of a 56 volume percent fiber content composite by nearly four times. The improved impact strength for the composite was related to improved matrix impact strength.

An additional objective of work conducted in Ref. 46 was to determine if the potential impact resistance of tungsten fiber/superalloy composites was sufficient to warrant their consideration as turbine blade or vane materials. Alloys with miniature Izod impact values less than 1.7 joules (15 in.-lb) have been successfully run as turbine blades (Refs. 48-49). Based on this, the value of 1.7 joules (15 in.-lb) was taken as the minimum value for Izod impact strength to indicate if a material has promise for further evaluation leading to turbine blade use. Figure 28 compares the tungsten fiber/superalloy impact strength values with the minimum standard. At room temperature as fabricated composites containing fiber contents greater than 35 percent did not meet the minimum requirement. Heat treatment and hot working, however, improved the impact strength so that high fiber content composites met the minimum requirement. At 760° C (1400° F), the higher fiber content as-fabricated composites have impact strengths distinctly above the minimum requirement. High Charpy impact strength values were obtained at 1093° C (2000° F), 37.3 J (330 in.-lb) for a 60 volume percent fiber content specimen, implying that most of this strength is maintained to at least 1093° C (2000° F). The impact strength potential for tungsten fiber/superalloy composites thus appears adequate for turbine blade and vane applications.

A further comparison of the room temperature and 760° C (1400° F) miniature Izod impact strength for some superalloys and other composite systems is plotted in Fig. 29. Values for the Inconel-713C, 25 percent W/Nichrome and WI-52 were obtained from Ref. 26 while data for the 25 percent W-1 percent ThO_2/Hastelloy X are from Ref. 47. Inconel-713C and Guy Alloy represent past turbine blade materials while the WI-52 alloy is representative of an older vane material. The room temperature impact strength values obtained for the composites all exceed the minimum standard, which is equal to the value for Guy Alloy, as well as that for the vane material WI-52. At 760° C (1400° F) the 25 percent W/Nichrome composite bent but did not fracture when impacted at 127 J (157 in.lb) at a velocity of 54 cm per second (136 in. per second). At 760° C (1400° F) the 25 percent W-1 percent ThO_2/Hastelloy X composites bent and cracked, but the crack did not extend or completely propagate through the cross section of the specimen. The impact strength of the composites were thus much higher than that obtained for Inconel 713C or WI-52 at 760° C (1400° F) and should be adequate for turbine blade or vane applications.

Oxidation and Corrosion

The gaseous environment in the gas turbine engine is highly oxidizing with oxygen partial pressures of the order of 2-4 atm, Ref. 50. However, this environment also contains significant amounts of combustion product impurities including sulfur from the fuel and alkali salts ingested with the intake air. Under these conditions an accelerated oxidation may be encountered, sometimes, but not always, accompanied by the formation of sulfides within the alloy: this is commonly referred to as hot corrosion.

The basic design of the composite material assumes that the superalloy matrix will provide oxidation resistance, including protection of the tungsten fibers. Superalloys that are used for hot section engine components are oxidation resistant for material operating temperatures up to about 980° C (1800° F). Above a material temperature of 980° C (1800° F) it is necessary to coat or clad the material to provide the required oxidation resistance. Claddings that are used for superalloy oxidation protection such as NiCrAlY and FeCrAlY are oxidation resistant to temperatures above 1090° C (2000° F). These materials may also be considered as the matrix for composites so that the composite would not have to be coated or clad for high temperature oxidation resistance. Preliminary oxidation studies were conducted in Ref. 27 on a nickel base superalloy reinforced with tungsten-1 percent ThO_2 fibers and clad with Inconel. The specimens were exposed in air at 1100° C (2010° F) for times up to 300 hours. Figure 30 is a transverse section of a clad composite specimen exposed for 50 hours at 1093° C (2000° F). The Inconel cladding was oxidized, and a coherent oxide scale formed on the Inconel. Oxidation had not progressed to the composite, and the surface fibers were not affected by the oxidation of the cladding. Composite specimens of W-1 percent ThO_2/FeCrAlY having completely matrix protected fibers were exposed to static air at 1038°, 1093° and 1149° C (1900°, 2000°, and 2100° F) for up to 1000 hours. The weight change in 1000 hours was 0.3 mg/cm^2 for 1038° C (1900° F) and 1.26 mg/cm^2 for 1149° C (2100° F). These values are in agreement with values obtained for the matrix material without any reinforcement, Ref. 51. Oxidation did not progress to the surface fibers.

The oxidation and corrosion resistance of composite materials having exposed fibers is also an important consideration. Although the fibers in the composite would not be designed to be exposed to the engine environment an understanding of the high temperature oxidation and corrosion behavior is desirable in the event of a coating, cladding, or matrix failure during service which could occur, for example, from impact due to foreign objects passing through the engine. Figure 31 illustrates the principal paths for oxidation and corrosion of exposed fibers. Oxidation proceeding perpendicular to the fibers (through the blade or vane thickness) would destroy the exposed fibers, but intervening matrix would prevent oxidation of subsequent layers. Thus only a partial loss of strength would result. Oxidation parallel to the fibers (along the blade or vane span) potentially is more severe, since all the exposed fibers in the cross section potentially could be oxidized along their entire length. However, studies conducted to evaluate oxidation along fibers showed only limited oxidation penetration along the fibers, Refs. 45, 50, and 52. In Ref. 45 it was found that after 10 hours exposure to static air at 1200° C (2190° F), the fibers in W-1 percent ThO$_2$/FeCrAlY were oxidized to a depth of only 2.5 mm (0.1 in.). After 100 hours exposure at 1100° C (2010° F), the fibers in a W/Nimocast 258 composite oxidized to a depth of 1.3 mm (0.05 in.) in static air and to 2.5 mm (0.1 in.) in a low velocity simulated engine exhaust gas stream moving at

1.8 m/s (6 ft/s), Ref. 52. Measurements of the depth of fiber oxidation for end exposed tungsten fibers in oxidized tungsten fiber reinforced Ni-20 percent Cr specimens obtained from photographs presented in Ref. 50 indicate similar values. In Ref. 50, however, considerable distortion and degradation of the matrix surrounding the oxidized tungsten fiber was observed. Typical weight gain-time curves for the oxidation of 40 volume percent tungsten fiber content nickel- 20 percent chromium material tested at 900° and 1000° C (1650° and 1830° F) in 1 atmosphere oxygen are shown in Fig. 32, Ref. 50. While tungsten can form volatile oxides, the large weight gains reflected in these data indicate the possible formation of complex matrix-tungsten oxides.

The hot corrosion behavior of tungsten fiber reinforced Ni-20 percent Cr composite specimens was also examined in Ref. 50, under the following exposure conditions: (a) sulfidation in H_2-10 percent H_2S; (b) pre-sulfidation in H_2-10 percent H_2S followed by oxidation in oxygen; and (c) oxidation in 1 atmosphere of oxygen after precoating with Na_2SO_4. During sulfidation, only the matrix formed sulfides and the fibers remained unaffected. Consequently, presulfidation, although having a dramatic effect on the oxidation of the matrix did not have a damaging effect on the fibers. The presence of sodium sulfate was also not critical. Thus, hot corrosion conditions were not harmful to the tungsten reinforced composites studied, and catastrophic loss of the exposed tungsten fibers did not occur upon exposure to a high-temperature oxidizing environment.

Thermal Conductivity

High thermal conductivity is desirable in a turbine blade material to reduce temperature gradients; this, in turn, results in reduced thermally induced strains that can cause cracks or distortion. In addition, higher thermal conductivity can reduce coolant flow requirements in some impingement cooled blades leading to greater engine efficiency or durability (Ref. 53). High conductivity for some turbine blade applications could result in unacceptable higher disc temperatures. However, proper design could alleviate the problem, Ref. 54. The conductivity of tungsten fiber/superalloy composites is markedly superior to that of superalloys. The thermal conductivity of tungsten is much higher that that for superalloys and the more tungsten added to a composite the greater the conductivity. Thermal conductivity of the composite is greatest in the direction of the fiber axes since there is a continuous path for conduction along the tungsten fibers. Conduction perpendicular to the fiber axes is lower because the heat cannot find a continuous path through tungsten. The thermal conductivity of some tungsten fiber/superalloy composites was determined as a function of temperature in Ref. 53. Figure 33 shows the thermal conductivity values obtained for a composite containing 65 volume percent fibers in a nickel base alloy and tested in the direction of the fiber axis and for a composite containng 50 volume percent fiber and tested in the direction perpendicular to the fiber axis. The longitudinal thermal conductivity is seen to be much higher than the transverse conductivity. Also shown in the figure are values for the matrix materials. The composites have much higher values for thermal conductivity over the entire temperature range. The results reported in Ref. 53 indicate that at blade conditions (30 to 60 volume percent fiber content and 730° C to 1130° C (1350° to 2060° F)) the transverse (through the wall) conductivity of the composite will be 35 to 50 W/mK. The longitudinal (spanwise) conductivity will be about 45 to 65 W/m-K. Typical superalloys have conductivities of about 25 W/m-K at 1030° C (1880° F).

Composite Turbine Blade Material Requirements

Relatively small increases in turbine component use temperature such as 6° C (10° F) are beneficial and result in cost savings in aircraft engine operation. Numerous modifications in engines, including material changes, have been made to further such gains. However, the introduction of a fiber reinforced superalloy component, such as a tungsten fiber reinforced super-alloy turbine blade would represent a large material change. Such a change would require a larger potential gain to justify the considerable effort required to make the application possible.

It has been assumed in the program at NASA-Lewis that a composition selected as a serious turbine component material candidate must achieve at least a 50° C (90° F) use temperature advantage over the best currently used superalloys and the goal of our program is to permit a 100° to 150° C (180° to 270° F) use temperature increase.

The properties of the composite are controlled by the fiber and matrix properties. Therefore, the properties needed in the composite dictate the fiber and matrix materials which may be used. Hence, selecting the composite involves selecting the best compromise in the combination of fiber and matrix alloys used. Specific composite fiber and matrix requirements are as follows:

(a) Fiber Property Requirements

Creep-rupture and mechanical fatigue strength must be adequate to permit at least a 50° C (90° F) composite blade metal temperature advantage over current superalloys.

Toughness must be adequate at operating temperatures to insure the needed foreign object damage (FOD) resistance.

Fiber costs must be low so that blade fabrication costs can be kept acceptably low.

Source: NASA Technical Memorandum 82590, January 1981, 68 pages

(b) <u>Matrix Property Requirements</u>

Compatibility with the fibers is required at fabrication and operating temperatures so that fiber strength is not excessively degraded by inter-diffusion.

Mechanical and thermal fatigue resistance is needed at operating temperatures. Thermal fatigue damage in the matrix can be initiated by the thermal expansion coefficient mismatch between fiber and matrix as well as by large temperature gradients.

Oxidation and hot corrosion resistance are needed in the matrix for use at temperatures up to 1100° C (2010° F). Not only is this a severe temperature in general, but at these levels the fibers lack oxidation resistance and must be protected by the matrix.

Density must be low to help offset the high density of the tungsten fibers. This is a particularly important consideration in aircraft engines where weight must be minimized because high blade densities can lead to high disk weights.

Toughness and ductility must be high at low temperatures because the matrix imparts impact damage resistance to the composite at low temperatures.

Shear creep strength must be adequate to allow fiber angle plying for airfoil chordwise strength and torsional strength.

First Generation Composite Turbine Blade Material Selection

Tungsten-fiber/FeCrAlY was identified as a promising first generation turbine blade material because of the excellent combination of complementary properties possible with this combination of fiber and matrix, Ref. 55. The matrix provides a high melting point, low density, excellent oxidation and hot corrosion resistance, limited fiber-matrix interdiffusion at proposed blade temperatures, as well as excellent ductility to aid in thermal fatigue resistance. The fiber provides high stress-rupture, creep, fatigue, and impact strength along with high thermal conductivity. Properties reported for this material indicate that it has adequate properties for turbine blade use and it could permit turbine blade operating temperatures over 50° C (90° F) greater than those of current directionally solidified (DS) super-alloy blades.

Composite Component Fabrication

Having demonstrated adequate properties for application as a turbine blade material, the next area of consideration is whether complex shapes such as hollow turbine blades can be designed and fabricated from such material and at reasonable cost. The selected composite fabrication techniques must result in a composite whose properties meet those required for application of the composite. The processes must be capable of first producing component shapes to required dimensions, second incorporating both uniaxial and off-axis fiber positioning; third providing uniform matrix cladding to prevent fiber oxidation, and fourth, providing if necessary, for cooling or weight reduction passages. The combined fabrication techniques must also be cost effective and reproducible.

Investment casting has been considered for fabrication of composite blades. However, there are two obstacles to overcome. A way must be found to hold the fibers in positive "angle plied" alignment during infiltration by the molten matrix. Also, small diameter fibers which must be used in hollow blades because of a space limitation are subject to damage caused by fiber-matrix interdiffusion and to displacement during casting. A W-2 per-cent ThO_2/MAR M322 composite JT9D turbine blade was fabricated by a casting technique reported in Ref. 19. Separation of the fiber bundle from the matrix occurred in the concave areas of the blade. Test specimens fabricated by the same process revealed that outer fibers are exposed to higher temperatures than inner fibers. Furthermore, the matrix composition because richer in tungsten which indicated dissolution of the fibers. Given the current state of the art, investment casting would appear to be more suitable for fabrication of uniaxially reinforced solid blades in which large fibers can be used.

Diffusion bonding of monolayer composite plies is currently the most promising, cost effective method of fabrication for a hollow blade. The composite plies consist of aligned tungsten fibers sandwiched between layers of matrix material. This approach has the capability for accurate fiber distribution and alignment; moreover, it limits fiber-matrix interdiffusion during fabrication. This approach also is capable of producing blade shapes that are close to final dimensions; hence only root machining and touch up grinding is needed. Solid B/Al and B/Ti fan and compressor blades have been fabricated using this approach. A solid W-2 percent ThO_2/conventionally cast MAR-M200 prototype airfoil for potential application in an advanced industrial gas turbine engine was also fabricated, Ref. 56, using a similar approach.

The feasibility of fabricating a composite hollow turbine blade was successfully demonstrated in work reported in Ref. 57. A JT9D-7F first stage, convection cooled blade was selected as the model from which a W-1 percent ThO_2/FeCrAlY composite blade was designed. The major purpose of the fabrication effort was not only to demonstrate the feasibility of fabricating a hollow blade but to also demonstrate that design requirements could be met in the fabricated blade. The design features incorporated into the fabricated blade are indicated in Figs. 34 and 35. The external airfoil was identical to that of the current MAR-M200 (D.S.), JT9D-7F blade. However, the airfoil walls were designed to be thinner to reduce the composite blade weight to within 10 percent of the current blade weight and to allow for more efficient air cooling. An impingement cooling insert was also added to improve cooling efficiency. The walls of the airfoil were built up of composite plies containing W-1 percent ThO_2 fibers. The inner and outer plies of the wall consisted of unreinforced FeCrAlY for oxidation resistance. The composite airfoil extends down into the blade root. The matrix used in the airfoil above the root was FeCrAlY for oxidation and thermal fatigue resistance. The matrix used in the airfoil within the root was an alloy optimized for shear strength and thermal fatigue resistance. The root requires high strength superalloy or a composite built-up of plies graded for differing thermal expansion to minimize thermal fatigue problems at the airfoil-root interface. The blade was designed to have a potential 50° C (90° F) use temperature advantage over the current MAR-M-200 (D.S.) blade.

Figure 36 shows the fabrication sequence used to produce the hollow composite turbine blade. A tungsten fiber mat was sandwiched between powder sheets of FeCrAlY which was subsequently hot pressed to form a monotape. The monotape was then cut into the plies necessary to arrive at the final blade dimension. The plies were then stacked around a steel core. Root inserts and outserts could also be stacked around the assembly or could be attached in a secondary fabrication step. The entire assembly was then placed in a refractory metal die, heated and pressed to arrive at the proper airfoil contour. After pressing the steel core was removed by leaching out with an acid. A tip cap was then welded on to the end of the airfoil and an impingement cooling insert was placed in the leached out cavity and brazed to the root of the blade.

Source: NASA Technical Memorandum 82590, January 1981, 68 pages

Figure 37 shows the as fabricated W-1 percent ThO_2/FeCrAlY composite hollow JT9D-7F airfoil containig a bonded on end cap and trailing-edge coolant slots. Figure 38 shows the composite airfoil which was brazed to a high strength superalloy arc root. A cross section of the composite airfoil is shown in Fig. 39. Excellent fiber alignment and fiber distribution was obtained and the fibers were fully protected by a layer of FeCrAlY in the interior and exterior of the airfoil.

Successful fabrication of a hollow composite airfoil has demonstrated that this material can be fabricated into the complex design shapes for hot turbine section components. While components such as vanes or combustion liners have not been fabricated these components are less complex than the blade and can be considered for future programs.

The fabrication process sequence used to produce a hollow composite blade was used in a fabrication cost study (Ref. 58). Fabrication costs were estimated for high technology turbine blades fabricated using three different materials. The same turbine blade configuration, a first stage JT9D-7F blade was used for each material. Directionally solidified eutectic (DSE), an oxide dispersion strengthened superalloy (ODSS), and W-1 percent ThO_2/FeCrAlY blade manufacturing costs were compared with the cost of producing the same blade from a DS superalloy, the current blade material. The relative costs are shown in Fig. 40. The study indicates that W/FeCrAlY manufacturing costs should be competitive with current manufacturing costs of manufacturing this blade by directional solidification of a superalloy – provided the projected manufacturing yields can be realized in actual commercial production of blades.

56

CONCLUDING REMARKS

Exploratory development and material property screening have indicated that tungsten fiber reinforced superalloy composites have considerable potential for application as advanced high temperature turbine engine component materials. A first generation TFRS composite has been selected to serve as a demonstration system to evaluate the merit and problems of this family of composites for turbine engine applications. Based on the data obtained from the development and property screening, thoriated tungsten wire in a FeCrAlY matrix has potential to become a viable candidate for application.

The choice of commercially available thoriated tungsten wire was made despite the appreciable difference in properties obtainable when stronger wire such as W-Hf-C were used because of current cost and availability. The manufacturing technology effort to make the stronger wire available can more readily be justified after TFRS has become better accepted for application. Test parameters can be evaluated equally well with the less expensive, readily available wire.

The choice of FeCrAlY as a matrix is based on the ductility, low strain hardening and oxidation and corrosion resistance of the alloy and good chemical compatibility with tungsten fibers. The oxidation-corrosion resistance of the composite can be achieved by a thin layer on the external and internal surfaces of a component. The ductility and low strain hardening rate of the matrix are vital to resist the plastic deformation that occurs from thermally induced strains from thermal gradient and thermal expansion mismatch between the fiber and matrix. FeCrAlY is not the ideal matrix because its thermal expansion coefficient is much higher than the fibers and its shear strength is low at elevated temperatures. As in the

Source: NASA Technical Memorandum 82590, January 1981, 68 pages

case of the thoriated tungsten wire, however, it is a reasonable choice as a first generation system to indicate that TFRS is a viable material for further improvement.

Use of tungsten/FeCrAlY composite material could permit engine turbine blade temperatures over 50° C (90° F) higher than those possible using conventional superalloys. Moreover, blade fabrication studies have demonstrated the feasibility of producing hollow W/FeCrAlY turbine blades at a cost competitive with DS superalloy blade costs. Still, a great deal of work remains to be done on this material to aid in its transition from laboratory feasibility to rig testing of prototype hardware and then on manufacturing technology and detailed design. Current Lewis Research Center efforts are addressing these problems.

REFERENCES

1. L. H. Amra, L. F. Chamberlin, F. R. Adams, J. G. Tavernelli and G. J. Polanka, "Development of Fabrication Process for Metallic Fibers of Refractory Metal Alloys," NASA CR-72654, 1970.

2. G. W. King, "Development of Wire-Drawing Processes for Refractory-Metal Fibers," NASA CR-120925, 1972.

3. D. W. Petrasek and R. A. Signorelli, "Stress-Rupture and Tensile Properties of Refractory-Metal Wires at 2000° and 2200° F (1093° and 204° C)," NASA TN D-5139, 1969.

4. D. W. Petrasek, "High-Temperature Strength of Refractory-Metal Wires and Consideration for Composite Applications," NASA TN D-6881, 1972.

5. D. W. Petrasek and J. W. Weeton, "Alloying Effect on Tensile Properties and Micro-Structure of Tungsten-Fiber-Reinforced Composites," NASA TN D-1568, 1963.

6. T. Montebano, J. Brett, L. Castleman and L. Seigle, "Nickel Induced Recrystallization of Doped Tungsten," Trans. Metall. Soc. AIME, 242, 1973-1979 (1968).

7. J. Hoffman, S. Hofmann and L. Tillmann, "Recrystallization of Tungsten Fibers in Nickel Matrix Composites," Z. Metallk., 65, 721-726 (1974).

8. H. Gruenling and G. Hofer, "Deferred Recrystallization of Tungsten Wire in Nickel and Nickel-Chromium Matrices, "Z. Werkstofftech., 5 (2), 69-72 (1974).

9. D. W. Petrasek, R. A. Signorelli and J. W. Weeton, "Refractory-Metal-Fiber-Nickel-Base-Alloy Composites for Use at High Temperatures," NASA TN D-4787, 1968.

10. D. M. Karpinos, L. I. Tuchinskii, 1. R. Vishnyakov, L. N. Pereselentseva and L. N. Klimenko, "Effect of Alloying a Nickel Matrix With The Metal Of Reinforcing Fibers On Structural Stability Of Nickel-Tungsten and Nickel-Molybdenum Composite Materials," Fiz. Khim. Obrab. Mater., 6, 107-113 (1972).

11. B. A. Klypin, A. M. Maslov and S. B. Maslenkov, "Effect of Alloying of the Structural Stability of Ni-W Co-W Composite Materials," Metalloved. Term. Obrab. Met., 5, 6-11 (1977). Transl. Met. Sci. Heat Treat. Met., 19 (5-6) 343-348 (1977).

12. V. S. Mirotvorskii and A. A. Ol'shevskii, "Reactions of Tungsten Fibers With Cobalt-Base Matrices," Poroshk. Metall., 7 (187), 57-64 (1978). Transl. Sov. Powder Metall. Met Ceram., 17 (7), 536-541 (1978).

13. V. S. Mirotvorskii and A. A. Ol'shevskii, "Reactions of Thoriated Tungsten Fibers with Iron-Base Powder Matrices," Poroshk. Metall., 7 (163), 46-52 (1976). Transl. Sov. Powder Metall. Met. Ceram., 15 (7), 534-540 (1976).

14. W. D. Brentnall, "Metal Matrix Composites for High Temperature Turbine Blades," TRW-ER-7790-F, TRW, Inc., Cleveland, Ohio, 1976 and NADC-76225-30, Naval Air Development Center, Warminster, Penn., 1976.

15. V. S. Mirotvorskii and A. A. Ol'shevskii,"Interaction of Thoriated Tungsten At 1200°-1600° C With Matrices Based On Various Metals," Metalloved. Term. Obrab. Met., 11, 12-15 (1979). Transl. Met Sci. Heat Treat. Met., 21 (11-12), 826-829 (1980).

16. A. W. H. Morris and A. Burwood-Smith, "Fiber Strengthened Nickel-Base Alloy," High Temperature Turbines, AGARD-CP-73-71, Jan. 1971.

17. R. A. Signorelli, "Review of Status and Potential of Tungsten Wire: Superalloy Composites for Advanced Gas Turbine Engine Blades," NASA TM X-2599, 1972.

18. R. J. E. Glenny, "Fibrous Composites with High Melting-Point Matrices," Proc. Roy. Soc., Ser. A, 319 (1536), 33-44 (1970).

19. I. Ahmad and J. M. Barranco, "Reinforced Cobalt Alloy Composite Eor Turbine Blade Application," SAMPE Q., 8, 38-49 (1977).

20. D. M. Karpinos, L. I. Tuchinskii, L. R. Vishnyakov and V. Ya. Fefer, "Expenditure of Energy in the Free Forging of Reinforced Metal Composites," Poroshk. Metall., 6 (150), 20-26 (1975). Transl. Sov. Powder Metall. Met. Ceram., 14 (6), 447-452 (1975).

21. V. P. Severdenko, A. S. Matusevich and A. E. Piskarev, "Production of Composite Sheet Materials Based On KhN77TYuR and VZh98 Alloys," Poroshk. Metall., 6 (138), 51-54, (1974). Transl. Sov. Powder Metall. Met. Ceram., 13 (6), 476-479 (1974).

22. W. D. Brentnall and I. J. Toth, "Fabrication of Tungsten Wire Reinforced Nickel-Base Alloy Composites. NASA CR-134664, 1974.

23. P. R. Sahm, "Eutectic and Artificial Composite Superalloys," pp. 73-114 in Proceedings of the Third Symposium On High-Temperature Materials In Gas Turbines, Edited by P. R. Sahm, Elsevier Scientific Publishing Co., Amsterdam, 1974.

24. R. J. E. Glenny and B. E. Hopkins, "Gas Turbine Requirements," Philios. Trans., R. Soc. (London), Series A, 282 (1307), 105-118 (1976).

25. W. Endres, "Design Principles of Gas Turbines," in pp. 1-14 in Proceedings of the Third Symposium on High-Temperature Materials in Gas Turbines, Edited by P. R. Sahm, Elsevier Scientific Publishing Co., Amsterdam, 1974.

26. A. R. Stetson, B. Ohnysty, R. J. Akins and W. A. Compton, "Evaluation Of Composite Materials For Gas Turbine Engines," AFML-TR-66-156, Part 1, Air Force Materials Lab., Wright-Patterson AFB, OH, June 28, 1966.

27. D. W. Petrasek and R. A. Signorelli, "Preliminary Evaluation of Tungsten-Alloy-Fiber-Nickel-Base Alloy Composites for Turbojet Engine Applications," NASA TN D-5575, 1970.

28. D. W. Petrasek and R. A. Signorelli, "Stress-Rupture Strength and Microstructural Stability of Tungsten-Hafnium-Carbon-Wire Reinforced Superalloy Composites," NASA TN D-7773, 1974.

29. A. V. Dean, "The Reinforcement of Nickel-Base Alloys with High-Strength Tungsten Wires," J. Inst. Met., 95, 79-86 (1967).

30. V. M. Chubarov, Yu V. Levinskii, S. E. Salibekov, A. F. Trefilov, L. V. Grachev, E. M. Rodin, M. Kh. Levinskaya and L. V. Dvoichenkova, "A Nickel Base Heat Resistant Composite Material," Probl. Prochn., 3, (7), 100-104 (1971). Transl. Strength Mater., 3 (7), 856-859 (1972).

31. A. W. H. Morris and A. Burwood-Smith, "Some Properties of a Fiber-Reinforced Nickel-Base Alloy," Fibre Sci. Technol., 3 (1), 53-78 (1970).

32. G. I. Friedman and J. N. Fleck, "Tungsten Wire-Reinforced Superalloys For 1093° C (2000° F) Turbine Blade Applications," NASA CR-159720, 1979.

33. R. H. Baskey, "Fiber-Reinforced Metallic Composite Materials," AFML-TR-67-196, Air Force Materials Lab., Wright Patterson AFB, OH, 1967.

34. A. Kannappan and H. F. Fischmeister, "High Temperature Stability of Tungsten Fiber Reinforced Nickel Composites," pp. 85-98 in Proceedings of the Fourth Nordic Symposium on High Temperature Materials Phenomena, Vol. II, Physical Metallurgy, Edited by M. Tilli, Helsinki University of Technology, Esbo, 1975.

35. M. O. Speidel, "Fatigue Crack Growth at High Temperatures," pp. 207-255 in Proceedings of the Third Symposium On High-Temperature Materials In Gas Turbines," Edited by P. R. Sahm, Elsevier Scientific Publishing Co., Amsterdam, 1974.

36. J. N. Fleck, "Fabrication of Tungsten-Wire/FeCrAlY - Matrix Composites Specimens," TRW, Inc., Cleveland, OH. TRW Report ER-8076, 1979.

37. C. T. Sims and W. C. Hagel, "The Superalloys. Wiley-Interscience, New York, 1972.

38. N. Nilsen and J. H. Sovik, "Fatigue Of Tungsten Fiber Reinforced Nickel," pp. B51-B54 in Practical Metallic Composites; Proceedings of the Spring Meeting, Institution of Metallurgists, London, 1974.

39. G. Garmong, "Elastic-Plastic Analysis of Deformation Induced By Thermal Stress in Eutectic Composites," Metall. Trans., 5, 2183-2205 (1974).

40. A. A. Baranov and E. V. Yakovleva, "Deformation Of a Composite Material During Thermocycling, II," Probl. Prochn., 8, 50-53 (1975). Transl. Strength Mater., 7 (8), 966-969 (1976).

41. V. V. Gaiduk, A. S. Lavrenko and Yu V. Sukhanov, "Dilatometric Method for Determining Thermal Stresses In Tungsten-Nichrome Composites," Probl. Prochn., 9, 108-111 (1972).

42. G. I. Dudnik, F. P. Banas and B. V. Aleksandrov, "Nature of Failure Of Reinforced Sheets Subjected To Thermal Cycling," Probl. Prochn., 5, 99-100 (1973). Transl. Strength Mater., 5 (1), 106-107 (1973).

43. F. P. Banas, A. A. Baranov and E. V. Yakovleva, "Deformation Of Composite Material During Alternate Heating and Cooling," Probl. Prochn., 6, 82-86 (1975). Transl. Strength Mater., 7 (6), 744-748 (1976).

44. W. D. Brentnall and D. J. Moracz, "Tungsten Wire-Nickel Base Alloy Composite Development," TRW, Inc., Cleveland, OH, TRW ER-7849, NASA CR-135021, 1976.

45. W. D. Brentnall and D. J. Moracz, "Tungsten Wire-Nickel Base Alloy Composite Development," TRW, Inc., Cleveland, OH, TRW ER-7849, NASA CR-135021, 1976.

46. E. A. Winsa and D. W. Petrasek, "Factors Affecting Miniature Izod Impact Strength Of Tungsten-Fiber-Metal-Matrix Composites," NASA TN D-7393, 1973.

47. B. Ohnysty and A. R. Stetson, "Evaluation of Composite Materials For Gas Turbine Engines," AFML-TR-66-156, Part 11, Dec. 1967.

48. R. A. Signorelli, J. R. Johnston and J. W. Weeton, "Preliminary Investigation of Guy Alloy as a Turbojet-Engine Bucket Material for Use at 1650° F," NASA RM E56I19, 1956.

49. W. J. Waters, R. A. Signorelli and J. R. Johnston, "Performance of Two Boron-Modified S-816 Alloys in a Turbojet Engine Operated at 1650° F. NASA Memo 3-3-59E, 1959.

50. M. E. El-Dahshan, D. P. Whittle and J. Stringer, "The Oxidation and Hot Corrosion Behavior of Tungsten-Fiber Reinforced Composites," Oxid. Met., 9, 45-67 (1975).

51. C. S. Wukusick, "The Physical Metallurgy and Oxidation Behavior of Fe-Cr-Al-Y Alloys," GEMP-414, General Electric Co., Cincinnati, OH, 1966.

52. A. V. Dean, "The Reinforcement of Nickel-Base Alloys with High Strength Tungsten Wires," NGTE-R-266, National Gas Turbine Establishment, Pyestock, England, 1965.

53. E. A. Winsa, L. J. Westfall and D. W. Petrasek, "Predicted Inlet Gas Temperatures For Tungsten Fiber Reinforced Superalloy Turbine Blades," NASA TM-73842, 1978.

54. H. J. Gladden, "Air Cooling of Disk of a Solid Integrally Cast Turbine Rotor for an Automotive Gas Turbine," NASA TM X-3471, 1977.

55. D. W. Petrasek, E. A. Winsa, L. J. Westfall and R. A. Signorelli, "Tungsten Fiber Reinforced FeCrAlY-A First Generation Composite Turbine Blade Material," NASA TM-79094, 1979.

56. P. J. Mazzei, G. Vandrunen and M. J. Hakim, "Powder Fabrication of Fibre-Reinforced Superalloy Turbine Blades," pp. SC 7.1-SC 7.16 in AGARD Conference Proceedings No. 200 on Advanced Fabrication Techniques in Powder Metallurgy and Their Economic Implications, Advisory Group for Aerospace Research and Development, Paris, France, 1976.

57. P. Melnyk and J. N. Fleck, "Tungsten Wire/FeCrAlY Matrix Turbine Blade Fabrication Study," NASA CR-159788, 1979.

58. C. F. Barth, D. W. Blake and T. S. Stelson, "Cost Analysis of Advanced Turbine Blade Manufacturing Processes," NASA CR-135203, 1977.

TABLE I. – CHEMICAL COMPOSITION OF WIRE MATERIALS (REFS. 3 AND 4)

Material	Weight percent of component									
	W	Ta	Mo	Nb	Re	Ti	Zr	Hf	ThO$_2$	C
Tungsten alloys										
218CS	99.9	-----	---	---	-----	----	-----	----	----	-----
W–1ThO$_2$	bal	-----	---	---	-----	----	-----	----	0.95	-----
W–2ThO$_2$		-----	---	---	-----	----	-----	----	1.6	-----
W–3Re		-----	---	---	2.79	----	-----	----	----	-----
W–5Re–2ThO$_2$		-----	---	---	4.89	----	-----	----	1.78	-----
W–24Re–2ThO$_2$		-----	---	---	22.54	----	-----	----	1.7	-----
W–Hf–C		-----	---	---	-----	----	-----	0.37	----	0.030
W–Re–Hf–C		-----	---	---	4.1	----	-----	.38	----	.021
Tantalum alloys										
ASTAR 811C	8.2	bal	---	---	-----	1.13	-----	.91	----	.027
Molybdenum alloys										
TZM	----	-----	bal	---	-----	.45	0.085	-----	----	.031
TZC	----	-----	bal	---	-----	1.18	.27	-----	----	.12
Niobium alloys										
FS85	10.44	27.95	---	bal	-----	----	.85	-----	----	.031
AS30	20	-----	---	bal	-----	----	1	-----	----	-----
B88	28.3	-----	---	bal	-----	----	-----	1.94	----	.58

TABLE II. – REPRESENTATIVE PROPERTIES OF REFRACTORY-ALLOY WIRES (REFS. 3 AND 4)

Alloys	Density, gm/cm^3	Wire diameter, mm	Ultimate tensile strength		Stress for 100-hr rupture		Stress/density for 100-hr rupture, cmx10^3
			ksi	MN/m^2	ksi	MN/m^2	
A. 1093° C (2000° F) Data							
Tungsten alloys							
218CS	19.1	0.20	126	869	63	434	234
W–1ThO$_2$	19.1	.20	142	979	77	531	282
W–2ThO$_2$	18.9	.38	173	1193	95	655	356
W–3Re	19.4	.20	214	1475	69	476	249
W–5Re–2ThO$_2$	19.1	.20	176	1213	70	483	254
W–24Re–2ThO$_2$	19.4	.20	211	1455	50	345	183
W–Hf–C	19.4	.38	207	1427	161	1110	584
W–Re–Hf–C	19.4	.38	314	2165	205	1413	744
Tantalum alloys							
ASTAR 811C	16.9	.51	108	745	84	579	351
Molybdenum alloys							
TZM	10.0	.38	113	779	42	290	295
TZC	10.0	.13	125	862	38	262	267
Niobium alloys							
FS85	10.5	.13	66	455	44	303	295
AS30	9.7	.13	61	421	31	214	224
B88	10.2	.51	77	531	48	331	328
B. 1204° C (2200° F) Data							
Tungsten alloys							
218CS	19.1	0.20	108	745	46	317	170
W–1ThO$_2$	19.1	.20	122	841	54	372	198
W–2ThO$_2$	18.9	.38	150	1034	70	483	257
W–3Re	19.4	.20	157	1082	46	317	168
W–5Re–2ThO$_2$	19.1	.20	148	1020	44	303	160
W–24Re–2ThO$_2$	19.4	.20	147	1014	28	193	102
W–Hf–C	19.4	.38	201	1386	111	765	404
W–Re–Hf–C	19.4	.38	281	1937	132	910	480
Tantalum alloys							
ASTAR 811C	16.9	.51	71	490	38	262	157
Molybdenum alloys							
TZM	10.0	.20	77	531	19	131	135
TZC	10.0	.13	79	545	18	124	127
Niobium alloys							
FS85	10.5	.13	40	276	23	159	155
AS30	9.7	.13	33	228	---	----	---
B88	10.2	.51	50	345	28	193	190

TABLE III. – COMPARISON OF FIBER-MATRIX REACTIONS FOR VARIOUS MATRIX MATERIALS (REF. 15)

Annealing temperature		Matrix	No. compositions investigated	Relative no. of cases, percent				
°C	°F			Recrystallization	Intermetallic compound	Diffusion penetration	No. recrystallization	No. reaction
1200	2190	Ni-base	27	93	55	--	7	4
		Co-base	29	10	83	12	90	10
		Fe-base	30	3	30	--	97	70
1300	2370	Ni-base	27	96	63	--	4	4
		Co-base	19	21	84	--	79	10
		Fe-base	30	20	80	3	80	13

TABLE IV. – RUPTURE STRENGTHS AND COMPOSITIONS FOR COMPOSITES AND SUPERALLOYS

A. 100 hr rupture strength at 1100° C (2010° F) for composites and superalloys

Ref.	Alloy	Wire	Wire diam		Vol. %	Density		100 hr rupture strength		Stress–density for 100 hr rupture	
			mm	in.		gm/cc	lb/in^3	MN/m^2	ksi	m	in.
30	ZhS6	VRN tungsten	0.3–0.5	0.012–0.020	40	12.5	0.45	138	20	1125	44 300
29	EPD–16	----------	-------	-----------	--	8.3	0.3	51	7.4	635	25 000
29		tungsten	0.25	0.010	40	12.7	.46	131	19	1040	41 000
31	Nimocast 713C	----------	-------	-----------	--	8.0	0.29	48	7	613	24 000
31		tungsten	1.27	0.050	20	10.3	.37	93	13.5	927	36 500
19	MARM322E	----------	-------	-----------	--	----	----	48	7	----	------
19		W–2%ThO$_2$	0.08	0.003	40	----	----	207	30	----	------
9	Ni,Cr,W,Al,Ti	----------	-------	-----------	--	9.15	0.33	23	3.3	254	10 000
9		218CS (tungsten)	0.38	0.015	40	13.3	.48	138	20	1058	41 700
27		W–2%ThO$_2$	0.38	0.015	40	13.0	0.47	193	28	1513	59 600
28		W–Hf–C	.38	.015	40	13.3	.48	324	47	2491	98 000
14	FeCrAlY	W–1%ThO$_2$	0.38	0.015	56	12.5	0.45	831 hr rupture strength–242 MN/m^2 (35 ksi)		1957	77 000
32	FeCrAlY	W–Hf–C	0.38	0.015	35	11.3	0.41	242	35	2147	84 500

B. Nominal composition of matrix alloys (weight %)

ZhS6	Ni–12.5Cr–4.8Mo–7W–2.5Ti–5Al
EPD–16	Ni–6Al–6Cr–2Mo–11W–1.5Nb
Nimocast 713C	Ni–12.5Cr–6Al–1Ti–4Mo–2Nb–2.5Fe
MARM322E	Co–21.5Cr–25W–10Ni–0.8Ti–3.5Ta
Ni,Cr,W,Ti,Al	Ni–15Cr–25W–2Ti–2Al
FeCrAlY	Fe–24Cr–5Al–1Y

TABLE V. – THERMAL CYCLING DATA FOR TUNGSTEN/SUPERALLOY COMPOSITES

Ref.	Composite material	Heat source	Cycle	No. of cycles	Remarks
28	40 v/o W/Nimocast 258	Fluidized bed	RT–1100° C (2010° F)	400	No apparent damage to interface
16	13 v/o W/Nimocast 713C	Fluidized bed	20°–600° C (70°–1110° F) 550°–1050° C (1020°–1920° F) 20°–1050° C (70°–1920° F)	200 2–12 2–25	No cracks Cracks at interface Cracks at interface
42	W/EI435 (14, 24, and 35 v/o)	Electric resistance furance	RT–1100° C (2010° F) 2.5 min to temp. Water quench	100	No. of cycles for fiber-matrix debonding 14 v/o – 90 to 100 cycles 24 v/o – 60 to 70 cycles 35 v/o – 35 to 50 cycles
43	W/EI435 (15 and 32 v/o)	Self resistance	30 sec to heat and cool 480°–700° C (900°–1290° F) 500°–800° C (930°–1470° F) 530°–900° C (980°–1650° F) 570°–1000° C(1050°–1830° F) 600°–1100° C (1110°–2010° F)	1000	All 15 v/o specimens warped and had a specimen length decrease. Cracks at interface. 35 v/o specimens did not deform externally but matrix cracks between fibers observed.
44	W/NiWCrAlTi (35 and 50 v/o)	Self resistance	1 min to heat 4 min to cool	100	35 v/o warpage and shrinkage 50 v/o no damage
	W/NiCrAlY (35 and 50 v/o)		RT–1093° C (2000° F)	100	35 v/o warpage 50 v/o no damage
	W/2IDA (35 and 50 v/o)			100	35 v/o warpage and shrinkage 50 v/o no damage
	50 v/o W/NiCrAlY		427°–1093° C (800°–2000° F)	1000	Internal microcracking
45	30 v/o W–1%ThO$_2$/FeCrAlY	Passage of electric current	1 min to heat 4 min to cool RT–1204° C (2200° F)	1000	No damage Surface roughening but no cracking

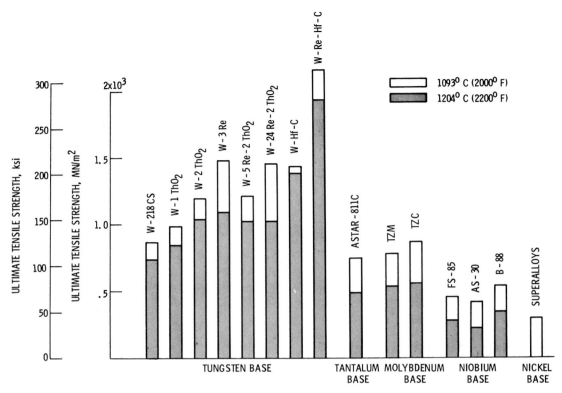

Figure 1. – Ultimate tensile strength for refractory metal wires and superalloys. (Refs. 3 and 4)

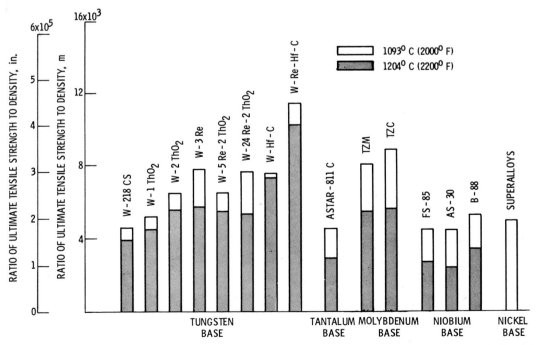

Figure 2. – Ratio of ultimate tensile strength to density for refractory metal wires and superalloys. (Refs. 3 and 4)

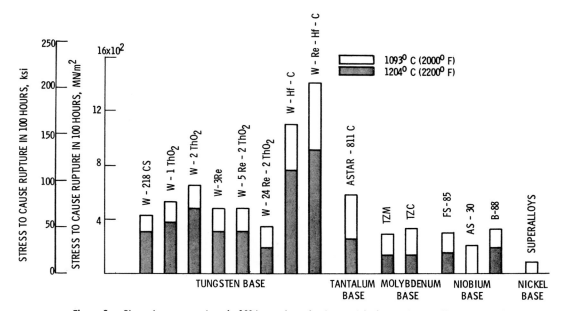

Figure 3. - Stress to cause rupture in 100 hours for refractory metal wires and superalloys. (Refs. 3 and 4)

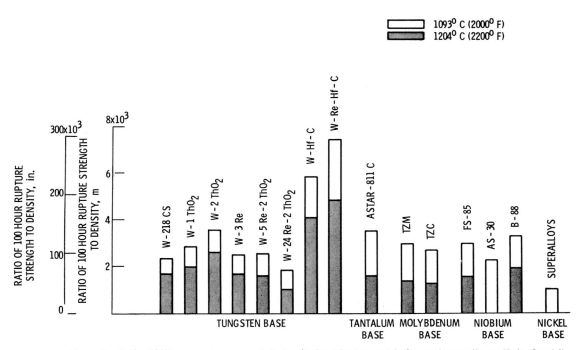

Figure 4. - Ratio of 100 hour rupture strength to density for refractory metal wires and superalloys. (Refs. 3 and 4)

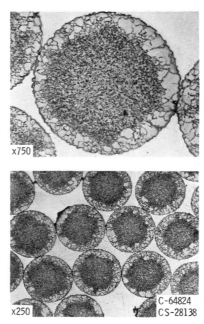

x750

x250

C-64824
CS-28138

Figure 5. - Recrystallization of tungsten
fibers in a copper plus 10 percent nickel
matrix. (Ref. 9)

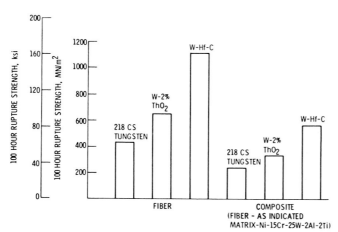

Figure 6. - Comparison of 100 hour rupture strength at 1093° C (2000° F) for fibers and
70 volume percent fiber composites.

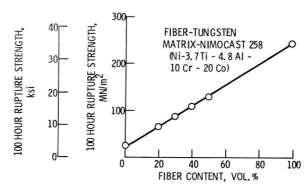

FIBER-TUNGSTEN
MATRIX-NIMOCAST 258
(Ni-3. 7 Ti - 4. 8 Al -
10 Cr - 20 Co)

Figure 7. - Effect of fiber content on 100 hour composite
rupture strength at 1100° C (2010° F). (Ref. 29)

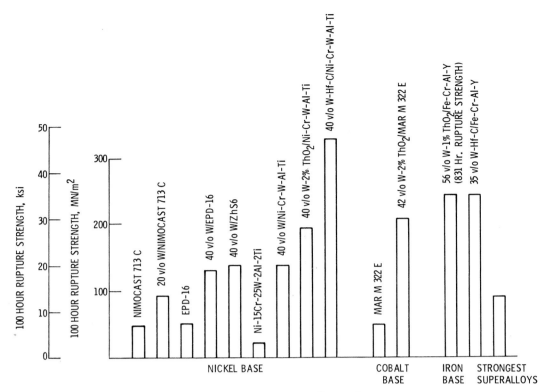

Figure 8. – Comparison of 100 hour rupture strength at 1093° C (2000° F) for composites and superalloys.

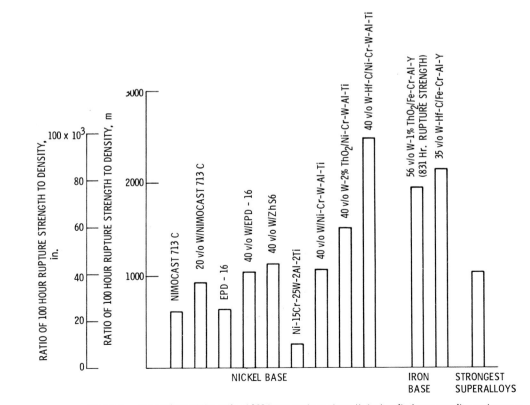

Figure 9. – Comparison of the ratio of 100 hour rupture strength to density for composites and super-alloys at 1093° C (2000° F).

Source: NASA Technical Memorandum 82590, January 1981, 68 pages

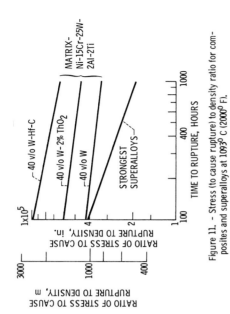

Figure 11. – Stress (to cause rupture) to density ratio for composites and superalloys at 1093° C (2000° F).

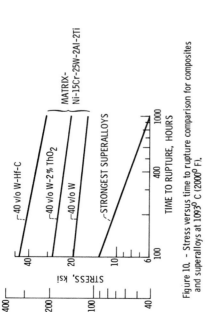

Figure 10. – Stress versus time to rupture comparison for composites and superalloys at 1093° C (2000° F).

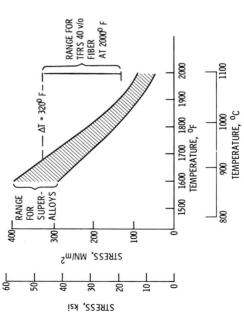

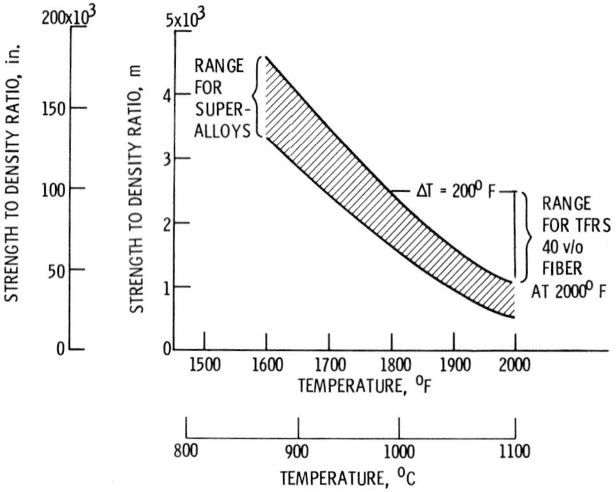

Figure 13. – Comparison of the ratio of 100 hour rupture strength to density for TFRS and superalloys.

Figure 12. – Comparison of 100 hour rupture strength for TFRS and superalloys.

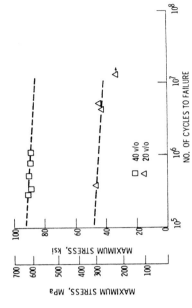

Figure 15. - Stress for failure in 1x10⁶ cycles for Hastelloy X and composite. (Ref. 33)

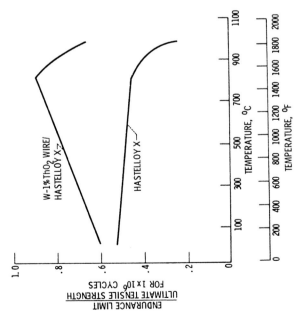

Figure 17. - Stress as a function of number of cycles to failure for W-1%ThO₂/FeCrAlY composites tested at 760° C (1400° F). (Ref. 36)

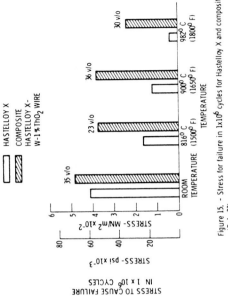

Figure 14. - Comparison of typical creep behaviour of Nimocast 713C with and without tungsten reinforcement at 1100° C (2010° F). (Ref. 31)

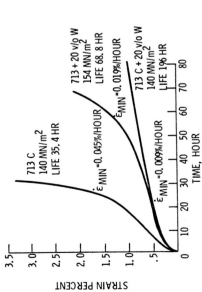

Figure 16. - Ratio of endurance limit to ultimate tensile strength for Hastelloy X and composite tested in axial tension-tension. (Ref. 33)

Source: NASA Technical Memorandum 82590, January 1981, 68 pages

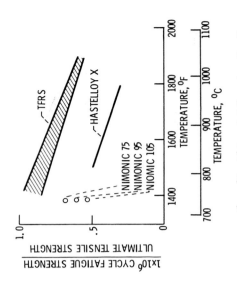

Figure 19. – High-cycle fatigue strength ratio comparison for TFRS and superalloys.

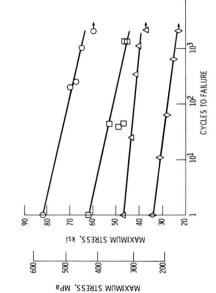

Figure 21. – Maximum stress versus cycles to failure for W-1%ThO$_2$/FeCrAlY composites. (Ref. 36)

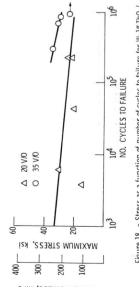
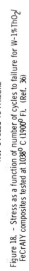
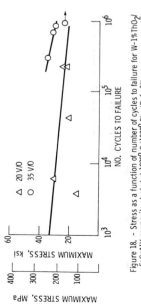

Figure 18. – Stress as a function of number of cycles to failure for W-1%ThO$_2$/FeCrAlY composites tested at 1038° C (1900° F). (Ref. 36)

Figure 20. – Ratio of fatigue strength (σ_F) to ultimate tensile strength (UTS) for Ni/W continuously reinforced composites. (Ref. 38)

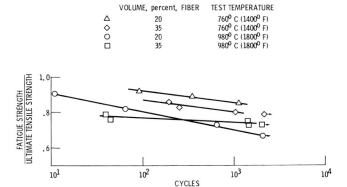

	VOLUME, percent, FIBER	TEST TEMPERATURE
△	20	760° C (1400° F)
◇	35	760° C (1400° F)
○	20	980° C (1800° F)
□	35	980° C (1800° F)

Figure 22. - Ratio of fatigue strength to ultimate tensile strength versus cycles to failure for W-1%ThO₂/ FeCrAlY composites. (Ref. 36)

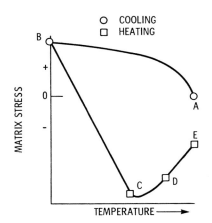

Figure 23. - Matrix stress as a function of heating and cooling.

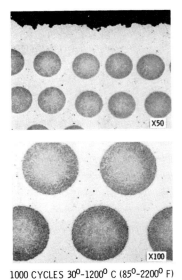

1000 CYCLES 30°-1200° C (85°-2200° F)

Figure 24. - Photomicrographs of thermally cycled tungsten wire reinforced FeCrAlY composite. (Photos courtesy of Irving Machlin). (Ref. 45)

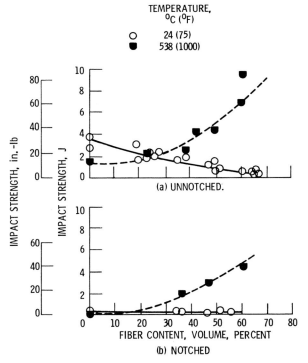

Figure 25. - Impact strength of unnotched and notched as-HIP tungsten/superalloy as a function of fiber content. (Ref. 53)

Source: NASA Technical Memorandum 82590, January 1981, 68 pages

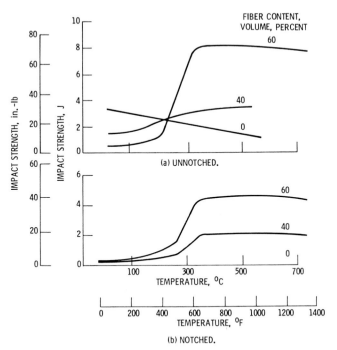

(a) UNNOTCHED.

(b) NOTCHED.

Figure 26. – Impact strength of unnotched and notched tungsten/superalloy as function of temperature and various fiber contents. (Ref. 53)

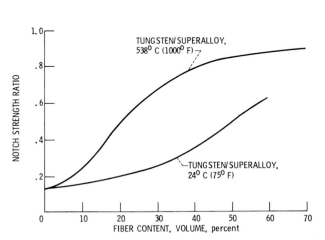

Figure 27. – Notch strength ratio as function of fiber content for tungsten/ metal composites tested at 24 and 538° C (75° and 1000° F). (Ref. 53)

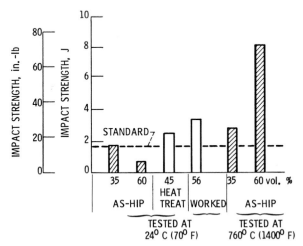

Figure 28. – Miniature Izod impact strengths of unnotched tungsten/nickel-base superalloys compared to minimum impact strength standard used to screen potential turbine blade and vane materials. (Ref. 53)

76

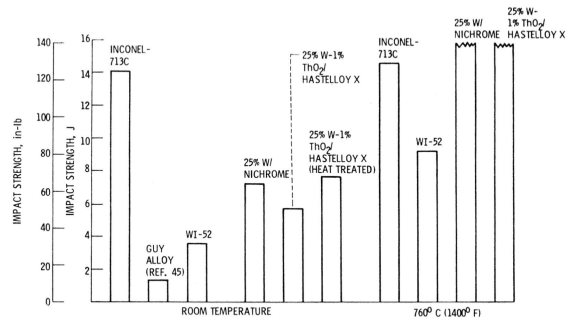

Figure 29. – Miniature Izod impact strength for superalloys and composites.

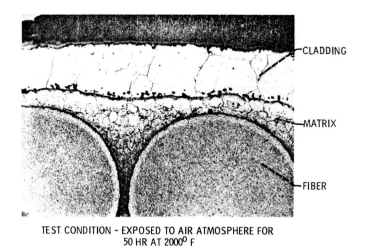

TEST CONDITION – EXPOSED TO AIR ATMOSPHERE FOR
50 HR AT 2000° F

Figure 30. – Transverse section of oxidized refractory fiber –
nickel alloy composite. (Ref. 27)

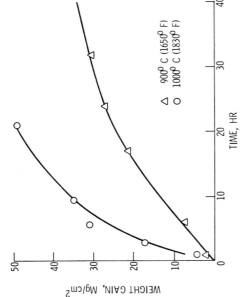

Figure 32. – Weight gain-time curves for the oxidation of W-reinforced Ni-20Cr at 900 and 1000° C (1650 and 1830° F) containing end exposed fibers. (Ref. 50)

Δ 900° C (1650° F)
O 1000° C (1830° F)

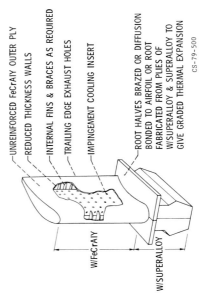

Figure 34. – Schematic of FRS JT9D blade designed for fabrication feasibility experiments.

UNREINFORCED FeCrAlY OUTER PLY
REDUCED THICKNESS WALLS
INTERNAL FINS & BRACES AS REQUIRED
TRAILING EDGE EXHAUST HOLES
IMPINGEMENT COOLING INSERT
ROOT HALVES BRAZED OR DIFFUSION BONDED TO AIRFOIL OR ROOT FABRICATED FROM PLIES OF W/SUPERALLOY & SUPERALLOY TO GIVE GRADED THERMAL EXPANSION

W/FeCrAlY
W/SUPERALLOY

CS-79-500

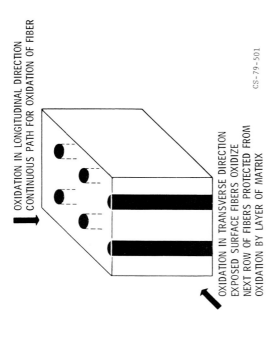

OXIDATION IN LONGITUDINAL DIRECTION
CONTINUOUS PATH FOR OXIDATION OF FIBER

OXIDATION IN TRANSVERSE DIRECTION
EXPOSED SURFACE FIBERS OXIDIZE
NEXT ROW OF FIBERS PROTECTED FROM
OXIDATION BY LAYER OF MATRIX

CS-79-501

Figure 31. – Principal paths for exposed fiber oxidation.

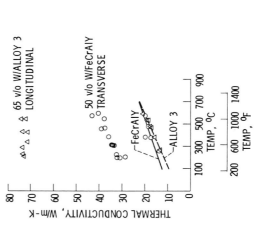

65 v/o W/ALLOY 3
Δ LONGITUDINAL

50 v/o W/FeCrAlY
O TRANSVERSE

FeCrAlY
ALLOY 3

Figure 33. – Measured thermal conductivity of TFRS composites and matrix alloys. (Ref. 53)

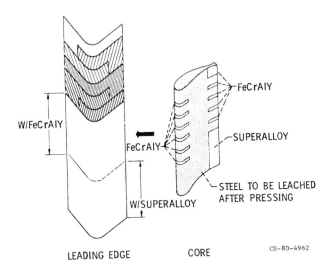

W/FeCrAlY

FeCrAlY

FeCrAlY

W/SUPERALLOY

SUPERALLOY

STEEL TO BE LEACHED
AFTER PRESSING

LEADING EDGE   CORE   CS-80-4962

Figure 35. - Schematic of airfoil ply configuration in JT9D blade.

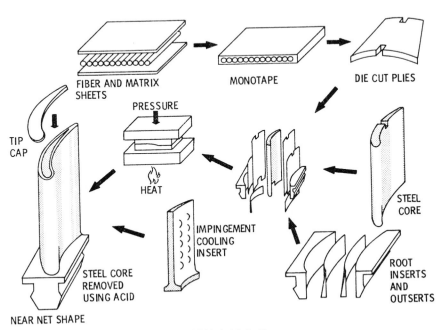

FIBER AND MATRIX
SHEETS

MONOTAPE

DIE CUT PLIES

PRESSURE

HEAT

TIP
CAP

IMPINGEMENT
COOLING
INSERT

STEEL
CORE

ROOT
INSERTS
AND
OUTSERTS

STEEL CORE
REMOVED
USING ACID

NEAR NET SHAPE

Figure 36. - TFRS blade fabrication process.

Source: NASA Technical Memorandum 82590, January 1981, 68 pages

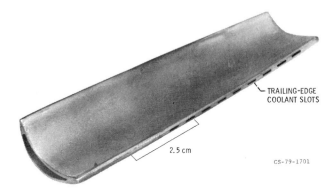

TRAILING-EDGE
COOLANT SLOTS

2.5 cm

CS-79-1701

Figure 37. - Composite hollow airfoil.

Figure 38. - Hollow composite blade.

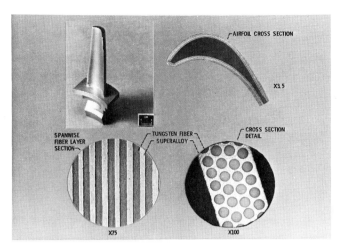

Figure 39. – Tungsten fiber/superalloy composite blade.

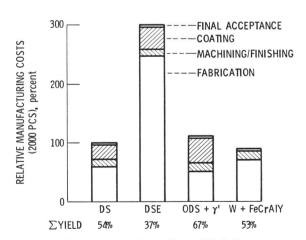

Figure 40. – Cost analysis for JT9D first stage turbine blade. (Ref. 58)

Source: NASA Technical Memorandum 82590, January 1981, 68 pages

HIGH-TEMPERATURE STRENGTH OF REFRACTORY-METAL WIRES

AND CONSIDERATION FOR COMPOSITE APPLICATIONS

by Donald W. Petrasek

Lewis Research Center

SUMMARY

Tensile and stress-rupture tests were conducted on refractory-metal alloy wires at room temperature, 1093° C (2000° F), and 1204° C (2200° F). Wires of W-Hf-C, W-Re-Hf-C, ASTAR 811C (a tantalum-base alloy), and B-88 (a columbium-base alloy) were tested in the 0.038- to 0.051-centimeter- (0.015- to 0.020-in.-) diameter range. The wires were tested in a vacuum of 1×10^{-6} to 5×10^{-5} torr in tension and for rupture. The wire specimens were examined after fracture, and reduction-in-area measurements were made. Metallographic examinations were also made of the wire microstructure after testing.

The highest ultimate tensile strength values obtained were 2170 meganewtons per square meter (314 000 psi) at 1093° C (2000° F) and 1940 meganewtons per square meter (281 000 psi) at 1204° C (2200° F) for W-Re-Hf-C wire. The ultimate tensile strength/density value obtained for the W-Re-Hf-C wire was 11.4×10^{3} meters (450×10^{3} in.) at 1093° C (2000° F), and 10.2×10^{3} meters (400×10^{3} in.) at 1204° C (2200° F). The best strength values obtained for a 100-hour rupture life were 1410 meganewtons per square meter (205 000 psi) at 1093° C (2000° F) and 910 meganewtons per square meter (132 000 psi) at 1204° C (2200° F) for W-Re-Hf-C wire.

The tensile and stress-rupture strengths of the wires investigated were superior to those reported for rod, bar, or sheet forms of refractory metals with the exception of B-88. The superior strengths obtained suggested that the wires studied showed promise as potential fiber reinforcement in the 1093° to 1204° C (2000° to 2200° F) temperature range. These results also suggest that it may be possible to produce W-Re-Hf-C fiber reinforced nickel or cobalt superalloys with over four times the tensile strength and up to ten times the 100-hour rupture strength at 1093° C (2000° F) of the strongest superalloys. Further, it may be possible to produce B-88 columbium-alloy-fiber - superalloy composites with a specific 100-hour rupture strength adequate for turbine blade service at 1093° C (2000° F). A solid turbine blade of B-88 wire-superalloy composite would be 10 to 18 percent heavier than a conventional superalloy blade but may permit a 111° C (200° F) increase in operating temperature.

Reprinted courtesy of NASA, Technical Note D-6881, August 1972, 40 pages

INTRODUCTION

The attractive high-strength potential of fiber composites is largely based on the properties of the fibers. As such, there is a need for fibers with improved properties and for fiber mechanical property data to aid in the selection and design of fiber-reinforced composite materials.

Refractory-metal alloy wires are of interest for fiber reinforcement of superalloy type matrix materials for use between 1093° C (2000° F) and 1204° C (2200° F) because of their high strength at these temperatures. In previous work (refs. 1 and 2) at the Lewis Research Center, composites of refractory-metal-fiber-reinforced, nickel-base alloys were produced that had stress-rupture strengths superior to conventional super-alloys at use temperatures of 1093° C (2000° F) and 1204° C (2200° F). Strength for 1000-hour rupture life as great as six times that for the strongest conventional super-alloys was obtained at 1093° C (2000° F). Even stronger composites would have been possible if higher strength fibers had been available. Research has been sponsored by the Lewis Research Center to fabricate stronger alloys into wire form. Alloys of tungsten, tantalum, and columbium having high creep resistance at temperatures of 1093° C (2000° F) and above were selected. The specific alloys were W-Hf-C, W-Re-Hf-C, ASTAR 811C (a tantalum alloy), and B-88 (a columbium alloy). These alloys were drawn into 0.051- and 0.038-centimeter- (0.020- and 0.015-in.-) diameter wire.

This investigation was conducted to determine the tensile and stress rupture properties of the materials cited above and to assess whether these materials could be considered for applications to metallic composite turbine blades of jet engines. Room-temperature tensile tests and elevated-temperature tensile and stress-rupture tests were conducted at 1093° and 1204° C (2000° and 2200° F). The wires were tested in a vacuum of 1×10^{-6} to 5×10^{-5} torr in tension and for rupture times up to 500 hours. The wire specimens were examined after fracture, and reduction-in-area measurements were made. Metallographic examinations were also made of the wire microstructure after testing to relate the observed structures to measured properties.

MATERIALS, APPARATUS AND PROCEDURE

Wire Material

Alloys of tungsten, tantalum, and columbium having high creep resistance at temperatures of 1093° C (2000° F) and above were selected to be drawn into wire form. The specific alloys selected were W-Re-Hf-C, W-Hf-C, ASTAR 811C (a tantalum alloy), and B-88 (a columbium alloy). These alloys were drawn into fine-diameter wires.

This work was done by two contractors. The fabrication procedure used by the contractors was not optimized to obtain the highest possible wire strength but, rather, was selected to successfully obtain the alloys in wire form.

A chemical analysis of the composition of each alloy is listed in table I. The W-Hf-C and W-Re-Hf-C alloys were drawn into wire in the fully hardened condition (no in-process anneals during working) by one contractor. Alloys ASTAR 811C, B-88, and an additional different lot of W-Hf-C were drawn by a different contractor using annealing treatments during the working process. The wire diameter tested for the W-Hf-C and W-Re-Hf-C alloys was 0.038 centimeter (0.015 in.), while that for B-88 and ASTAR 811C was 0.051 and 0.038 centimeter (0.020 and 0.015 in.). All wire was tested in the as-drawn, cleaned, and straightened condition.

Tensile Tests

The wire was tested in tension in a vacuum chamber at a vacuum of 1×10^{-6} to 5×10^{-5} torr with a constant-strain, screw-driven tensile machine. The strain rate used for all tests was 0.25 centimeter (0.1 in.) per minute. Tensile tests were conducted at room temperature, 1093° C (2000° F), and 1204° C (2200° F).

Stress-Rupture Tests

The equipment used to conduct constant-load, stress-rupture tests is shown in figure 1 and described in detail in reference 3. Wire specimens were cut to 38-centimeter (15-in.) lengths. Each specimen was clamped to a fixed mount, strung through a tantalum-wound resistance furnace, passed over a pulley, and attached to a weight which applied the required load. The weights were supported by retractable supports, while the furnaces and wire test specimens were heated to the test temperature and were stabilized. Microswitches were actuated by fallen weights as each specimen broke, thereby disconnecting power to the furnace and recording the time to fracture. The entire assembly was covered by a cooled metal bell jar. Testing was conducted in a vacuum of 1×10^{-6} to 5×10^{-5} torr. The test temperature was monitored with platinum - platinum-13-percent-rhodium thermocouples. The test temperature did not vary more than $\pm3^{\circ}$ C ($\pm5^{\circ}$ F) during the course of the tests.

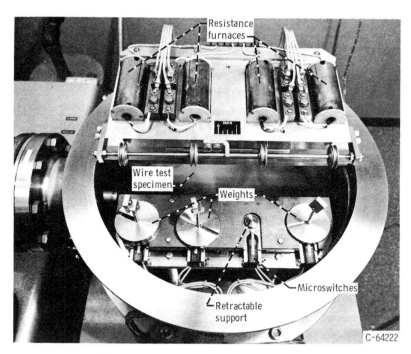

Figure 1. – Fiber stress-rupture testing apparatus.

Reduction-in-Area Measurements

After testing, the fracture area of each tensile and stress-rupture specimen was examined with a microscope at a magnification of 100. Reduction-in-area calculations were based on the difference between the known original wire area and the final area calculated from the average of two diametral measurements taken after testing at a magnification of 100.

Metallographic Analysis

After testing, a longitudinal section of the fracture edges of wire specimens were mounted in epoxy resin and metallographically polished with successively finer grit abrasive papers (to 600 grit size). The specimens were next polished successively with 3- and 0.5-micrometer diamonds, and a slurry of 0.3-micrometer alumina powder and water on a silk cloth.

The specimens were then etched with the solutions listed in the following table:

Source: NASA Technical Note D-6881, August 1972, 40 pages

Wire material	Etch	Method of apply-ing etch
W-Hf-C W-Re-Hf-C	30 cm^3 Lactic 10 cm^3 HF 10 cm^3 HNO_3	Swabbing
ASTAR 811C	60 cm^3 Lactic 40 cm^3 HNO_3 20 cm^3 HF	Immersion
B-88	90 cm^3 Lactic 30 cm^3 HNO_3 15 cm^3 HF $25 \text{ g } NH_4F$	Immersion

After etching, the samples were cleaned for several minutes in an ultrasonic bath of water and were given a final cleaning for 1 minute in an ultrasonic bath of ethyl alcohol. The specimens were then dried in a warm air blast, and the sample surfaces were further cleaned by dry-stripping with a replicating medium.

A two-step technique was used to replicate the wires. The samples were first replicated with 0.025 percent Mowital dissolved in chloroform and, after drying, they were reinforced with 1.5 percent Parlodion in amyl acetate. The two plastic layers were then dry-stripped with pressure-sensitive cellulose tape, shadowed with platinum carbon, and reinforced with a 0.01 micrometer layer of carbon. The replica was cut into grid-size squares and placed into an amyl acetate solution to remove the pressure-sensitive cellulose tape and to dissolve the Parlodion. The Mowital-carbon replica was then viewed in an electron microscope at magnifications of 9000 to 58 000, and photographs were taken at magnifications of 9000 to 12 000.

MECHANICAL PROPERTY DATA

Tensile Properties

The room-temperature, 1093° C (2000° F), and 1204° C (2200° F) tensile properties of the wire materials investigated are listed in table II. The tensile strength of tungsten wire containing 2 percent thoria was previously determined (ref. 4) and is included in the table for comparison.

Room temperature. - The tungsten-alloy wires were superior in tension at room temperature to the tantalum- and columbium-base alloys investigated. The hard-drawn W-Re-Hf-C wire material was the strongest wire material investigated, having a room-temperature tensile strength of 3160 meganewtons per square meter (458 000 psi). It is also interesting to note that the hard-drawn W-Re-Hf-C alloy wire is much more ductile at room temperature than the hard-drawn W-Hf-C alloy wire, having a reduction in area of 27.5 percent as compared with 1.9 percent for the hard-drawn W-Hf-C alloy wire. The tantalum-base alloy (ASTAR 811C) wire was weaker than the tungsten-alloy wires but slightly stronger than the columbium-base alloy (B-88) wire. The room-temperature tensile strength of the tungsten - 2-percent-thoria wire was slightly lower than that of the hard-drawn W-Re-Hf-C wire and the in-process-annealed W-Hf-C wire, and it was higher than that of the other wires investigated.

Elevated temperature. - The tungsten-base alloy wires were much stronger at 1093° and 1204° C (2000° and 2200° F) than either the tantalum- or columbium-base alloy wires studied. The tensile strength of the tantalum-base alloy (ASTAR 811C) wire also compares favorably with the tensile strength obtained for conventional tungsten lamp filament wire in the work of reference 4 (869 MN/m^2 (126 000 psi) at 1093° C (2000° F)). The W-Re-Hf-C wire had the highest tensile strengths of all the wires tested at 1093° and 1204° C (2000° and 2200° F), with 2170 and 1940 meganewtons per square meter (314 000 and 281 000 psi), respectively.

Figure 2 shows the percentage of the room-temperature tensile-strength retention of the various wire materials. The percentage of strength retention was calculated for two test temperatures, 1093° C (2000° F) and 1204° C (2200° F). The tungsten-alloy wires showed the best strength retention, while the columbium-alloy wires showed the poorest strength retention. The strength retention of the ASTAR 811C alloy wire appears good up to 1093° C (2000° F) and compares favorably with some of the tungsten-alloy wires. At 1204° C (2200° F), however, the strength retention of the tantalum-alloy wire was much lower than that of any of the tungsten-alloy wires. The columbium-alloy (B-88) strength retention was the least of the materials tested.

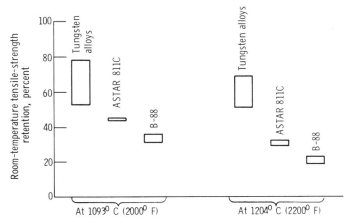

Figure 2. - Percentage of room-temperature tensile-strength retention for wires at two test temperatures.

Source: NASA Technical Note D-6881, August 1972, 40 pages

Stress-Rupture Properties

Stress-rupture strength. - Results of stress-rupture tests on the wire materials studied are presented in table III and plotted as stress to cause rupture against rupture life in figures 3 to 6. Figure 3 is a plot of stress to cause rupture against rupture life for hard-drawn and in-process annealed W-Hf-C wire tested at 1093° and 1204° C (2000° and 2200° F). The rupture strength of the hard-drawn wire is greater than that of the in-process annealed wire at both temperatures. At 1204° C (2200° F), however, there is little superiority in rupture strength. The strongest fiber of all those tested in stress rupture was the hard-drawn W-Re-Hf-C wire. Figure 4 is a plot of the stress to cause rupture against rupture life for the W-Re-Hf-C wire tested at 1093° and 1204° C (2000° and 2200° F). The ASTAR 811C wire had lower stress-rupture strength values than the tungsten wires at both temperatures. Figure 5 is a plot of the stress to cause rupture

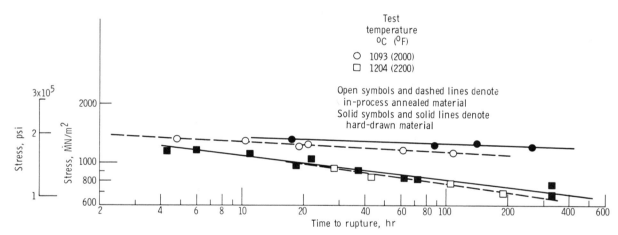

Figure 3. - Time to rupture as function of stress for W-Hf-C wire.

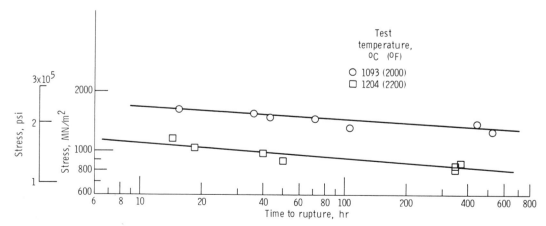

Figure 4. - Time to rupture as function of stress for W-Re-Hf-C wire.

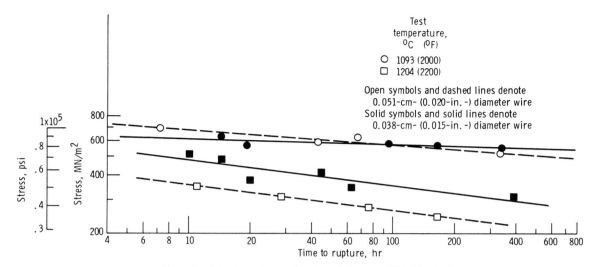

Figure 5. - Time to rupture as function of stress for ASTAR 811C wire.

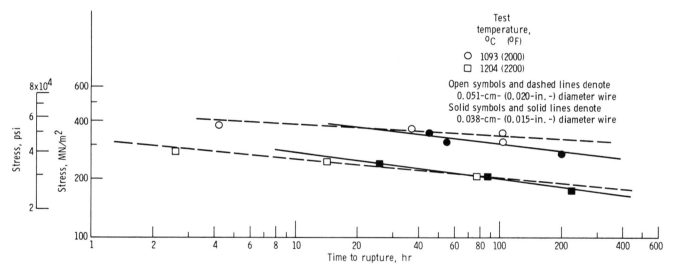

Figure 6. - Time to rupture as function of stress for B-88 wire.

against rupture life for both the 0.51- and 0.038-centimeter- (0.020- and 0.015-in.-) diameter ASTAR 811C wire tested at 1093° and 1204° C (2000° and 2200° F). At 1093° C (2000° F), the stress-rupture strength was nearly equivalent for both wire diameters. At 1204° C (2200° F), the 0.038-centimeter- (0.015-in.-) diameter wire was stronger in stress-rupture than the 0.051-centimeter- (0.020-in.-) diameter wire. A plot of stress to cause rupture against rupture life for 0.051- and 0.038-centimeter- (0.020- and 0.015-in.-) diameter B-88 wire is shown in figure 6. The B-88 wire was the weakest in stress-rupture of all the alloys studied. The 0.051- and 0.038-centimeter- (0.020- and 0.015-in.-) diameter wires had nearly equivalent strength in rupture at both temperatures.

Source: NASA Technical Note D-6881, August 1972, 40 pages

In figure 7, the stress to cause rupture is plotted as a function of rupture life for the wires of all alloys tested at 1093° C (2000° F) so that their relative strengths and the slopes of the curves can be compared. With the exception of the 0.051-centimeter- (0.020-in.-) diameter ASTAR 811C wire and the 0.038-centimeter- (0.015-in.-) diameter B-88 wire, the alloy wire materials appear to have equivalent slopes for the stress-rupture-time curves at 1093° C (2000° F). The same type of plot for the 1204° C (2200° F) stress-rupture tests is shown in figure 8. The W-Re-Hf-C alloy and the 0.51-centimeter- (0.020-in.-) diameter B-88 wire stress-rupture against time curves have the lowest slopes at this temperature.

Stress-rupture ductility. - The elevated-temperature ductility of the wire specimens tested in this investigation was determined by reduction-in-area measurements (table III). Although considerable decrease in stress rupture ductility was obtained for some tungsten alloy specimens, there was no consistent trend. However, a fairly consistent decrease in ductility with time was observed for both the tantalum- and columbium-base alloys.

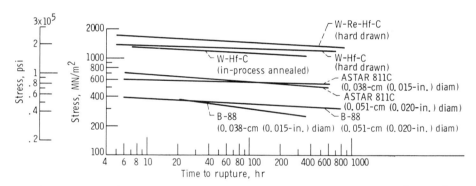

Figure 7. - Comparison of time to rupture as function of stress for wires tested at 1093° C (2000° F).

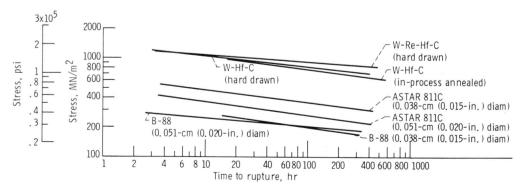

Figure 8. - Comparison of time to rupture as function of stress for wires tested at 1204° C (2200° F).

MICROSTRUCTURAL STUDY

Tungsten-Base Alloys

Table IV provides a summary of the wire microstructure data as a function of exposure time and temperature.

Hard-drawn W-Hf-C. - Figure 9(a) is an electron photomicrograph of a hard-drawn W-Hf-C wire in the as-drawn condition. Heavily worked elongated grains are visible. Occasional large particles were observed. The largest particle observed was 0.05 micrometer. Small particles were evenly distributed throughout the wire and were approximately 0.015 to 0.040 micrometer in size and occupied approximately 1 percent of the area of the wire. The small particles are believed to be precipitates of HfC. The grains increased slightly in width, and a small degree of subgrain formation occurred for the specimen which failed after a long-time exposure at 1093° C (2000° F), as shown in figure 9(b). The small particles observed in this specimen were between 0.020 and 0.060 micrometer in size. Large particles were also observed in this specimen. The largest particle observed was 0.20 micrometer. Grain broadening was also observed for specimens exposed at 1204° C (2200° F), as shown in figures 9(c) and (d). The grains also were not as distinct as those of the specimens in the as-drawn condition and those exposed at 1093° C (2000° F).

In-process annealed W-Hf-C. - Figure 10(a) is an electron photomicrograph of an in-process annealed W-Hf-C wire in the as-drawn condition. Heavily worked elongated grains are visible. Small particles were very sparsely distributed throughout the wire. The particle size ranged from 0.030 to 0.100 micrometer. The particles are believed to be HfC. Increased temperature and time of exposure resulted in grain broadening and a small degree of subgrain formation, as shown in figures 10(b) to (e).

Hard-drawn W-Re-Hf-C. - The microstructure of a hard-drawn W-Re-Hf-C wire in the as-drawn condition is shown in figure 11(a). Small particles were evenly distributed throughout the wire and ranged in size from 0.010 to 0.120 micrometer. Heavily worked elongated grains are visible. Figure 11(b) shows the microstructure of a specimen exposed at 1093° C (2000° F) for approximately 15 hours. The microstructure is similar to the microstructure of the as-drawn specimen. The microstructure of a specimen exposed at 1093° C (2000° F) for approximately 442 hours is shown in figure 11(c). Grain broadening, as well as some subgrain formation, has occurred. The small particles have not increased in size and range between 0.010 to 0.120 micrometer. Figures 11(d) and (e) show the microstructure of specimens exposed at 1204° C (2200° F). Grain broadening and partial recrystallization have occurred.

Source: NASA Technical Note D-6881, August 1972, 40 pages

(a) As drawn.

(b) After 327.4 hours at 1093° C (2000° F).

(c) After 11.1 hours at 1204° C (2200° F).

(d) After 334.1 hours at 1204° C (2200° F).

Figure 9. – Replica electron micrographs of hard-drawn W-Hf-C wire tested at various temperatures. X11 000. (Reduced 50 percent in printing.)

(a) As drawn.

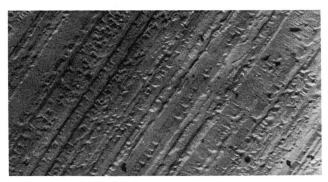

(b) After 4.4 hours at 1093° C (2000° F).

1 μm

(c) After 108.3 hours at 1093° C (2000° F).

Figure 10. – Replica electron micrographs of in-process annealed W-Hf-C wire tested at various temperatures. X12 000. (Reduced 50 percent in printing.)

Source: NASA Technical Note D-6881, August 1972, 40 pages

93

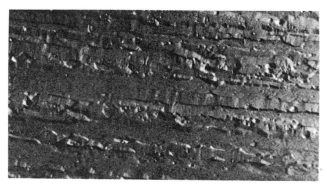

(d) After 28.3 hours at 1204° C (2200° F).

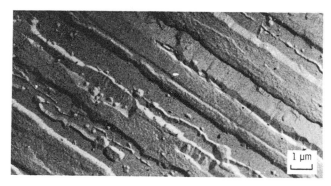

(e) After 188.4 hours at 1204° C (2200° F).

Figure 10. – Concluded.

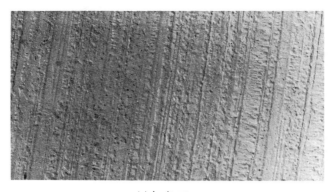

(a) As drawn.

(b) After 15.6 hours at 1093° C (2000° F).

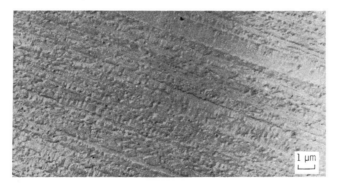

1 μm

(c) After 442.6 hours at 1093° C (2000° F).

Figure 11. – Replica electron micrographs of hard-drawn W-Re-Hf-C wire tested at various temperatures. X9 000. (Reduced 50 percent in printing.)

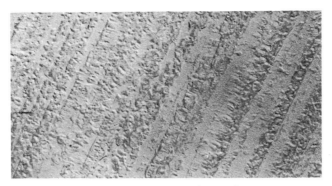

(d) After 18.4 hours at 1204° C (2200° F).

(e) After 365.5 hours at 1204° C (2200° F).

Figure 11. – Concluded.

Tantalum-Base Alloy (ASTAR 811C)

Figures 12(a) to (e) are electron photomicrographs of 0.051-centimeter- (0.020-in.-) diameter ASTAR 811C wire exposed to the indicated temperatures and times. The as-drawn structure of the wire is shown in figure 12(a). Pronounced particle alinement is observed. The dispersed particles range in size from 0.015 to 1.0 micrometer. The precipitate particles are believed to be tantalum dimetal carbide, Ta_2C. Increased temperature and time of exposure resulted in coarsening of precipitates and volume-percent increase of precipitates. Major differences were noticed between the periphery and the center portions of the wires exposed to elevated temperatures. Particles of larger size and in greater abundance were observed at the periphery of the wires.

Figures 13(a) to (e) are electron photomicrographs of 0.038-centimeter- (0.015-in.-) diameter ASTAR 811C wire exposed to the indicated temperatures and times. The microstructures observed were similar to those obtained for the larger diameter wire except for the specimen exposed at 1093° C (2000° F) for 338.2 hours, which had a lower particle content than the large-diameter wire.

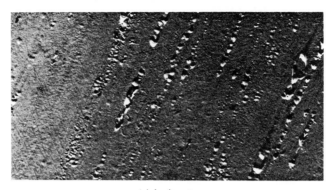

(a) As drawn.

(b) After 7.3 hours at 1093° C (2000° F).

1 μm

(c) After 338.2 hours at 1093° C (2000° F).

Figure 12. – Replica electron micrographs of 0.051-centimeter-
(0.020-in.-) diameter ASTAR 811C wire tested at various temperatures.
X12 000. (Reduced 50 percent in printing.)

Source: NASA Technical Note D-6881, August 1972, 40 pages

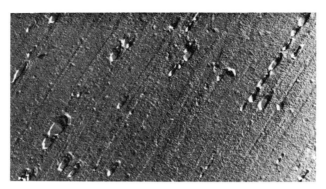

(d) After 10.8 hours at 1204° C (2200° F).

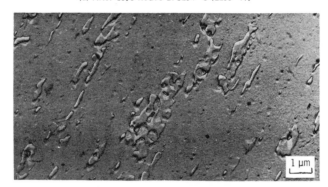

(e) After 166.7 hours at 1204° C (2200° F).

Figure 12. – Concluded.

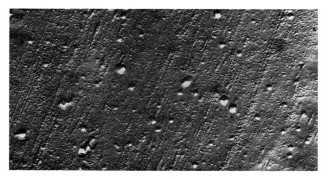

(a) As drawn.

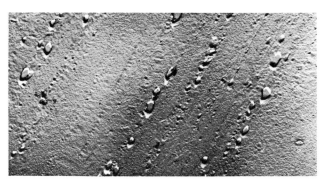

(b) After 14.6 hours at 1093° C (2000° F).

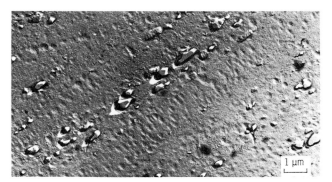

1 μm

(c) After 338.2 hours at 1093° C (2000° F).

Figure 13. – Replica electron micrographs of 0.038-centimeter-
(0.015-in.-) diameter ASTAR 811C wire tested at various temperatures.
X12 000. (Reduced 50 percent in printing.)

(d) After 10.2 hours at 1204° C (2200° F).

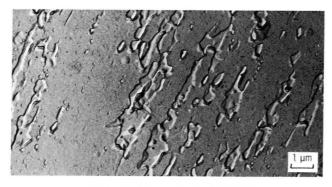

1 μm

(e) After 391.9 hours at 1204° C (2200° F).

Figure 13. – Concluded.

Columbium-Base Alloy (B-88)

Figure 14(a) is an electron photomicrograph of a 0.051-centimeter- (0.020-in.-) diameter B-88 as-drawn wire. Microporosity resulting from what is assumed to be a preferential etching attack of highly worked wires is observed. The microstructure of a specimen exposed to 1093° C (2000° F) for 4.1 hours is shown in figure 14(b). Particles ranging in size from 0.030 to 2 micrometers are seen. The larger particles had a tendency to aline themselves in the longitudinal direction of the wire. The particles are believed to be the face-centered cubic monocarbide, $(Cb, Hf) C_{1-x}$. Figure 14(c) is an electron photomicrograph of a wire specimen exposed to 1093° C (2000° F) for 101.3 hours. The particles were slightly larger and more numerous than in the short-time exposure at 1093° C (2000° F). Structures of specimens exposed at 1204° C (2200° F) for 2.6 and 78.5 hours are shown in figures 14(d) and (e).

Figure 15(a) to (e) are electron photomicrographs of 0.038-centimeter- (0.015-in.-) diameter B-88 wire exposed to the indicated temperatures and times. The microstructures were similar to those obtained for the larger diameter wire with the exception that at 1093° C (2000° F) an apparent loss of fibrous structure is observed and much larger subgrains are formed after long time exposure at 1204° C (2200° F).

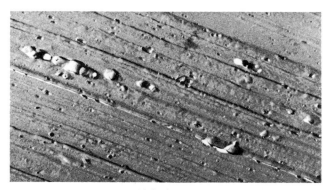

(a) As drawn.

(b) After 4.1 hours at 1093° C (2000° F).

1 μm

(c) After 101.3 hours at 1093° C (2000° F).

Figure 14. - Replica electron micrographs of 0.051-centimeter-
(0.020-in.-) diameter B-88 wire tested at various temperatures.
X12 000. (Reduced 50 percent in printing.)

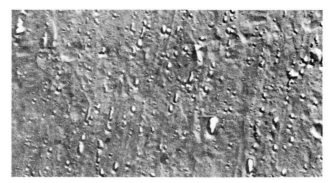

(d) After 2.6 hours at 1204° C (2200° F).

(e) After 78.5 hours at 1204° C (2200° F).

Figure 14. – Concluded.

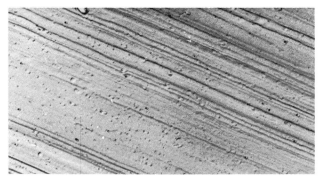

(a) As drawn.

(b) After 44.8 hours at 1093° C (2000° F).

1 μm

(c) After 199.3 hours at 1093° C (2000° F).

Figure 15. – Replica electron micrographs of 0.038-centimeter-
(0.015-in.-) diameter B-88 wire tested at various temperatures.
X12 000. (Reduced 50 percent in printing.)

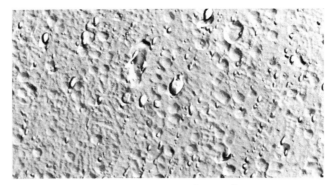

(d) After 25.4 hours at 1204° C (2200° F).

(e) After 224.1 hours at 1204° C (2200° F).

Figure 15. - Concluded.

DISCUSSION

Mechanical Properties

The W-Re-Hf-C wire used in this investigation had a 100-hour rupture strength at 1093° C (2000° F) over twice that of the W - 2-percent-ThO_2 wire which has successfully been utilized to reinforce superalloys (ref. 1). The W-Re-Hf-C wire was found to be the strongest wire material of those tested in stress-rupture at both 1093° and 1204° C (2000° and 2200° F). Table V compares the 100-hour rupture strengths of the wires tested in this investigation with those of the W - 2-percent-ThO_2 wire. This comparison is also shown in figure 16. All of the tungsten alloy wires used in the present investigation were stronger in stress-rupture than the W - 2-percent-ThO_2 wire. The 100-hour rupture strength for the ASTAR 811C wire compares favorably with that of the W - 2-percent-ThO_2 wire while the stress to cause rupture in 100 hours for the columbium-base alloy (B-88) was much lower. However, the densities of the tantalum and columbium alloys are lower than those of the tungsten alloys. Therefore, where the strength-to-density ratio (specific strength) is a criterion for material selection, the

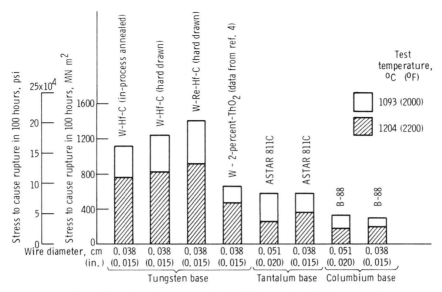

Figure 16. – Stress to cause rupture in 100 hours for refractory metal wires.

tantalum and columbium alloys would compare more favorably with the tungsten alloys. Table V also compares the 100-hour rupture strength divided by the wire density for the wires studied in this investigation with that of W - 2-percent-ThO_2. The strength-to-density values in this report were calculated by using weight-density data and are reported in units of meters (in.). A bar chart showing this comparison is provided in figure 17. The tantalum- and columbium-base wire materials have about the same 100-hour specific rupture strengths as that of the W - 2-percent-ThO_2 wire at 1093° C

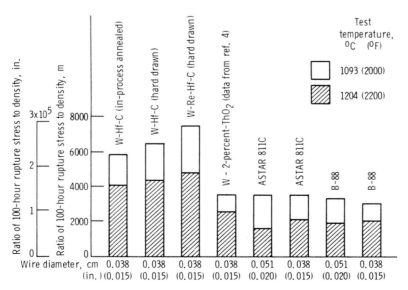

Figure 17. – Ratio of 100-hour rupture strength to density for refractory metal wires.

Source: NASA Technical Note D-6881, August 1972, 40 pages

(2000° F), and they have slightly lower strengths at 1204° C (2200° F). Even when density is taken into account, however, the W-Re-Hf-C and W-Hf-C wires are much superior to the other wire materials.

The ultimate tensile-strength values obtained for the wire materials were found to be much higher than those reported for the same materials in rod, bar, or sheet form. The only unexpected exceptions were the 1093° and 1204° C (2000° and 2200° F) tensile strengths of the B-88 wire, which were lower than the tensile strengths reported for this material in rod form. The ultimate tensile strengths obtained for the wire materials investigated are shown in figure 18. Also shown in figure 18 are the tensile-

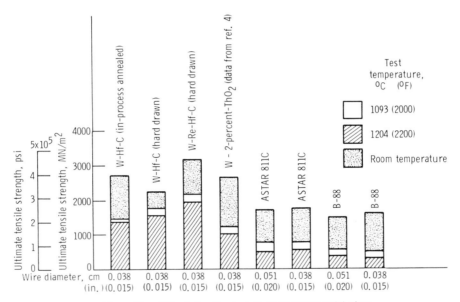

Figure 18. - Ultimate tensile strength of refractory metal wires.

strength values reported in reference 4 for W - 2-percent-ThO$_2$ wire, which, as stated previously, was successfully used as a reinforcement material in composites reported in reference 1. The tungsten-base alloys had the highest ultimate tensile strengths at all the temperatures investigated.

When density is taken into account, the tantalum- and columbium-base wire materials compare considerably more favorably with the tungsten-base alloys, particularly at room temperature, as indicated in figure 19. At elevated temperatures, however, the tungsten-base alloys have superior specific tensile strengths compared to the other wire materials studied.

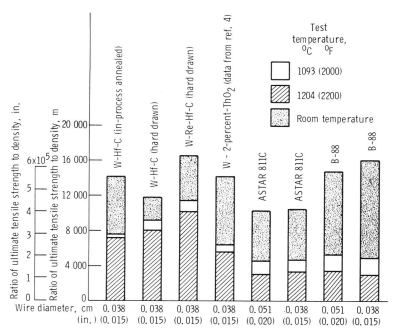

Figure 19. – Ratio of ultimate tensile strength to density of refractory metal wires.

Microstructure

Generally, all wire materials studied had similar microstructural stability and had about the same stabilities in stress rupture, as seen by their equivalent slopes in the plots of stress-to-rupture as a function of time-to-rupture of figures 7 and 8. However, minor concurrent variations in both microstructural and stress-rupture stability were observed for the following few cases. The 0.051-centimeter- (0.020-in.-) diameter ASTAR 811C wire had a steeper slope for stress to rupture against rupture time at 1093° C (2000° F) than did the 0.038-centimeter- (0.015-in.-) diameter wire. At 1093° C (2000° F), the larger diameter ASTAR 811C wire had a less stable microstructure than did the smaller diameter wire, as evidenced by the higher particle content for the 0.051-centimeter- (0.020-in.-) diameter wire for long-time exposure at this temperature. At 1204° C (2200° F), the ASTAR 811C material of both diameters had similar stress-rupture stability and microstructural stability. A difference in stress rupture and microstructural stability was also noted for the B-88 wire material, at both 1093° and 1204° C (2000° and 2200° F). The 0.038-centimeter- (0.015-in.-) diameter B-88 wire was less stable in both stress-rupture and microstructure at both temperatures than the larger diameter wire. At 1093° C (2000° F), the smaller diameter B-88 wire showed an apparent loss of fibrous structure with exposure time, while the larger diameter B-88 wire did not. At 1204° C (2200° F), the 0.038-centimeter- (0.015-in.-)

Source: NASA Technical Note D-6881, August 1972, 40 pages

diameter B-88 wire formed much larger subgrains after long-time exposure than did the larger diameter B-88 wire.

Potential of Refractory Metal Fiber in Superalloy Composites

Refractory metal-alloy wires are of interest for fiber reinforcement of superalloy matrix materials for use between 1093° and 1204° C (2000° and 2200° F). Previous experimental work at Lewis (refs. 1 and 2) has shown that composites of superalloys reinforced with available refractory metal wires can be produced that have stress-rupture strengths superior to conventional superalloys at use temperatures of 1093° and 1204° C (2000° and 2200° F). Composite strength was found to be dependent on fiber properties and on the compatibility of the fiber with the matrix. The stress-rupture properties of the composite could be approximated if the fiber stress-rupture properties were known. One of the objectives of this investigation was to determine the potential of the wire materials investigated as reinforcing fibers for superalloy composites.

Tensile strength. - The potential ultimate tensile strengths of superalloy composites containing wires studied in this investigation were calculated for 1093° C (2000° F) by using the rule-of-mixtures relation. It was assumed that the composite contained 70 volume percent wire and that negligible reaction would occur between the wire and matrix during fabrication or exposure for short times at 1093° C (2000° F). The strongest wire was selected from each wire alloy group. Figures 20 and 21 show the predicted

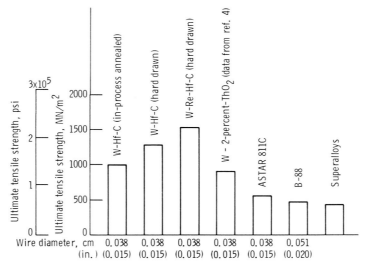

Figure 20. - Potential ultimate tensile strengths of refractory metal wire and superalloy composites at 1093° C (2000° F). Fiber content of composite, 70 volume percent.

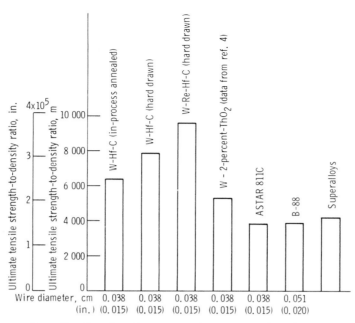

Figure 21. - Potential ultimate tensile strength-to-density ratio of wire-and-superalloy composites at 1093° C (2000° F). Fiber content of composite, 70 volume percent.

1093° C (2000° F) ultimate tensile strength and the specific ultimate tensile strength of composites compared with those of the strongest superalloys. An ultimate tensile strength of over 1380 meganewtons per square meter (200 000 psi) may be projected for a composite containing 70 volume percent W-Re-Hf-C wire. The strongest superalloys have an ultimate tensile strength of approximately 350 meganewtons per square meter (50 000 psi) at this temperature.

Tensile strength-to-density ratio. - The tungsten-alloy wire composite appears to offer the most potential, even when density is taken into consideration, as shown in figure 21. The W-Re-Hf-C wire composite has a potential specific ultimate tensile strength of 9630 meters (379 000 in.) at 1093° C (2000° F) compared to 4240 meters (167 000 in.) for the strongest superalloys. The superalloys have a specific ultimate tensile strength of 9630 meters (379 000 in.) at 871° to 927° C (1600° to 1700° F). The potential specific ultimate tensile strength of the W-Re-Hf-C wire reinforced superalloy composite represents a use-temperature advantage of 165° to 222° C (300° to 400° F) over superalloys. The ASTAR 811C and B-88 wire reinforced superalloy composites have potential ultimate tensile strengths equivalent to those of the strongest superalloys at 1093° C (2000° F).

Stress-rupture strength. - The potential 100-hour rupture strengths at 1093° C (2000° F) were also determined for superalloy composites containing 70 volume percent of refractory metal alloy wire. The potential rupture strengths of the composites were determined for two conditions. The first condition assumed that reaction with the wire and matrix does not occur, which would represent a composite system in which the matrix is insoluble in the wire or in which the wires are coated with a diffusion barrier. The second condition assumed that reaction between the matrix and wire does occur but that 80 percent of the wire strength is retained after exposure at 1093° C (2000° F) for 100 hours. Seventy-four percent of the wire strength was actually retained in stress-rupture for the W - 2-percent-ThO_2 composites investigated in reference 1. When tungsten lamp filament wire was used as a reinforcement, 90 percent of the wire strength in stress-rupture was retained after exposure at 1093° C (2000° F), as reported in reference 2. The 80-percent value used in our calculations thus appears reasonable. The potential 100-hour rupture strength for wire-reinforced superalloy composites is given in figure 22. Also shown in figure 22 are the actual 100-hour rupture strengths

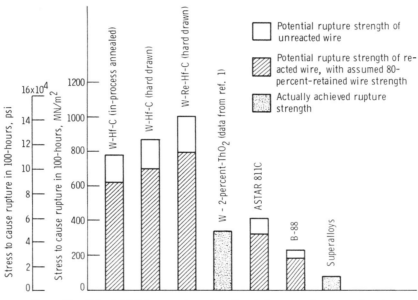

Figure 22. - Potential 100-hour rupture strengths of wire-superalloy composites at 1093° C (2000° F). Fiber content of composites, 70 volume percent.

for superalloys and for a W - 2-percent-ThO_2 wire-and-superalloy composite which was investigated in a previous program at Lewis and which is also reported in reference 1. A 100-hour rupture strength at 1093° C (2000° F) of 993 meganewtons per square meter (144 000 psi) might be obtained for a composite containing W-Re-Hf-C wire if reaction between the wire and superalloy matrix could be avoided. Composites containing either ASTAR 811C or B-88 as a reinforcement would have a 100-hour rupture strength much lower than that obtained with W-Re-Hf-C or W-Hf-C wire even if reaction could be avoided. The 100-hour rupture strength for all the potential wire material would be greater than that obtained for the strongest superalloys at this temperature. If it is assured that reaction with the wire cannot be prevented, the value for the 100-hour rupture life of the composite would be approximated by that shown in figure 22. The W-Re-Hf-C wire composite would have a 100-hour rupture strength of 793 meganewtons per square meter (115 000 psi) compared to a value of 340 meganewtons per square meter (49 000 psi) reported in reference 1 for a tungsten - 2-percent-thoria wire composite. Composites reinforced with ASTAR 811C or B-88 wire would have 100-hour rupture strengths lower than those obtainable for the W-Re-Hf-C or W-Hf-C wire composites but higher than those obtained for superalloys. A potential increase in the 100-hour rupture strength of over 100 percent exists for composites containing W-Re-Hf-C wire as compared with the composites produced in reference 1, which contained tungsten - 2-percent-thoria wire. The potential 100-hour rupture strength of the W-Re-Hf-C wire composite is also approximately 10 times that of the strongest conventional superalloys at this temperature. The composite has a 100-hour stress for rupture value at 1093° C (2000° F) equivalent to that of conventional superalloys at 649° C (1200° F), or a 444° C (800° F) use-temperature advantage.

Stress-rupture strength-to-density ratio. - The densities of the composite materials, however, are much greater than those of superalloys and should be taken into consideration for applications such as turbojet engines where weight is important. The potential 100-hour specific rupture strength for the wire-superalloy composites is presented in figure 23. The two assumptions (unreacted wire and reacted wire) used in plotting figure 22 were also used to construct this figure. When density is taken into consideration, the tantalum- and columbium-base wire materials appear more promising. The potential specific 100-hour rupture strength for the W-Re-Hf-C wire composite, if reaction is assumed, is 5030 meters (198 000 in.) as compared to 2110 meters (83 000 in.) for

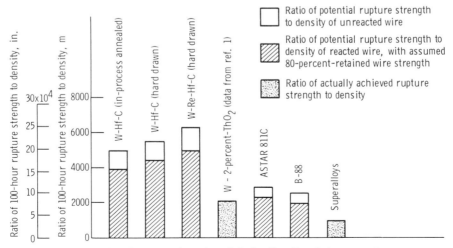

Figure 23. - Potential 100-hour rupture strength to density ratios of wire-superalloy composites at 1093° C (2000° F). Fiber content of composites, 70 volume percent.

Source: NASA Technical Note D-6881, August 1972, 40 pages

the tungsten - 2-percent-thoria wire and superalloy composite reported in reference 1 and 1000 meters (39 000 in.) for the strongest superalloys. The W-Re-Hf-C composite has a potential 100-hour specific rupture strength at 1093° C (2000° F) equivalent to that of conventional superalloys at 871° C (1600° F), or a 222° C (400° F) use-temperature advantage. The ASTAR 811C and B-88 wire composites are equivalent to the tungsten - 2-percent-thoria wire composite on a strength-to-density basis for rupture in 100 hours at 1093° C (2000° F) and have a use-temperature advantage of 111° C (200° F) over the strongest conventional superalloys.

The tungsten - 2-percent-thoria wire and superalloy composite has previously been shown to have (ref. 1) a potential 111° C (200° F) turbine-blade use-temperature advantage over superalloys. The composite contains 70 volume percent fibers, however, and is quite dense. This composite density of 16.1 grams per cubic centimeter (0.58 lb/in.^{3}) is considerably above the 8.30 to 9.13 grams per cubic centimeter (0.30 to 0.33 lb/in.^{3}) of conventional superalloys used for turbine blades. The superior stress-to-density properties of the composite can be used to increase the turbine-blade operating temperature by 111° C (200° F) without increasing blade weight, with an appropriate hollow blade design. However, if a direct substitution of a fixed-external-geometry solid-blade design is used, the 111° C (200° F) increase in temperature for the composite blade will be achieved with about $1\frac{1}{2}$ times as heavy a blade. The density penalty can be reduced for substitutional solid blades by using stronger tungsten alloy fibers and reducing the fiber content of the composite or by using lower density fibers. Figure 24 is a plot of the potential densities for composites having a 100-hour specific rupture strength of 2100 meters (83 000 in.) at 1093° C (2000° F) and the specific 100-hour rupture strength actually obtained for composites containing 70 volume percent of W - 2-percent-thoria

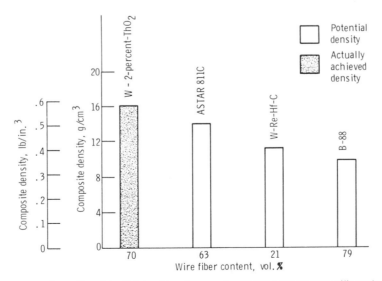

Figure 24. - Potential densities of composites having a 100-hour specific rupture strength of 2100 meters (83 000 in.) at 1093° C (2000° F). Reacted wire with 80-percent-retained strength assumed.

fibers. Fiber reaction is assumed, so that 80 percent of the fiber properties are retained after exposure for 100-hours at 1093° C (2000° F). The fiber content necessary to obtain this rupture strength was determined by the rule-of-mixtures relation assuming that the fibers carry all the load. It is also indicated in the plot. The tungsten - 2-percent-thoria fiber composite has a density of 16.1 grams per cubic centimeter (0.58 lb/in.^{3}). By use of the B-88 fibers, the density of the composite can be lowered to 9.8 grams per cubic centimeter (0.35 lb/in.^{3}) while maintaining the same specific rupture strength for 100 hours as obtained for the tungsten - 2-percent-thoria fiber composites.

Based on the potential composite strengths and densities indicated above, it may be possible to achieve a B-88 columbium alloy wire reinforced superalloy which has the strength-to-density ratio required for 1093° C (2000° F) turbine-blade operation. A solid blade of this composite would permit a 111° C (200° F) service-temperature increase over conventional superalloys, with only a 10- to 18-percent weight increase.

SUMMARY OF RESULTS

The tensile properties of refractory-metal alloy wires of W-Re-Hf-C, W-Hf-C, ASTAR 811C, and B-88 were determined at room temperature, 1093° C (2000° F), and 1204° C (2200° F). Stress-rupture properties were also determined at 1093° C (2000° F) and 1204° C (2200° F) for rupture times up to 100 hours. The rupture properties were correlated to microstructure, and the potential of the wires as a reinforcement for superalloys was determined. The following results were obtained:

1. Tungsten-base alloy wire had a 100-hour rupture strength over twice that of either the tantalum-base wire (ASTAR 811C) or the columbium-base wire (B-88). The W-Re-Hf-C wire had the highest 100-hour rupture strength at both 1093° and 1204° C (2000° and 2200° F), with 1410 and 910 meganewtons per square meter (205 000 and 132 000 psi), respectively. Even when density was taken into consideration, the tungsten-base alloy wire materials were stronger in stress-rupture than either the tantalum- or columbium-base alloy wire materials investigated. Specific 100-hour rupture strengths of 7440 meters (293 000 in.) at 1093° C (2000° F) and 4801 meters (189 000 in.) at 1204° C (2200° F) were obtained for W-Re-Hf-C wire.

2. The ultimate tensile strength values obtained for the wire materials at room temperature, 1093° C (2000° F), and 1204° C (2200° F) were much higher than those reported for rod, bar, or sheet forms of the materials, except for the B-88 wire material.

3. Of the several refractory alloy materials tested, the tungsten-base alloy wire materials studied had the highest ultimate tensile strengths at 1093° C (2000° F) and 1204° C (2200° F). Ultimate tensile strengths of 2170 meganewtons per square meter

Source: NASA Technical Note D-6881, August 1972, 40 pages

(314 000 psi) for W-Re-Hf-C tested at 1093° C (2000° F) and 1440 meganewtons per square meter (281 000 psi) for W-Re-Hf-C wire tested at 1204° C (2200° F) were obtained.

4. All of the wire materials investigated were microstructurally stable with time at temperature. Only minor differences in microstructure were observed. Where minor differences in microstructure were observed, there was a corresponding minor difference in the slope of the rupture stress against time curves.

CONCLUDING REMARKS

The attractive strengths indicated for these wires offer the potential for increased strength if they can be applied to metallic composites. The strength of superalloy matrix composites reinforced with these refractory alloy fibers were calculated. The following potentialities may be realized if these composites can be fabricated:

It may be possible to produce W-Re-Hf-C fiber reinforced nickel- or cobalt-base superalloys with over four times the tensile strength and up to ten times the 100-hour rupture strength at 1093° C (2000° F) of the strongest conventional superalloys. Such a composite material may be applied to hollow turbine blades for operating temperatures in the 1093° to 1204° C (2000° to 2200° F) range.

It may be possible to produce B-88 columbium alloy fiber and superalloy composites with a specific 100-hour rupture strength adequate for turbine blade service at 1093° C (2000° F). A solid turbine blade of B-88 and superalloy composite would be 10 to 18 percent heavier than a conventional superalloy blade but would permit a 111° C (200° F) increase in operating temperature.

Lewis Research Center,
National Aeronautics and Space Administration,
Cleveland, Ohio, May 4, 1972,
134-03.

REFERENCES

1. Petrasek, Donald W.; and Signorelli, Robert A.: Preliminary Evaluation of Tungsten Alloy Fiber - Nickel-Base Alloy Composites for Turbojet Engine Applications. NASA TN D-5575, 1970.

2. Petrasek, Donald W.; Signorelli, Robert A.; and Weeton, John W.: Refractory-Metal-Fiber-Nickel-Base-Alloy Composites for Use at High Temperatures. NASA TN D-4787, 1968.

3. McDanels, David L.; and Signorelli, Robert A.: Stress-Rupture Properties of Tungsten-Wire from 1200° to 2500° F. NASA TN D-3467, 1966.

4. Petrasek, Donald W.; and Signorelli, Robert A.: Stress-Rupture and Tensile Properties of Refractory-Metal Wires at 2000° and 2200° F (1093° and 1204° C). NASA TN D-5139, 1969.

TABLE I. - CHEMICAL ANALYSIS OF MATERIAL COMPOSITION

Wire material	Component element, wt. %							
	C	Cb	Hf	N	O	Re	Ta	W
W-Hf-C (hard drawn)	0.042		0.33					Balance
W-Hf-C (in-process annealed)	.030		.37					Balance
W-Re-Hf-C	.021		.38			4.1		Balance
ASTAR 811C	.027		.91	0.0026	0.0058	1.13	Balance	8.2
B-88	.058	Balance	1.94	.0029	.010			28.3

TABLE II. - TENSILE PROPERTIES OF REFRACTORY METAL WIRES

Wire material	Wire diameter		Test temperature		Ultimate tensile strength		Percent elongation in 2.5 cm (1 in.)	Reduction in area, percent
	cm	in.	°C	°F	MN/m^2	psi		
W-Hf-C (in-process annealed)	0.038	0.015	(a)	(a)	2700	392 000	5.4	21.1
			1093	2000	1430	207 000	---	67.8
			1204	2200	1390	201 000	---	70.9
W-Hf-C (hard drawn)	0.038	0.015	(a)	(a)	2250	326 000	2.8	1.9
			1093	2000	1740	253 000	---	44.2
			1204	2200	1540	224 000	---	46.9
W-Re-Hf-C (hard drawn)	0.038	0.015	(a)	(a)	3160	458 000	4.8	27.5
			1093	2000	2160	314 000	---	24.7
			1204	2200	1940	281 000	---	37.6
ASTAR 811C	0.051	0.020	(a)	(a)	1700	247 000	6.9	51.0
			1093	2000	744	108 000	---	80.8
			1204	2200	490	71 000	---	89.8
	0.038	0.015	(a)	(a)	1740	253 000	5.3	42.9
			1093	2000	779	113 000	---	66.4
			1204	2200	550	80 000	---	66.9
B-88	0.051	0.020	(a)	(a)	1480	215 000	4.8	26.5
			1093	2000	530	77 000	---	87.4
			1204	2200	350	50 000	---	97.9
	0.038	0.015	(a)	(a)	1620	235 000	7.7	54.8
			1093	2000	490	71 000	---	94.5
			1204	2200	310	45 000	---	95.7
W - 2-percent-ThO$_2$ (data from ref. 4)	0.038	0.015	(a)	(a)	2650	384 000	5.5	14.2
			1093	2000	1190	173 000	---	50.2
			1204	2200	1030	150 000	---	51.0

[a]Room temperature.

TABLE III. - STRESS-RUPTURE PROPERTIES OF WIRE MATERIALS

(a) Tungsten-base alloys

Wire material	Wire diameter		Test temperature		Stress		Rupture time, hr	Reduction in area, percent
	cm	in.	^{o}C	^{o}F	MN/m^2	psi		
W-Hf-C (thermally annealed during drawing)	0.038	0.015	1093	2000	1300	189 000	4.4	44.2
					1290	187 000	10.3	58.4
					1230	178 000	21.1	23.2
					1210	175 000	19.1	35.0
					1150	167 000	61.5	44.5
					1110	161 000	108.5	18.0
	0.038	0.015	1204	2200	918	133 000	28.3	15.3
					841	122 000	42.9	21.9
					765	111 000	104.3	11.5
					689	100 000	188.4	28.5
W-Hf-C (hard drawn)	0.038	0.015	1093	2000	1310	190 000	17.7	44.2
					1240	180 000	139.4	37.0
					1230	178 000	88.6	22.6
					1210	175 000	262.0	57.3
					1100	160 000	(a)	----
	0.038	0.015	1204	2200	1170	170 000	6.0	29.4
					1140	165 000	4.3	30.6
					1100	160 000	11.1	20.2
					1040	150 000	22.5	24.9
					965	140 000	18.6	20.2
					896	130 000	37.4	16.6
					827	120 000	63.0	22.6
					793	115 000	74.3	17.8
					758	110 000	334.1	50.2
					689	100 000	329.6	65.8
W-Re-Hf-C (hard drawn)	0.038	0.015	1093	2000	1590	230 000	15.6	15.7
					1520	220 000	36.2	19.5
					1480	215 000	42.8	34.0
					1450	210 000	72.1	27.1
					1380	200 000	442.6	34.7
					1310	190 000	104.2	16.7
					1240	180 000	522.3	37.3
	0.038	0.015	1204	2200	1140	165 000	14.4	32.0
					1040	150 000	18.4	43.2
					965	140 000	39.7	23.0
					896	130 000	49.8	32.8
					862	125 000	365.5	33.7
					827	120 000	345.5	44.3
					793	115 000	342.2	43.2

[a] Test stopped at 233.4 hr.

Source: NASA Technical Note D-6881, August 1972, 40 pages

TABLE III. - Concluded. STRESS-RUPTURE PROPERTIES OF WIRE MATERIALS

(b) Tantalum-base alloy

Wire material	Wire diameter		Test temperature		Stress		Rupture time, hr	Reduction in area, percent
	cm	in.	°C	°F	MN/m^2	psi		
ASTAR 811C	0.051	0.020	1093	2000	690	100 000	7.3	17.5
					620	90 000	68.5	8.7
					590	85 000	43.0	7.0
					520	75 000	338.2	3.8
	0.038	0.015	1093	2000	620	90 000	14.6	29.8
					590	85 000	94.6	4.2
					570	82 000	19.1	19.3
					570	82 000	162.8	6.6
					550	80 000	338.2	7.4
					550	80 000	(b)	----
	0.051	0.020	1204	2200	350	50 000	10.8	8.8
					310	45 000	28.5	9.7
					280	40 000	78.3	8.3
					240	35 000	166.7	7.3
	0.038	0.015	1204	2200	520	75 000	10.2	15.3
					480	70 000	14.7	10.3
					410	60 000	45.4	5.2
					380	55 000	20.1	6.5
					350	50 000	62.7	2.8
					310	45 000	391.9	<1.0

(c) Columbium-base alloy

Wire material	Wire diameter		Test temperature		Stress		Rupture time, hr	Reduction in area, percent
B-88	0.051	0.020	1093	2000	380	55 000	4.1	32.8
					370	53 000	37.1	15.8
					350	50 000	101.3	18.6
					310	45 000	102.1	16.6
	0.038	0.015	1093	2000	350	50 000	44.8	22.1
					310	45 000	55.4	23.3
					280	40 000	199.3	20.7
	0.051	0.020	1204	2200	280	40 000	2.6	34.7
					240	35 000	14.4	39.9
					210	30 000	78.5	20.9
	0.038	0.015	1204	2200	240	35 000	25.4	32.0
					210	30 000	86.8	28.6
					170	25 000	224.1	26.1

[b]Test stopped at 348.9 hr.

TABLE IV. - SUMMARY OF WIRE MICROSTRUCTURE OBSERVATIONS AS FUNCTIONS OF EXPOSURE TIME AND TEMPERATURE

Wire material	As drawn	Condition			
		Exposed at 1093°C (2000°F)		Exposed at 1204°C (2200°F)	
		Short time, <50 hr	Long time, >80 hr	Short time, <50 hr	Long time, >80 hr
W-Hf-C (hard drawn)	Heavily worked elongated grains; HfC particles, 0.015 to 0.040 micrometer in size	-------------	Grain-width increase and subgrain formation	Grain-width increase	Grain-width increase
W-Re-Hf-C (hard drawn)	Heavily worked elongated grains; small particles, 0.010 to 0.120 micrometer in size	Structure similar to as-drawn condition	Grain-width increase and some subgrain formation	Grain-width increase and subgrain formation	Grain-width increase
W-Hf-C (annealed during drawing)	Heavily worked elongated grains; small particles of HfC, 0.030 to 0.100 micrometer in size	Structure similar to as-drawn condition	Grain-width increase	Grain-width increase and subgrain formation	Formation of long wide grains
ASTAR 811C	Pronounced particle alinement; particles range in size from 0.015 to 1.0 micrometer	Structure similar to as-drawn condition	Particle coarsening; higher particle content than for short-time exposure	Particle coarsening	Particle coarsening and higher particle content than for short-time exposure; apparent loss of fibrous structure
B-88	Heavily worked elongated grains; particles range in size from 0.030 to 2 micrometers	Particle coarsening	Higher particle content than for short-time exposure; particle coarsening	Particle-content increase and subgrain formation	Particle coarsening and subgrain formation

Source: NASA Technical Note D-6881, August 1972, 40 pages

TABLE V. - 100-HOUR RUPTURE STRENGTH AND SPECIFIC 100-HOUR RUPTURE STRENGTH OF REFRACTORY METAL WIRES AT 1093° AND 1204° C (2000° AND 2200° F)

(a) Test temperature, 1093° C (2000° F)

Wire material	Approximate wire density		Wire diameter		100-Hour rupture strength		100-Hour rupture strength-to-density	
	gm/cm³	lb/in.³	cm	in.	MN/m²	psi	m	in.
W-Hf-C (hard drawn)	19.37	0.7	0.038	0.015	1240	180 000	6450	257 000
W-Re-Hf-C (hard drawn)	19.37	0.7	0.038	0.015	1410	205 000	7440	293 000
W-Hf-C (in-process annealed)	19.37	0.7	0.038	0.015	1110	161 000	5840	230 000
ASTAR 811C	16.9	0.61	0.051 / 0.038	0.020 / 0.015	580 / 580	84 000 / 84 000	3500 / 3500	138 000 / 138 000
B-88	10.3	0.373	0.051 / 0.038	0.020 / 0.015	330 / 300	48 000 / 44 000	3280 / 3000	129 000 / 118 000
W - 2-percent-ThO₂ (data from ref. 4)	18.91	0.68	0.038	0.015	660	96 000	3560	141 000

(b) Test temperature, 1204° C (2200° F)

Wire material	Approximate wire density		Wire diameter		100-Hour rupture strength		100-Hour rupture strength-to-density	
	gm/cm³	lb/in.³	cm	in.	MN/m²	psi	m	in.
W-Hf-C (hard drawn)	19.37	0.7	0.038	0.015	827	120 000	4340	171 000
W-Re-Hf-C (hard drawn)	19.37	0.7	0.038	0.015	910	132 000	4800	189 000
W-Hf-C (in-process annealed)	19.37	0.7	0.038	0.015	765	111 000	4040	159 000
ASTAR 811C	16.9	0.61	0.051 / 0.038	0.020 / 0.015	260 / 355	38 000 / 51 500	1600 / 2100	62 000 / 84 000
B-88	10.3	0.373	0.051 / 0.038	0.020 / 0.015	190 / 200	28 000 / 29 000	1900 / 2000	75 000 / 78 000
W - 2-percent-ThO₂ (data from ref. 4)	18.91	0.68	0.038	0.015	480	69 000	2570	101 000

SUPER-HYBRID COMPOSITES -- AN EMERGING STRUCTURAL MATERIAL

By C. C. Chamis, R. F. Lark, T. L. Sullivan

NASA Lewis Research Center

SUMMARY

Specimens of super-hybrids and advanced fiber composites were tested for smooth and center notch tensile strength, flexural strength, and Izod impact strength along the fiber direction and in the transverse direction. Specimens were also subjected to thermal fatigue and then tested for possible degradation at room temperature. The smooth tensile specimens were instrumented to obtain data for stress-strain curves. Laminate analysis was used to analyze the super-hybrid specimens with respect to elastic and thermal properties, residual stresses, and ply stresses at the specimen fracture stress condition.

The results show that the super-hybrid composites exhibit superior resistance to Izod impact compared with other hybrid and advanced fiber composites, are only slightly degraded by thermal fatigue, and have transverse flexural strengths about three times that of diffusion bonded boron/aluminum.

INTRODUCTION

The national need for the conservation of natural resources provides a strong motivation for the efficient utilization of materials. In the area of materials for flight vehicle applications, the super-hybrid composite concept provides a means of utilizing advanced composite materials efficiently while meeting diverse design requirements.

Advanced fiber/resin and fiber/aluminum matrix composites are used effectively when the fiber and load directions are coincident. To provide strength or stiffness in more than one direction, composites with fibers oriented in several directions are necessary. Orienting fibers in more than one direction in the same composite, however, reduces their efficiency substantially and introduces lamination residual stresses comparable to the transverse and shear strength properties of the unidirectional composite. These lamination residual stresses can severely limit the resistance of composite components in flight vehicle structures to cyclic loads. In addition, current commercially available advanced fiber composites are weak in impact and erosion resistance. Also graphite-fiber resin composites are susceptible to moisture degradation.

The aforementioned difficulties may be overcome to a significant extent via the super-hybrid composite concept. Briefly, superhybrids combine the best characteristics of fiber/resin, fiber/metal, and metal foil materials using adhesive bonding. Assembly of a super-hybrid is schematically illustrated in figure 1. Preliminary studies on fabrication feasibility and potential of super-hybrid composites as an aerospace material were reported in reference 1. In that study, one specific super-hybrid composite was made and tested for mechanical properties and Izod impact strength. The resistance of super-hybrids to thermal fatigue was not investigated in the study reported in reference 1.

To further evaluate the super-hybrid composite concept and to further illustrate their distinct advantages over other forms of advanced composites, an investigation was conducted wherein super-hybrids were made from various constituents and evaluated as to their mechanical properties, including impact and thermal fatigue. Also this paper describes the fabrication and test procedures used in making super-hybrids, and provides comparisons of theoretical predictions of mechanical properties and experimental results.

DESCRIPTION OF COMPOSITE SYSTEMS

In the following section the various types of laminates investigated are discussed.

Constituent Plies and Materials

Four types of laminates were made, two boron/aluminum (B/Al) and two super-hybrids. The types of laminates, laminate designations, constituent materials, and material suppliers are listed in table I(a). The laminate configurations are shown in table I(b). The numbering system for the laminate types is consistant with that of reference 1.

Thermal, physical and mechanical properties of the constituent materials are summarized in reference 1, table II. Note the thermal and mechanical properties of the boron/1100-Al alloy composites are approximately the same as those for the boron/6061-Al alloy.

Laminate Fabrication

Type II-A. – Eight plies of 5.6 mil-B/1100-Al were diffusion bonded by the manufacturer, Amercom, Inc. to provide a unidirectional laminate. The diffusion bonding conditions consisted of 4500 psi pressure at 950 °F for 1/2 hour.

Type II-B. – Six plies of 8.0 mil-B/1100-Al were diffusion bonded by the manufacturer, Amercom, Inc. to produce a unidirectional laminate. The diffusion bonding conditions were the same as those for Type II-A.

Type VI. - Five sheets of titanium foil and 10 plies of graphite/epoxy (Gr/Ep) were adhesively bonded using FM 1000 structural adhesive to produce a unidirectional laminate. The foil was so oriented that its primary rolling direction was parallel to the fiber direction. Before bonding, the titanium foil plies were degreased and treated with a 5-percent hydrogen fluoride solution for 30 seconds at room temperature. This was followed by a water and methyl alcohol rinse and then by drying.

The graphite/epoxy plies were bonded using PR 288 epoxy matrix resin supplied by the Minnesota Mining and Manufacturing Co. (3 M Co.). The time-pressure-temperature curing cycle was selected to initially cure the graphite/epoxy plies and then to cure the FM 1000 interfaces. The procedure was as follows: The various components of the laminate were assembled in a metal mold. A laminating press was then preheated to 350 °F. The cold mold was placed in the press and 15 psig contact pressure was applied. Contact pressure was maintained for 3.6 minutes. A pressure of 600 psig was then applied. Pressure and temperature were maintained for 2 hours to complete curing of the epoxy resin matrix and the FM 1000 adhesive. Upon completion of curing, the press pressure was reduced to zero and the mold was removed from the press in a hot condition. The laminate was then removed from the mold after cooling.

Type VII. - Four sheets of titanium foil, two plies of B/Al, and seven plies of graphite/epoxy were adhesively bonded to produce a unidirectional laminate. Super-hybrid Type VII is the same as Type V, (ref. 1), without the mid titanium foil layer. The bonding and curing procedures for Type VII were the same as for Type VI. It is noted here that the fabrication procedure for super-hybrids VI and VII is similar to that for super-hybrid Type V in reference 1, except that 3501 epoxy resin matrix was used for the Type V laminate.

Typical cross sections of the laminates are shown in the photomicrographs of figure 2. The materials and various plies in these laminates are also indicated in this figure. The detailed arrangement of the materials, plies and their corresponding thicknesses are given in table I(b).

DESCRIPTION OF TEST PROGRAM

In this section the specimen preparation, instrumentation, types of tests, and procedures are described.

Specimen Preparation

Unidirectional laminates ranging from 0.058 to 0.064 in. thick were cut into 0.500 in. wide specimens by using a precision wafer cutting machine equipped with a diamond cutting wheel. A specimen layout plan is shown in figure 3. The ends of all specimens used to determine longitudinal smooth tensile properties were reinforced with adhesively bonded aluminum tabs.

To determine the notch sensitivity of the laminates, through-the-thickness center notches were placed in specimens using electrical discharge machining. In all cases the notch root radius was 0.003 in. or less. A notch length of 0.17 in. was used.

The flexural specimens were used for the thermal fatigue and subsequent residual strength studies. The flexural specimens were selected because flexure bending is the most convenient way to assess the structural integrity retention of a material, as measured by its mechanical properties since this test subjects the material to tension, compression and shear simultaneously.

Specimen Instrumentation

The specimens used to determine smooth tensile properties were instrumented with strain gages to measure longitudinal and transverse strain. A photograph of an instrumented specimen is shown in figure 4.

Types of Tests and Procedures

Composite density. - Samples of each of the four types of laminates were evaluated for density by using the ASTM D-792 test method for "Specific Gravity and Density of Plastics by Displacement."

Smooth and notch tensile strengths. - The smooth and notch tensile specimens were loaded to failure using a hydraulically actuated universal testing machine. Longitudinal specimens had a test section that was about 3 in. long, and the transverse specimens had a test section that was about 2 in. long. The notched specimens were loaded to failure at a loading rate of approximately 0.01 in./in./minute, and the maximum load noted. Loading was halted at convenient intervals when testing the smooth specimens so that strain gage data (using a digital strain recorder) could be obtained.

Flexural strengths. - Test specimens having a length of 3 in. were tested for flexural strength in an Instron testing machine. A three-point loading system was used with a span of 2 in.

Izod impact strengths. - Non-standard unnotched thin specimens were subjected to Izod impact strength tests using a TMI impact tester equipped with a 2-lb hammer. The velocity of the hammer was 11 ft/sec. The specimen dimensions were 0.50 in. wide by 2.5 in. long.

Thermal fatigue tests. - Test specimens having a length of 3 in. and a width of 0.50 in. were subjected to a thermal fatigue test using a hot-cold shock chamber. A photograph of the thermal fatigue specimen magazine is shown in figure 5. The thermal fatigue chamber is shown in figure 6. Samples were cycled (without applied external stress) over a temperature range from -100° F to 300° F for 1000 cycles. A typical cycle consisted of a two-minute cooling and two-minute heating period. Samples were periodically withdrawn from thermal fatigue testing and were subjected to flexural strength tests using a three-point loading system with a span of 2 inches. All specimens, before and after thermal fatigue testing, were examined optically, at 30X magnification, for possible cracks or delaminations.

EXPERIMENTAL TEST PROGRAM RESULTS

In this section results obtained for density, tensile (smooth and notched), flexural, flexural after thermal fatigue, and Izod impact tests are summarized and discussed. Data for super-hybrid V are from reference 1. They are included here for comparison.

Density

The measured densities of the laminates tested are given in the third column of table II. Note that all the super-hybrids have about the same density which is the same as that of glass/epoxy, 0.074 lb/in.^{3}, and about 25 percent less than B/1100-Al.

Smooth Tensile Tests

Table II summarizes the test data obtained from smooth specimens (specimens without notches). This table includes laminate longitudinal (load applied parallel to fibers) and transverse (load applied normal to fibers) tensile properties. The initial tangent moduli and Poisson's ratios are given. As can be seen, inclusion of titanium foil layers in the super-hybrids improves the transverse strength properties relative to the B/Al unidirectional materials. The longitudinal and transverse fracture strains of the super-hybrids are approximately equal.

Note in table II that the longitudinal fracture stress of super-hybrid Type VI is about the same as that for the B/Al composites. The corresponding strain is greater by about 50-percent. Also, note that both the longitudinal fracture stress and strain of the 5.6-B/Al (II-A) are about 10-percent greater than those of the 8.0 B/Al (II-B). The transverse fracture stress of the 5.6 and 8.0 B/1100-Al is only about 50-percent of that of 5.6 B/6061-Al (ref. 1).

Stress-strain curves for all types of laminates are shown in figure 7(a) for loads parallel to fibers and in figure 7(b) for loads transverse to the fibers. The stress-strain curves are linear to fracture, or nearly so, for specimens loaded parallel to the fibers. However, specimens loaded transverse to the fibers exhibit considerable nonlinearity (fig. 7(b)). Curves of Poisson's strain versus axial strain are shown in figure 8.

One interesting result (also mentioned in ref. 1) is the failure mode of super-hybrids V and VII tested in longitudinal tension. The boron/aluminum plies failed when the tensile stress produced strain about equal to the fracture strains of the boron fibers. The Gr/Ep plies remained intact and were therefore still capable of carrying mechanical load. The authors believe this failure mode to be significant because these hybrids can be designed to have inherent fail-safe design characteristics.

Notch Tensile Tests

The test data obtained from notched specimens are summarized in table III. Two interesting points to be observed from the data in table III are:

(1) The notch effects are small and about the same for both the longitudinal and transverse directions in the super-hybrid composites.

(2) Notch strengthening for the transverse tensile specimens was observed in both of the diffusion bonded B/Al laminates. This strengthening may be attributed, in part, to the transverse restraining effects of the fibers at the notch ends, and to possible changes in the aluminum matrix strength due to machining the notch.

Flexural Tests

The test data obtained from subjecting test specimens to three-point flexural loading are summarized in table IV. The important points to be observed from the data are:

(1) The super-hybrid composites exhibit significant improvement in transverse strength compared with the B/1100-Al composites.

(2) The super-hybrid composites containing B/Al exhibit a decrease in the longitudinal flexural strength compared with other composites.

(3) The super-hybrid composite flexural longitudinal modulus is about 70-percent that of the corresponding tensile modulus (table II).

(4) The flexural transverse modulus of the super-hybrid composites is smaller than the corresponding tensile modulus (table II).

Impact Tests

Unnotched Izod impact test data are summarized in table V. To compare Izod impact resistance of the different laminates, the data were normalized with respect to the cross sectional area of the composite. In table V the low and high Izod impact strengths and the number of specimens for each composite or material are given.

The important point to be observed from the data in table V is the following: Using the super-hybrid composite concept, composite materials may be designed with Izod impact resistance approaching that of 6061 aluminum

(800 in.-lb/in.^{2}) (ref. 1). In addition, when the Izod impact values are normalized with respect to density, the longitudinal impact resistance of the type V hybrid is about 70-percent that of titanium (16,000 in.^{3}).

Another important point to be observed from the data in table V is that the super-hybrids have Izod longitudinal impact resistance about two times that of the B/1100-Al, while the transverse is about the same. This is significant because B/1100-Al is considered to be a relatively high energy absorbing material. It is noted, however, that the B/1100-Al unidirectional composite provided by the material supplier may not have been processed for optimum impact resistance.

Flexural Tests After Thermal Fatigue

The thermal fatigue effects on the flexural strength of super-hybrids are summarized in table VI. Prior to testing these specimens in flexure, they were subjected to 1000 thermal cycles from -100 °F to 300 °F. Optical examination, at 30X magnification, did not reveal cracks or delaminations. In order to make the thermal fatigue effects more pronounced, the flexural fracture stress after thermal cycling is compared to the as-fabricated highest value. As can be observed from the data in table VI, the thermal fatigue effects on the longitudinal flexural strength of the super-hybrids is less than 10-percent, while on the transverse, it is about 25-percent.

The important point to be noted is that the transverse flexural residual strength, after thermal fatigue, of the super-hybrids is still substantial when compared to the transverse strength of advanced unidirectional composites, both metal and resin-matrix.

THEORETICAL RESULTS

The theoretical results described below were obtained using the laminate theory and computer code described in reference 3. The results are for:

(1) mechanical and thermal properties

(2) ply residual stresses

(3) ply-stress and composite-stress influence coefficients.

Mechanical and Thermal Properties

The theoretically predicted mechanical properties are summarized in table VII. In this table results are given for density, membrane (in-plane) moduli and thermal coefficients of expansion, and bending moduli. Nominal measured properties for 6061-Al and Ti (6 Al-4V) are given for baseline material comparisons.

As can be observed from the data in table VII, the bending moduli of the super-hybrids are comparable to those of the titanium alloy in magnitude and the super-hybrids have less than one-half its density.

Lamination Residual Stresses

Lamination residual stresses are induced in the constituent material layers of the metal and resin matrix composites and the adhesive as a result of the lamination process because of:

(1) The mismatch of the thermal coefficient of expansion of the constituents

(2) The temperature difference between the cure and room temperatures (ref. 2).

The lamination residual stresses were computed as described in reference 1.

Selected results are summarized in table VIII. Note that the safety margins (S.M.) are also given; these were computed using the failure criterion in reference 4.

The ply residual stresses in table VIII are for a particular ply type as it is first encountered progressing inward from either surface. Several important observations from the data in table VIII are:

1. The S.M.'s for all plies is 0.66 or greater (zero denotes onset of fracture). Therefore, considerable capacity remains to resist mechanical load.

2. The longitudinal and transverse residual stresses in the adhesive are approximately equal and appear to be insensitive to the composition of the super-hybrid investigated. The adhesive residual stress is about 3.5 ksi which is about 50-percent of the bulk-state fracture stress, (ref. 1).

3. The longitudinal residual stress in the Gr/Ep plies is compressive.

4. The transverse residual stress in the B/Al plies is relatively small; less than 2.5 ksi.

Ply Stress Influence Coefficients

The concept of super-hybrid composites involves the strategic location of the titanium foil and B/Al plies to provide maximum resistance to transverse and shear forces. A direct way to assess whether this was achieved in the super-hybrids considered herein is to compute the ply stress influence coefficients due to uniaxial membrane and bending composite stresses. These influence coefficients were computed using the linear laminate analysis of reference 3. Selected results obtained for super-hybrid V are summarized in table IX. These results are for a particular ply type as it is first encountered progressing inward from the surface. Note that to obtain the ply stress, the influence coefficients must be multiplied by the membrane (bending) stress taken with the correct sign.

As can be observed from the data in table IX, the titanium foil and B/Al plies have large ply stress influence coefficients for uniaxial transverse and shear composite stresses. Therefore, the titanium foils and the B/Al in the super-hybrids provide practically all the resistance for transverse and shear forces. This verifies their role in the super-hybrid concept.

The other points to be observed from the data in table IX are:

1. The ply stress influence coefficients of the adhesive are negligible for all uniaxial composite stresses. Therefore, fracture will occur first in one of the non-adhesive constituents as desired in the super-hybrid concept.

2. The transverse and shear ply stress influence coefficients of the Gr/Ep due to uniaxial transverse and shear composite stresses are about 10-percent of the corresponding coefficients for the titanium and B/Al. This is another desired feature of the super-hybrid concept.

3. The transverse and shear ply stress influence coefficients for the titanium foils and the B/Al plies are approximately equal. This means that super-hybrids tested in the transverse direction, in in-plane shear, or in twisting will exhibit nonlinear stress-strain to fracture. The B/Al plies will fail first followed by yielding and finally fracture of the titanium. The super-hybrid stress-strain curves in figure 7(b) and the Poisson's strain curves in figure 8(b) are consistent with this observation.

The previous discussion leads to the following conclusion. The transverse and shear fracture modes of super-hybrids will be governed by the titanium foils in general. The transverse fracture strains for the super-hybrids in table II are about 1-percent which is approximately equal to the yield strain of titanium and which is in accord with the conclusion just stated. In-plane or twisting shear fracture strains need yet to be determined.

The linear stresses in the various plies resulting from the combined residual and applied loads may be obtained from the influence coefficients in table IX and by adding algebraicly the corresponding residual stress from table VIII.

Photographs of fractured super-hybrid specimens from various tests in this program are shown in figure 9. Note the well defined fracture surfaces on all these specimens.

Since the stress-strain curves for the super-hybrids tested in the longitudinal direction are linear to fracture, the influence coefficients in table IX can be used to compute the ply longitudinal fracture stress due to both membrane and bending loads. For example, the longitudinal fracture stresses in the B/Al plies are (204 ksi) due to tensile load and (216 ksi) due to flexural load. These values are about the same as those for laminate II-A, table II.

COMPARISON OF PREDICTED AND MEASURED DATA

Comparing corresponding data from table VII (in-plane) with those in table II for the super-hybrids, it is seen that all but one of the predicted values are within 10-percent of the measured values. Note, the measured values for longitudinal modulus for the B/Al composites are less than the predicted ones by an amount equivalent to that corresponding to approximately the aluminum modulus contribution (as determined using the rule-of-mixtures). This is probably the case because the 1100-Al is so soft that it was probably already stressed nonlinearly due to microresidual strains.

Comparing corresponding data for the longitudinal and transverse moduli in table VII (bending) with those in table IV, it is seen that they are in reasonably good agreement. These comparisons further substantiate that linear laminate theory is adequate for predicting elastic properties of super-hybrids.

SUMMARY OF RESULTS AND CONCLUSIONS

The key results from this investigation are:

1. Super-hybrids subjected to 1000 cycles of thermal fatigue from -100° to 300 °F retained over 90-percent of their longitudinal flexural strength and over 75-percent of their transverse flexural strength.

2. The transverse flexural strength of super-hybrids may be as high as eight-times that of the commerically supplied boron/1100-Al composite. The longitudinal stress in the boron/aluminum plies of the super-hybrids at fracture is about the same as that for the boron/1100-Al composite.

3. The thin specimen Izod longitudinal impact resistance of the super-hybrids is about twice that of the commercially supplied boron/1100-Al, while the transverse impact strength is about 100 to 150-percent of that of boron/1100-Al.

4. Linear laminate analysis is adequate for predicting initial membrane (in-plane) and bending elastic properties of super-hybrids.

5. Super-hybrids subjected to transverse tensile loads exhibit nonlinear stress-strain relationships.

The data obtained and analyzed in this investigation further substantiate the practicality and utility of the super-hybrid composite concept for attaining superior impact, transverse and shear strength properties and notch insensitivity. Since the titanium foils are on the surface, it may further be concluded that super-hybrids should have good erosion and moisture resistance.

REFERENCES

1. Chamis, Christos C.; Lark, Raymond F.; and Sullivan, Timothy L.: Boron/Aluminum-Graphite/Resin Advanced Fiber Composite Hybrids. NASA TN D-7879, 1975.

2. Chamis, Christos C.: Lamination Residual Stresses in Multilayered Fiber Composites. NASA TN D-6146, 1971.

3. Chamis, Christos C.: Computer Code for the Analysis of Multilayered Fiber Composites - User's Manual. NASA TN D-7013, 1971.

4. Chamis, Christos C.: Failure Criteria for Filamentary Composites. NASA TN D-5367, 1969.

TABLE I. - LAMINATE DESCRIPTIONS

(a) Layer materials and source

Laminate		Materials	Source
Type	Composition		
II-A	B/Al	Diffusion-bonded unidirectional layers 5.6-mil diam. boron fibers in 1100 aluminum alloy matrix	Amercom, Inc.
II-B	B/Al	Diffusion-bonded unidirectional layers 8-mil diam. boron fibers in 1100 aluminum alloy matrix	Amercom, Inc.
VI	Ti, A-S/Ep Superhybrid	Titanium (6Al-4V) foil, 0.0015-in. thick as rolled	Teledyne Rodney Metals
		Type A-S graphite fiber/PR 288 prepreg	3 M Company
		FM 1000 structural adhesive in film form	American Cyanamid Co.
VII, V(ref. 1)	Ti, B/Al, A-S/Ep Superhybrid	Titanium (6Al-4V) foil, 0.0015-in. thick as rolled	Teledyne Rodney Metals
		Individual monotape layers 5.6-mil diam. boron fibers in 6061 aluminum alloy matrix	Amercom, Inc.
		Type A-S graphite fiber/PR 288 prepreg	3 M Company
		FM 1000 structural adhesive in film form	American Cyanamid Co.

TABLE I. – LAMINATE DESCRIPTIONS

(b) Laminate

LAMINATE

COMPOSITION Dif. Bonded Boron/1100-Al (B/Al) Type-II-A			COMPOSITION Dif. Bonded Boron/1100-Al (B/Al) Type-II-B			COMPOSITION Titanium, Graphite/Epoxy, (Ti/(A-S/E)) Type-VI			COMPOSITION Titanium,Boron/6061-Al, Graphite/Epoxy (Ti/(B/Al)/(A-S/E)) Type-VII		
Layer no.	Material	t, in. [a]	Layer	Material	t, in.	Layer	Material	t, in.	Layer	Material	t, in.
1	B/Al, (5.6 mil, 1100)	0.0075	1	B/Al, (8.0 mil, 1100)	0.0107	1	Ti (6-4)	0.0015	1	Ti (6-4)	0.0015
2			2			2	FM 1000	0.0001	2	FM 1000	0.001
3			3			3	Ti (6-4)	0.0015	3	Ti (6-4)	0.0015
4			4			4	FM 1000	0.0001	4	FM 1000	0.001
5			5			5	A-S/E	0.005	5	B/Al, (5.6 mil, 6061)	0.0074
6			6	(total thickness, 0.064)		6	A-S/E	0.005	6	FM 1000	0.001
7						7	A-S/E	0.005	7	A-S/E	0.005
8	(total thickness, 0.060)					8	A-S/E	0.005	8	A-S/E	0.005
						9	A-S/E	0.005	9	A-S/E	0.005
						10	FM 1000	0.0001	10 [b]	A-S/E	0.005
						11	Ti (6-4)	0.0015	11	A-S/E	0.005
						12	FM 1000	0.0001	12	A-S/E	0.005
						13	A-S/E	0.005	13	A-S/E	0.005
						14	A-S/E	0.005	14	FM 1000	0.001
						15	A-S/E	0.005	15	B/Al, (5.6 mil, 6061)	0.0074
						16	A-S/E	0.005	16	FM 1000	0.001
						17	A-S/E	0.005	17	Ti (6-4)	0.0015
						18	FM 1000	0.0001	18	FM 1000	0.001
						19	Ti (6-4)	0.0015	19	Ti (6-4)	0.0015
						20	FM 1000	0.0001	(total thickness, 0.062)		
						21	Ti (6-4)	0.0015			
						(total thickness, 0.058)					

[a] t denotes layer thickness.

[b] For laminate Type V (ref. 1) this layer was Ti (6-4) with FM 1000 on each side.

TABLE II. – SMOOTH TENSILE TEST DATA

Laminate type	Constituents	Density, lb/in.^{3}	Fracture strength, 10^3 psi		Fracture strain, percent		Initial modulus of elasticity, 10^6 psi		Initial Poisson's ratio	
			Longi-tudinal	Trans-verse	Longi-tudinal	Trans-verse	Longi-tudinal	Trans-verse	Longi-tudinal	Trans-verse
II-A	B/Al (5.6)[a]	0.094	196	9.63	0.78	0.18	27	17	0.23	0.09
II-B	B/Al (8.0)[a]	0.093	183	10.6	0.67	0.12	28	17	0.23	0.15
V	Ti, B/Al,[b,c] Gr/Ep	0.074	125	31.5	0.73	1.01	18	8.5	0.25	0.11
VI	Ti, Gr/Ep[b]	0.068	182	25.7	1.2	0.91	15	3.5	0.32	0.07
VII	Ti, B/Al,[b] Gr/Ep	0.073	133	23.1	0.75	0.66	18	8.1	0.25	0.12

[a] Diffusion-bonded.

[b] Adhesive-bonded.

[c] Reference 1.

TABLE III. – NOTCHED TENSILE TEST DATA

Laminate type	Constituents	Notch length, in.	Net fracture strength, 10^3 psi		Notch strength ÷ unnotched strength	
			Longi-tudinal	Trans-verse	Longi-tudinal	Trans-verse
II-A	B/Al (5.6/1100)[a]	0.17	138	11.6	0.70	1.20
II-B	B/Al (8.0/1100)[a]	0.17	142	12.6	0.77	1.19
V	Ti,B/Al,Gr/Ep[b,c]	0.17	98	24.0	0.78	0.76
VI	Ti,Gr/Ep[b]	0.17	150	19.4	0.82	0.75
VII	Ti,B/Al,Gr/Ep[b]	0.17	107	15.6	0.80	0.68

[a] Diffusion-bonded.

[b] Adhesive-bonded.

[c] Reference 1.

TABLE IV. - RESULTS OF FLEXURAL TESTS

Laminate type	Constituents	Fracture stress 10^3 psi				Modulus, [a] 10^6 psi			
		Longitudinal		Transverse		Longitudinal		Transverse	
		Low	High	Low	High	Low	High	Low	High
II-A	B/Al (5.6) [b]	241	252	18.9	23.3	29.4	31.4	---	---
II-B	B/Al (8.0) [b]	204	204	16.6	24.2	21.0	23.2	---	---
V	Ti, B/Al [c,d] Gr/Ep	185	---	83.0	---	22.0	---	11.0	---
VI	Ti, Gr/Ep [c]	217	245	42.5	70.1	24.1	---	8.8	---
VII	Ti, B/Al, [c] Gr/Ep	161	188	71.3	77.3	20.6	---	12.5	---

[a] Modulus was computed using chart deflection and a calibration factor to account for Instron compliance. These values are only approximate.

[b] Diffusion-bonded.

[c] Adhesive-bonded.

[d] Reference 1.

Source: NASA Technical Memorandum X-71836, 1977, 24 pages

TABLE V. – SUMMARY OF THIN-SPECIMEN IZOD IMPACT STRENGTH RESULTS

Laminate type	Constituents	Test direction	IZOD impact strength in.-lb/in.^{2}		Number of specimens
			Low	High	
II-A	B/Al (5.6)[a]	Longitudinal	331	335	3
		Transverse	135	167	2
II-B	B/Al (8.0)[a]	Longitudinal	319	338	3
		Transverse	129	147	2
V	Ti, B/Al,[b] Gr/Ep[c]	Longitudinal	634	720	2
		Transverse	186	202	2
VI	Ti, Gr/Ep[b]	Longitudinal	573	734	3
		Transverse	142	171	3
VII	Ti, B/Al,[b] Gr/Ep	Longitudinal	454	658	6
		Transverse	129	143	2

[a] Diffusion-bonded.
[b] Adhesive-bonded.
[c] Reference 1.

TABLE VI. – COMPARISON OF FLEXURAL STRENGTH PROPERTIES OF AS-FABRICATED AND

THERMALLY-CYCLED SPECIMENS (1000 CYCLES FROM -100° TO 300 °F)

Laminate type	Constituents	Fracture stress 10^3 psi			
		As-fabricated		Thermally-cycled	
		Longitudinal	Transverse	Longitudinal	Transverse
V	Ti, B/Al, Gr/Ep[a]	185 (high)	83 (high)	169	62.4
VI	Ti, Gr/Ep	245 (high)	70.1 (high)	247	57.9
VII	Ti, B/Al, Gr/Ep	188 (high)	77.3 (high)	177	56.5

[a] Laminate reported in reference 1.

TABLE VII. – PREDICTED MECHANICAL AND THERMAL PROPERTIES

OF METAL-MATRIX AND SUPER-HYBRID COMPOSITES

Composite	Density, lb/in.³	Membrane (In Plane) Modulus, 10⁶ psi Longi-tudinal	Membrane Trans-verse	Membrane Shear	Membrane Poisson's Ratio Major	Membrane Poisson's Ratio Minor	Thermal Coefficient of Expansion 10⁻⁶ in./in./°F Longi-tudinal	Thermal Trans-verse	Bending Modulus, 10⁶ psi Longi-tudinal	Bending Trans-verse	Bending Shear	Bending Poisson's Ratio Major	Bending Poisson's Ratio Minor
II-A[a]	0.095	33.0	21.0	7.2	0.25	0.16	3.3	10.7	33.0	21.0	7.2	0.25	0.16
II-B[b]	0.095	33.0	21.0	7.2	0.25	0.16	3.3	10.7	33.0	21.0	7.2	0.25	0.16
V[b]	0.079	20.2	8.7	3.0	0.26	0.11	5.6	15.8	21.2	13.4	4.7	0.27	0.17
VI[b]	0.070	18.1	3.9	1.3	0.28	0.06	1.4	15.1	17.6	6.1	2.1	0.29	0.10
VII[b]	0.075	20.0	7.9	2.7	0.26	0.10	5.4	16.2	20.0	12.5	4.4	0.27	0.17
Al (6061)[c]	0.098	10.0	10.0	3.6	0.33	0.33	12.6	12.6	10.0	10.0	3.6	0.33	0.33
Ti (6Al-4V)[c]	0.160	16.0	16.0	6.2	0.30	0.30	3.2	3.2	16.0	16.0	6.2	0.30	0.30

[a] Diffusion bonded.

[b] Super-hybrid.

[c] Nominal properties included for comparison.

TABLE VIII. – PLY RESIDUAL (THERMAL) STRESSES IN SUPERHYBRIDS

DUE TO COOLING FROM 370 °F TO ROOM TEMPERATURE

PLY	SUPER-HYBRID V Stress, 10³ psi Longitudinal	SUPER-HYBRID V Transverse	SUPER-HYBRID V S.M.[a]	SUPER-HYBRID VI Stress, 10³ psi Longitudinal	SUPER-HYBRID VI Transverse	SUPER-HYBRID VI S.M.[a]	SUPER-HYBRID VII Stress, 10³ psi Longitudinal	SUPER-HYBRID VII Transverse	SUPER-HYBRID VII S.M.[a]
Titanium	12.0	-19.6	0.95	17.9	-20.2	0.92	12.6	-21.1	0.94
Adhesive	3.6	3.2	0.67	3.7	3.1	0.66	3.6	3.1	0.67
B/Al	11.7	2.4	0.99	---	---	---	13.2	0.4	0.99
Gr/Ep	-9.5	3.1	0.83	-2.7	2.9	0.85	-8.4	2.9	0.85

[a] Safety margin based on the failure criterion programmed in reference 2 and described in reference 3.

Source: NASA Technical Memorandum X-71836, 1977, 24 pages

TABLE IX. - PREDICTED PLY STRESS INFLUENCE COEFFICIENTS DUE TO UNIT

UNIAXIAL COMPOSITE STRESS FOR SUPER-HYBRID V (Ti, B/Al, Gr/Ep)

Ply	Uniaxial Membrane Stress					Uniaxial Flexural Stress[a]				
	Longitudinal		Transverse		Shear	Longitudinal		Transverse		Shear
	Longitudinal	Transverse	Longitudinal	Transverse		Longitudinal	Transverse	Longitudinal	Transverse	
Top Titanium alloy	0.824	0.032	0.373	1.95	2.08	0.768	0.026	0.161	1.22	1.28
Adhesive	0.011	0.002	0.009	0.028	0.024	0.010	0.002	0.004	0.016	0.014
B/Al	1.63	-0.014	0.184	2.45	2.43	1.12	-0.013	-0.017	1.13	1.10
Gr/Ep	0.912	-0.002	-0.184	0.226	0.206	0.420	0	-0.095	0.070	0.063
Center Gr/Ep	0.912	-0.002	-0.184	0.226	0.206	-0.420	0	0.095	-0.070	-0.063
B/Al	1.63	-0.014	0.184	2.45	2.43	-1.12	0.013	0.017	-1.13	-1.10
Adhesive	0.011	0.002	0.009	0.028	0.024	-0.010	-0.002	-0.004	-0.016	-0.014
Bottom Titanium alloy	0.824	0.032	0.373	1.95	2.08	-0.768	-0.026	-0.161	-1.22	-1.28

[a]To obtain ply stress, multiply influence coefficient by the flexural stress with the correct sign.

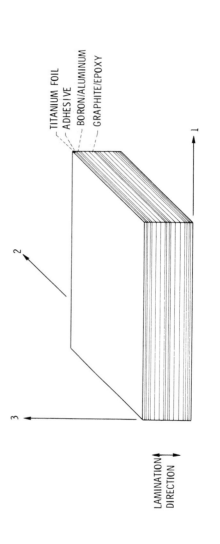

LAMINATION DIRECTION

TITANIUM FOIL
ADHESIVE
BORON/ALUMINUM
GRAPHITE/EPOXY

FIBER DIRECTION

Figure 1. - Schematic of adhesively bonded metal matrix and resin matrix fiber composite hybrid.

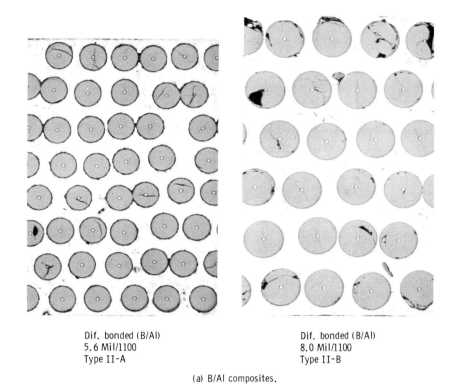

Dif. bonded (B/Al)
5.6 Mil/1100
Type II-A

Dif. bonded (B/Al)
8.0 Mil/1100
Type II-B

(a) B/Al composites.

Figure 2. - Photomicrographs of composite specimen cross sections. X50.

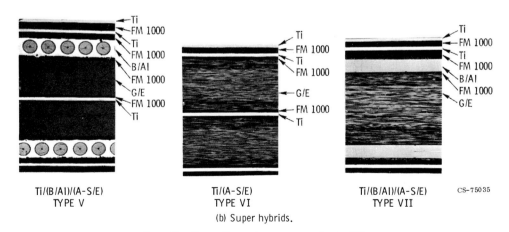

Ti/(B/Al)/(A-S/E)
TYPE V

Ti/(A-S/E)
TYPE VI

Ti/(B/Al)/(A-S/E)
TYPE VII

CS-75035

(b) Super hybrids.

Figure 2. - Composite specimen cross sections. X50.

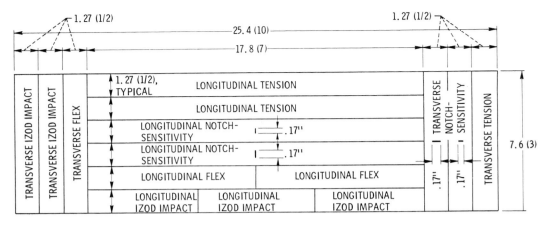

			1.27 (1/2), TYPICAL	LONGITUDINAL TENSION		TRANSVERSE NOTCH-SENSITIVITY		TRANSVERSE TENSION
TRANSVERSE IZOD IMPACT	TRANSVERSE IZOD IMPACT	TRANSVERSE FLEX		LONGITUDINAL TENSION				
				LONGITUDINAL NOTCH-SENSITIVITY	.17"			
				LONGITUDINAL NOTCH-SENSITIVITY	.17"			
				LONGITUDINAL FLEX	LONGITUDINAL FLEX			
			LONGITUDINAL IZOD IMPACT	LONGITUDINAL IZOD IMPACT	LONGITUDINAL IZOD IMPACT	.17"	.17"	

1.27 (1/2)   25.4 (10)   17.8 (7)   1.27 (1/2)   7.6 (3)

Figure 3. - Typical specimen layout plan. (Nominal values. All dimensions are in cm (in.).)

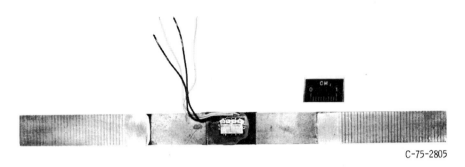

C-75-2805

Figure 4. - Instrumented super-hybrid specimen.

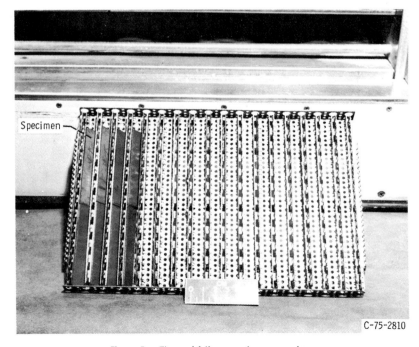

Specimen

C-75-2810

Figure 5. - Thermal fatigue specimen magazine.

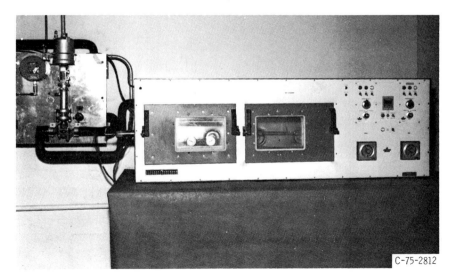

Figure 6. – Thermal fatigue chamber, closed.

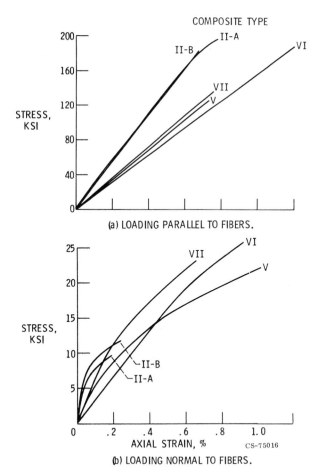

(a) LOADING PARALLEL TO FIBERS.

(b) LOADING NORMAL TO FIBERS.

Figure 7. – Stress-strain curves for smooth tensile specimens.

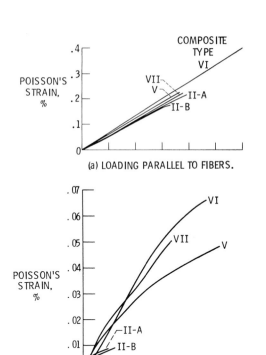

(a) LOADING PARALLEL TO FIBERS.

(b) LOADING NORMAL TO FIBERS.

Figure 8. – Poisson's strain for smooth tensile specimens.

Source: NASA Technical Memorandum X-71836, 1977, 24 pages

Longitudinal smooth and notch tensile

Transverse smooth and notch tensile

Transverse and longitudinal flexural

Transverse izod impact

Longitudinal izod impact   C-74-2121

Figure 9. – Various fractured composite specimens.

EVALUATION OF SILICON CARBIDE FIBER/TITANIUM COMPOSITES

R. W. Jech and R. A. Signorelli

SUMMARY

The potential usefulness of a composite composed of silicon carbide fiber in a titanium matrix (SiC/Ti) as a substitute for titanium alloys or stainless steel was evaluated. Composites were fabricated by hot pressing alternating layers of collimated silicon carbide fiber and unalloyed titanium foil.

Composites were evaluated for tensile strength and modulus of elasticity at room temperature and elevated temperature. Room temperature impact strength was determined for composites in the as-fabricated condition and after exposure in air at elevated temperature up to 1000 hours. The composites contained about 36 volume percent fiber.

Notched miniature Izod impact strength of SiC/A-70 Ti composites, in the as-fabricated condition, averaged 130 kJ/m^2 (60 ft-lb/in^2). This was increased to 190 kJ/m^2 (90 ft-lb/in^2) by changing the matrix to A-40 unalloyed titanium which was lower in oxygen content and higher in ductility. The impact strength of the A-40 titanium composites was equal to the titanium-6 aluminum-4 vanadium (6-4) alloy, 195 kJ/m^2 (95 ft-lb/in^2), but less than the unreinforced unalloyed (A-40) titanium, 515 kJ/m^2 (245 ft-lb/in^2).

SiC/A-70 Ti composites had a higher room temperature modulus of elasticity, 2.40 GPa (35.0×10^6 psi), than either stainless steel, 2.10 GPa (30×10^6 psi), or titanium, 1.20 GPa (18×10^6 psi). This advantage persisted at test temperatures up to, and including, 870 K (1110° F).

The average room temperature tensile strength of the SiC/A-70 Ti composites was 650 MPa (95 ksi), which was nearly identical to the unreinforced A-70 titanium, 655 MPa (95 ksi). At temperatures above 700 K (800° F), the composites were superior in tensile strength to unreinforced unalloyed (A-70) titanium, but had lower tensile strength than the titanium-6 aluminum-4 vanadium (6-4) alloy. Although about 10 percent lower in density, the composites were not competitive with 6-4 titanium, either on the basis of absolute strength or specific strength.

The data emphasize both the promise and the problems of SiC fiber reinforced unalloyed titanium composites. The composites exhibit low density, high stiffness, and moderate tensile strength, but at the price of decreased ductility (<2%) and impact strength to the extent that their impact strength is approximately equal to currently available titanium alloys.

INTRODUCTION

Temperatures encountered in gas turbine engines cover a wide range. Because of this, designers require a variety of materials for the various engine components, each compatible with its operating environment, in order to achieve powerful, lightweight, engines economical to operate. In the low temperature regions of the engine, up to 590 K (600° F), there is potential for the use of polymere matrix components and boron/aluminum composites. For application at temperatures in the intermediate range, to 810 K (1000° F), titanium alloys or stainless steel are used.

It is in the intermediate temperature region that silicon carbide/titanium (SiC/Ti) composites might be employed. In theory, SiC/Ti composites would be lower in weight and have higher tensile strength (especially at elevated temperature), greater creep resistance, and improved stiffness compared to either titanium alloys or stainless steel. Calculations indicate that titanium reinforced with 30 volume percent SiC fiber would have a density of about 4.1 grams/cc (0.148 lb/in^3) and a room temperature modulus of elasticity of about 2.10 GPa (30×10^6 psi). Offsetting these advantages is the possibility of lowered impact resistance and decreased ductility.

The problem is to maximize the desirable properties of the composite while minimizing the undesirable ones. Past work on the SiC/Ti system (ref. 1) has shown that composites containing high strength SiC fibers in a matrix of titanium-6 aluminum-4 vanadium (6-4) alloy exhibited modest improvements in tensile strength compared to the unreinforced alloy. However, this alloy is less ductile than unalloyed titanium, 8 percent elongation versus 15 to 20 percent, and therefore lower in impact strength. Thus, when a brittle, high strength fiber, such as SiC, was used as reinforcement, the resulting composite had high tensile strength and stiffness in the major reinforcement direction, but at the cost of lowered impact strength (ref. 2).

It was believed possible to increase the impact strength of SiC/Ti composites by using a more ductile matrix than those previously employed. This assumption was based upon the results obtained in reference 3 where changing the matrix in a boron/aluminum composite from 5052 aluminum alloy to the more ductile 1100 aluminum alloy increased the notched Charpy impact energy from 24 to 64 J (18 to 47 ft-lb). This was accomplished with no decrease in modulus of elasticity in either the longitudinal or transverse direction.

The purpose of the present investigation was to evaluate the potential usefulness of composites composed of SiC fiber in a titanium matrix as a substitute for unreinforced titanium or stainless steel for certain stiffness critical applications. The requirements were: adaquate modulus, tensile strength and impact strength in both the as-fabricated condition and after exposure in air at 920 K (1200° F) for up to 1000 hours.

SiC/Ti composites containing 36 volume percent fiber were fabricated using the hot pressed foil-fiber mat technique. Notched Izod impact tests were conducted on composites in the as-fabricated condition and after exposure at 920 K (1200° F) in air for times of 100 and 1000 hours. Tensile data and modulus of clasticity measurements were obtained at room temperature and elevated temperature.

MATERIALS, APPARATUS, AND PROCEDURE

Matrix Selection

Four titanium alloys in bar form were screened for use as possible matrix materials. They were: unalloyed titanium (A-75), titanium-6 aluminum-4 vanadium (6-4), titanium-6 aluminum-2 tin-4 zirconium-2 molybdenum (6242), and titanium-5 aluminum-6 tin-2 zirconium-1 molybdenum-0.5 silicon (5621 S). Their complete chemical compositions are given in table I(a).

Screening was done by testing notched miniature Izod impact specimens machined from bar stock, figure 1(a), for impact strength using a modified Bell Telephone Laboratory type miniature Izod impact machine. The specimens were tested in the as-received condition and after exposure in air at 865 K (1100° F) or 980 K (1300° F) for 10, 100, or 1000 hours.

Fiber

Commercially obtained SiC fiber made by chemical vapor deposition of SiC on a tungsten core was used in this investigation. The fiber diameter was 0.010 cm (0.004 in.).

Process Selection

Processing variables such as hot pressing temperature and hot pressing pressure were investigated for a range of temperatures from 1090 K (1500° F) to 1200 K (1700° F) and pressures from 70 MPa (10 ksi) to 210 MPa (30 ksi). The composites were made of alternating layers of unalloyed titanium (A-70) foil, 0.008 cm (0.003 in.) thick and collimated SiC fiber mat using polystyrene as a binder to hold the fibers in position. The assembled layers of fiber mat and foil, 8 and 9 in number, were encapsulated in type 304 stainless steel cans. The polystyrene binder was removed by heating to 700 K (800° F) in vacuum ($\sim 10^{-3}$ torr). The composites were preheated for 15 minutes while a light clamping load was applied, and hot pressed in a channel die for 5 minutes.

The composites were air cooled to room temperature and decanned by etching away the stainless steel with H_2O-HNO_3-HCL. These particular composites were evaluated as nonstandard, thin-sheet, unnotched Izod impact specimens (fig. 1(b)) to determine an appropriate specimen fabrication process. The impact tests were conducted by striking the specimens normal to the pressing surface.

Composite Fabrication

Fabrication of notched miniature Izod impact specimens was carried out using the appropriate parameters determined in the "Process Selection" portion of this paper. The hot pressing pressures and temperatures used were either 1200 K (1700° F)-210 MPa (30 ksi) or 1145 K (1600° F)-140 MPa (20 ksi). The notched Izod composite specimen blanks were thicker, 0.64 cm (0.25 in.) and contained more layers of fiber (71) and titanium foil (72) than the thin-sheet, unnotched specimens.

SiC/Ti notched miniature Izod specimens containing from 12 to 84 volume percent fiber oriented parallel to the longitudinal axis of the specimen were fabricated. Similar specimens, which contained 36 volume percent fiber angle-plied at ±10°, ±20°, or ±30° to the longitudinal axis were also fabricated.

Unreinforced titanium "control" specimens were made by hot pressing 90 layers of titanium foil encapsulated in a type 304 stainless steel can which had been evacuated and electron beam welded. The hot pressing pressure and temperature used was the same as that used to hot press the SiC/Ti composites.

Testing

Tensile tests and sonic modulus of elasticity tests were conducted on SiC/Ti composites in the as-fabricated condition. The tensile tests were carried out in a standard constant crosshead speed tensile machine at a crosshead speed of 0.13 cm/min (0.050 in./min). Elevated temperature tests were performed in a vacuum chamber.

Modulus of elasticity measurements were made using the sonic method. The method employed and equations used are described in reference 4.

Izod impact tests of the SiC/Ti composites and unreinforced titanium control specimens were made using notched miniature specimens in the as-fabricated condition or after exposure in air at 920 K (1200° F) for 100 or 1000 hours. Both the composites and the unreinforced titanium specimens were notched in the surface normal to the pressing direction. The specimen configuration and notch orientation is shown in figure 1(a). The notch was transverse to the longitudinal axis of the specimen.

RESULTS AND DISCUSSION

Matrix Selection

Room temperature notched Izod impact test results for the titanium alloys considered as possible matrix materials are given in table II. As shown in figure 2, the impact strength generally decreased as the time at temperature increased. Increasing the exposure temperature also decreased the impact strength. The material with the highest impact strength was unalloyed titanium (A-75), followed by the 6-4, 6242, and 5621S alloys, respectively.

Selection of unalloyed titanium was predication upon its notched impact strength which was more than twice that of the next best titanium alloy, 6-4. This was true for samples in both the as-received condition and after air exposure at 865 K (1100° F) or 980 K (1300° F) for up to 1000 hours. The 6-4 alloy has higher tensile strength than the unalloyed titanium, it was anticipated that tensile strength could be "built into" composites made with unalloyed titanium by utilizing the high strength of the SiC fibers.

Process Selection

A number of variables were considered when selecting suitable processing conditions for SiC/Ti composites. These variables were: hot pressing temperature, hot pressing pressure, preheat and hot pressing time, and the effect of possible contamination of the fiber and/or the matrix by the polystyrene binder used in the fiber mat.

Preheat and hot pressing time were arbitrarily set at 15 minutes and 5 minutes, respectively. Nothing occurred during the course of the investigation to require changing these times.

Contamination of the SiC fiber and A-70 titanium foil was investigated by tensile testing samples of each in the as-received condition, after polystyrene coating, and after coating and burn-off of the polystyrene. No change in the tensile strength of the SiC fiber, 3460 MPa (502 ksi) average, or the titanium, 702 MPa (102 ksi) average, was observed.

Chemical analyses showed no significant difference in the levels of oxygen, carbon, or nitrogen among the as-received foil; the hot pressed foil or the SiC/Ti composites (tables I(a) and (b)).

Hot pressing temperature and pressure were found to be important in their effect on the impact strength of the composites. The temperature had to be high enough to soften the titanium thus allowing it to flow around the fibers, but not so high as to cause thermal degradation of the fiber or reaction between the fiber and matrix. Hot pressing pressure had to be high enough to cause matrix flow, but not so high as to cause fiber breakage.

Hot pressing temperature and pressure were selected based upon the results presented in table III. These data are plotted in figure 3 and are the results of Izod impact tests on unnotched thin-sheet specimens. The results show that composites hot pressed at pressures of 70, 105, or 140 MPa (10, 15, or 20 ksi) exhibited their maximum impact strength when the hot pressing temperature was 1145 K (1600° F). As the hot pressing temperature was increased, the impact strength decreased. At a hot pressing pressure of 170 MPa (25 ksi), the maximum impact strength occurred when hot pressing temperature of 1170 K

Source: NASA Technical Memorandum 79232, July 1979, 26 pages

(1650° F) was used. When the hot pressing pressure was 210 MPa (30 ksi), the impact strength was essentially constant regardless of the temperature. The two highest average impact strengths were achieved when composites were hot pressed at 1145 K (1600° F)-140 MPa (20 ksi) and at 1170 K (1650° F)-170 MPa (25 ksi). Because of the greater risk of reaction or fiber degradation at 1170 K (1650° F), the lower temperature was used in making most of the specimens subsequently fabricated. The specific hot pressing temperature-pressure combination utilized was 1145 K (1600° F) and 140 MPa (20 ksi).

Impact Tests

In this investigation, major attention was directed toward determining the effects of processing parameters, matrix ductility, elevated temperature exposure, fiber content, and fiber orientation on the impact strength of SiC/Ti composites. These factors were examined in order to identify those areas which could be exploited to produce composites having good resistance to foreign object damage. Results of Izod impact tests on notched 36 v/o SiC/Ti composites are given in table IV(a), while table IV(b) contains impact data for unreinforced unalloyed titanium made from foil.

Influence of process parameters. - The influence of differing processing parameters on the impact strength of SiC/Ti composites was demonstrated using notched miniature Izod impact specimens. Previous results, described in the "Process Selection" section of this report, where unnotched thin sheet Izod impact specimens were used, showed that specimens made using different processing parameters had different impact strength. This was re-emphasized by the results shown in figure 4. Unalloyed (A-70) titanium, both reinforced with 36 v/o SiC fiber and unreinforced, exhibited greater impact strength when produced using 140 MPa (20 ksi) at 1145 K (1600° F), than when using 210 MPa (30 ksi) at 1200 K (1700° F).

Influence of matrix ductility. - SiC/Ti composites were made using either A-70 or A-40 unalloyed titanium as the matrix. The A-70 titanium contained 0.34 percent total interstitial impurities (carbon, oxygen, hydrogen, and nitrogen) and exhibited about 15 percent elongation when tested in tension at room temperature. Unreinforced notched miniature Izod impact specimens made from pressed A-70 foil, using 140 MPa (20 ksi) at 1145 K (1600° F), had an average impact strength of 245 kJ/m^2 (115 ft-lb/in^2). Incorporation of 36 volume percent SiC fiber into this matrix, using the same processing conditions, resulted in a composite whose average impact strength was 135 kJ/m^2 (65 ft-lb/in^2).

Composites made using the more ductile (20 percent elongation) unalloyed (A-40) titanium foil and the same processing conditions, had an average impact strength of 225 kJ/m^2 (105 ft-lb/in^2). Utilization of the higher purity (0.17 percent interstitials) unalloyed (A-40) titanium resulted in a 42 percent increase in impact strength. As was previously shown for boron/aluminum (ref. 3), the use of a more ductile matrix greatly improved the impact strength of composites. The impact strength of 36 v/o SiC/A-40 Ti composites was equal to unreinforced 6-4 titanium, 195 kJ/m^2 (90 ft-lb/in^2).

Influence of elevated temperature air exposure. - Exposure in air at 920 K (1200° F) for up to 1000 hours generally caused a decrease in the impact strength of the SiC/Ti composites (fig. 5). This was true for the composites made with either A-70 unalloyed titanium or the lower interstitial, more ductile A-40 unalloyed titanium. The decrease in impact strength was probably due to interstitial contamination, primarily oxygen, which embrittled the titanium. This is supported by the fact that the oxygen content of the two composites, after exposure for 1000 hours, was similar (table I(b)). The A-70 composite contained 0.65 percent oxygen, an increase of 0.38 percent, while the A-40 composite contained 0.53 percent oxygen, an increase of 0.40 percent. However, after exposure for 1000 hours, the impact strengths of both the A-70 and A-40 composites were very nearly the same, averaging 75 kJ/m^2 (35 ft-lb/in^2) for the A-70 composites and 80 kJ/m^2 (40 ft-lb/in^2) for the A-40 composites.

Influence of fiber content. - Notched impact strength of SiC/A-40 Ti composites as a function of fiber content was also examined. These data, table V and figure 6, show a decrease in the impact strength of the composites as the fiber content increased. It would appear that at some fiber content between 42 and 74 volume percent, the impact strength reached a minimum. Such behavior agrees with the equations proposed in references 2 and 5 for fiber reinforced composites composed of a ductile, impact resistant matrix and brittle fibers.

The specimens containing greater than 40 volume percent fiber failed in shear, as shown by the series of steps along the ends of the specimen (fig. 7). The reason for the change in failure mode was not investigated.

Influence of ply angle. - Angleply specimens of 36 v/o SiC fiber in A-40 titanium were tested for impact strength using specimens containing fibers oriented at 0°, ±10°, ±20°, or ±30° to the major axis of the composite. These results are presented in table VI. As shown in figure 8, the notched impact strength decreased from 225 kJ/m^2 (120 ft-lb/in^2) to 110 kJ/m^2 (55 ft-lb/in^2) when the ply-angle changed from 0° to ±10°. Increasing the plyangle resulted in further decreases in the impact strength although not nearly as drastically as the change in the first 10°.

Tensile Strength

Modest increases in tensile strength at elevated temperature were achieved when A-70 unalloyed titanium was reinforced with 36 volume percent SiC fiber. Results of tensile tests carried out on 36 v/o SiC/A-70 Ti composites; unreinforced unalloyed titanium (A-70) hot pressed foil, A-75 bar stock, and 6-4 bar stock are listed in table VII. As shown in figure 9, the room temperature tensile strength of the SiC/Ti composites was about the same as for unreinforced unalloyed (A-70) titanium made from pressed foil. At test temperatures above room temperature, the tensile strength of the composites was greater than the unreinforced unalloyed titanium, either A-70 ro A-75, but lower than the 6-4 titanium alloy. In no case was rule-of-mixtures strength attained. The measured tensile strength at room temperature, 650 MPa (94.5 ksi), was less than one-half that expected from rule-of-mixtures, 1550 MPa (225 ksi). As the test temperature increased the discrepancy between calculated and observed tensile strength became greater. This was probably due to inadaquate bonding between the fiber and the matrix. An indication of this is seen in figure 10 which shows the appearance of the fracture surface of specimens tested at room and elevated temperature. Fiber pull-out at 920 K (1200° F) and 1035 K (1400° F) was much more pronounced than at room temperature which indicated a weak bond between fiber and matrix. As the test temperature was increased the bond and titanium lost strength more rapidly than the fiber, resulting in increased fiber pull-out.

Modulus of Elasticity

Results of the sonic modulus of elasticity tests on the composite containing 36 v/o SiC in A-40 unalloyed titanium are presented in table VIII and figure 11. Figure 11 shows a plot of the ratio of the modulus at the test temperature (E_T) to the modulus at room temperature (E_{RT}). The room temperature modulus of elasticity of the composite was 2.41 GPa (35.0×10^6 psi). This is 14 percent greater than type 403 stainless steel (ref. 6) and 48 percent greater than unalloyed (A-75) titanium bar (ref. 7).

The composite exhibited good retention of elastic modulus up to the final test temperature, 870 K (1110° F). At 870 K (1110° F), the composite had a modulus of 1.95 GPa (28.5×10^6 psi), which was 0.81 of the room temperature modulus. In contrast, unalloyed titanium (A-75), at the same test temperature had a modulus of 1.24 GPa (12×10^6 psi) (ref. 7). The composite exhibited rule-of-mixtures behavior over the entire range of test temperatures. Calculations indicate that at 810 K (1000° F) the modulus of elasticity would be 1.95 GPa (28.0×10^6 psi), while the observed value was, in fact, 2.00 GPa (29.0×10^6 psi). Unlike the tensile strength, the modulus of elasticity of the composite demonstrated rule-of-mixtures behavior over the entire range of test temperatures.

SUMMARY OF RESULTS AND CONCLUSIONS

The purpose of this investigation was to evaluate the potential usefulness of SiC/Ti composites as a substitute for titanium alloys or stainless steel for stiffness critical applications in the compressor section of aircraft turbine engines. The major results and conclusions were:

1. Composites of 36 v/o SiC/A-40 Ti exhibited notched Izod impact strength equal to unreinforced titanium-6 aluminum-4 vanadium (6-4) alloy. The composites averaged 190 kJ/m^2 (90 ft-lb/in^2) in the as-fabricated condition compared to an average of 195 kJ/m^2 (95 ft-lb/in^2) for the 6-4 alloy. Unreinforced unalloyed (A-40) titanium averaged 515 kJ/m^2 (245 ft-lb/in^2).

2. SiC/A-40 Ti composites containing 36 volume percent fiber had a higher room temperature modulus of elasticity than either stainless steel or unreinforced unalloyed titanium. At room temperature, the modulus of elasticity of the composite was 2.40 GPa (35.0×10^6 psi) compared to 2.05 GPa (30×10^6 psi) for stainless steel and 1.15 GPa (17×10^6 psi) for A-75 unalloyed titanium. The composite retained a high fraction of its modulus at elevated temperature. At 870 K (1110$^{\rm o}$ F), the modulus was 1.95 GPa (28.5×10^6 psi), 81 percent of the room temperature modulus of elasticity compared to 66 percent for unreinforced unalloyed titanium.

3. Increasing the ductility of the unalloyed titanium used as the matrix in SiC/Ti composites resulted in an increase in the impact strength of the composites. Composites made with A-70 unalloyed titanium (15 percent elongation) had an average impact strength of 130 kJ/m^2 (60 ft-lb/in^2). When A-40 unalloyed titanium (20 percent elongation) was used as the matrix, the average impact strength was raised to 190 kJ/m^2 (90 ft-lb/in^2).

4. Hot pressing temperature and pressure affected the impact strength of the SiC/Ti composites. For example, composites of SiC/A-70 unalloyed titanium hot pressed at 140 MPa (20 ksi)- 1145 K (1600$^{\rm o}$ F) averaged 115 kJ/m^2 (55 ft-lb/in^2) while composites of the same composition hot pressed at 210 MPa (30 ksi)- 1200 K (1700$^{\rm o}$ F) had an average impact strength of 55 kJ/m^2 (30 ft-lb/in^2).

5. Use of 36 v/o SiC fiber to reinforce an unalloyed (A-70) titanium matrix resulted in composites having improved tensile strength at elevated temperature. The tensile strength of the composite was 405 MPa (58.5 ksi) at 810 K (1000$^{\rm o}$ F) compared to 225 MPa (33 ksi) for unreinforced unalloyed (A-70) titanium. At room temperature the tensile strengths were 650 MPa (94.5 ksi) and 655 K (95 ksi), respectively. However, compared to unreinforced 6-4 titanium, the composite is lower in tensile strength over the entire range of test temperatures.

6. Exposure at elevated temperature in still air decreased the impact strength of the SiC/Ti composites. After exposure for 1000 hours at 920 K (1200$^{\rm o}$ F), the SiC/A-40 Ti composites had an average impact strength of 80 kJ/m^2 (40 ft-lb/in^2) compared to the original 190 kJ/m^2 (90 ft-lb/in^2) for the as-fabricated composites.

CONCLUDING REMARKS

Impact results obtained in this investigation were encouraging from the standpoint that those trends observed in boron/aluminum composites were also observed in the SiC/Ti composites. The variation in impact properties of SiC/Ti due to both matrix ductility and processing variables were similar to boron/aluminum. Other changes such as increasing fiber diameter and hybridizing which have been success-

Source: NASA Technical Memorandum 79232, July 1979, 26 pages

ful in boron/aluminum, were not investigated. However, there is reason to expect that the results of such changes would also be similar in the SiC/Ti system. Furthermore, the impact properties observed for the SiC/Ti composites in this investigation may not be the best which can ultimately be achieved. It is therefore suggested that further investigation of fabrication variables would be useful for achieving optimum properties.

Although the processing of the SiC/Ti composites may not have been optimum, their properties compare well, in certain areas, to those of 403 stainless steel and 6-4 titanium. The composites had higher moduli of elasticity and these moduli were retained to a greater degree at elevated temperature than the monolithic materials. The composites exhibited room temperature impact strengths equal to that of unreinforced titanium 6-4 alloy. Composite tensile strength was lower than that of 6-4 titanium both at room temperature and elevated temperature. In order to provide a full fledged comparison with currently used monolithic materials, additional data would be required in other areas such as high and low cycle fatigue, erosion, and oxidation resistance.

REFERENCES

1. Ahmad, I.; et. al.: Metal Matrix Composites for High Temperature Applications. WVT-7266, Watervliet Arsenal, 1972. (AD-756867)

2. Winsa, E. W.; and Petrasek, D. W.: Factors Affecting Miniature Izod Impact Strength of Tungsten- Fiber- Metal-Matrix Composites. NASA TN D-7393, 1973.

3. McDanels, D. L.; and Signorelli, R. A.: Effect of Fiber Diameter and Matrix Alloy on Impact Resistant Boron-Aluminum Composites. NASA TN D-8204, 1976.

4. Roberts, M. H.; and Nortcliffe, J.: Measurement of Young's Modulus at High Temperatures. J. Iron Steel Inst., vol. 157, pt. 3, Nov. 1947, pp. 345-348.

5. Cooper, R. E.: The Work-to-Fracture of Brittle Fibre Ductile-Matrix Composites J. Mech. Phys. Solids, vol. 18, 1970, pp. 179-187.

6. Metals Handbook. Vol. 1. Eighth ed. American Society for Metals, 1961.

7. Jech, R. W.; and Weber, E. P.: Development of Titanium Alloys for Elevated Temperature Service by Powder Metallurgical Techniques, Clevite Research Center, Contract NOas-55-953-c, Final Report, July 15, 1957.

8. Crane, R. L.; and Krukonis, V. J.: Strength and Fracture Properties of Silicon Carbide Filament. Am. Ceram. Soc. Bull., Vol. 54, Feb. 1975, pp. 184-188.

TABLE I. – CHEMICAL COMPOSITION AND INTERSTITIAL CONTENT OF SiC/Ti COMPOSITES, HOT PRESSED TITANIUM FOIL, TITANIUM BAR, AND TITANIUM FOIL

(a) Titanium Alloys – Supplier's Analysis

Designation	Form	Composition, wt %									
		Al	V	Sn	Zr	Mo	C	H	O	N	Other
6-4	Bar	6.3	4.1	---	---	---	0.026	0.003	0.18	0.012	0.17 Fe
6242	↓	5.9	---	2.2	3.9	2.0	.019	.006	.09	.008	.06 Fe
5621S		4.8	---	6.1	1.9	1.0	.024	.003	.13	.010	.5 Si
											.13 Fe
A-75	↓	---	---	---	---	---	.028	.002	.17	.010	.15 Fe
A-70	Foil	---	---	---	---	---	.026	.003	.30	.012	.12 Fe
A-40	Foil	---	---	---	---	---	.022	.004	.13	.015	.07 Fe

(b) Pressed Foil and SiC/Ti Composites

Type	Condition	Composition, wt %			
		C	H	O	N
Pressed foil (A-70)	As fab - 1200 K (1700° F)	0.019	0.009	0.35	0.015
	Exposed 100 hr at 920 K (1200° F)	.008	.007	.50	.023
	Exposed 1000 hr at 920 K (1200° F)	.008	.005	.67	.021
SiC/A-70	As fab - 1200 K (1700° F)	[a]ND	[a]ND	.28	[a]ND
	Exposed 100 hr at 920 K (1200° F)	[a]ND	[a]ND	.35	[a]ND
	Exposed 1000 hr at 920 K (1200° F)	[a]ND	[a]ND	.65	[a]ND
SiC/A-40	As fab - 1145 K (1600° F)	[a]ND	[a]ND	.13	[a]ND
	Exposed 100 hr at 920 K (1200° F)	[a]ND	[a]ND	.24	[a]ND
	Exposed 1000 hr at 920 K (1200° F)	[a]ND	[a]ND	.53	[a]ND

[a]ND - Not determined.

Source: NASA Technical Memorandum 79232, July 1979, 26 pages

TABLE II. – ROOM TEMPERATURE NOTCHED IZOD IMPACT STRENGTH OF UNREINFORCED TITANIUM AND TITANIUM ALLOYS AFTER EXPOSURE IN AIR AT 865 K (1100° F) OR 980 K (1300° F)

Alloy	Exposure			Impact energy		Impact strength	
	Time	Temperature		J	in-lb	kJ/m^2	$ft\text{-}lb/in^2$
	Hr	K	°F				
A-75	As received			6.55	58.0	365	170
	As received			5.94	52.5	330	155
	10	865	1100	7.92	70.0	340	210
	100	865	1100	7.04	62.5	390	185
	1000	865	1100	7.47	66.0	415	195
	10	980	1300	7.65	67.5	425	200
	100	980	1300	8.85	78.5	490	230
	1000	980	1300	6.99	62.0	390	185
6-4	As received			3.27	29.0	180	85
	As received			3.72	33.0	210	100
	10	865	1100	3.47	30.5	190	90
	100	865	1100	3.27	29.0	180	85
	1000	865	1100	3.22	28.5	180	85
	10	980	1300	3.32	29.5	180	85
	100	980	1300	3.83	34.0	215	100
	1000	980	1300	2.78	24.5	155	75
6242	As received			3.42	30.0	190	90
	As received			3.02	26.5	165	80
	10	865	1100	2.93	26.0	160	80
	100	865	1100	2.54	22.5	140	65
	1000	865	1100	2.03	18.0	110	55
	10	980	1300	3.02	26.5	165	80
	100	980	1300	2.83	25.0	155	75
	1000	980	1300	1.94	17.0	110	50
5621S	As received			1.89	16.5	110	50
	As received			2.19	19.5	120	60
	10	865	1100	1.37	12.0	75	35
	100	865	1100	1.76	15.5	95	45
	1000	865	1100	0.63	5.5	35	15
	10	980	1300	1.59	14.0	90	40
	100	980	1300	1.01	9.0	55	25
	1000	980	1300	0.89	8.0	50	25

TABLE III. - ROOM TEMPERATURE IZOD IMPACT STRENGTH OF UNNOTCHED THIN-SHEET 36 V/O SiC/A-70 TITANIUM COMPOSITES

Hot pressing				Impact energy		Impact strength	
Pressure		Temperature					
MPa	ksi	K	°F	J	in-lb	kJ/m^2	ft-lb/in^2
70	10	1090	1500	0.34	3.0	95	45
				.23	2.0	65	30
		1115	1550	.23	2.0	65	30
				.23	2.0	65	30
		1145	1600	.28	2.5	80	35
				.40	3.5	110	55
		1170	1650	.23	2.0	65	30
				.17	1.5	45	25
		1200	1700	.17	1.5	45	25
				.17	1.5	45	25
105	15	1115	1550	.28	2.5	80	35
				.34	3.0	95	45
		1145	1600	.28	2.5	80	35
				.34	3.0	95	45
		1170	1650	.17	1.5	45	25
				.23	2.0	65	30
140	20	1090	1500	.34	3.0	95	45
				.28	2.5	80	35
		1145	1600	.34	3.0	95	45
				.40	3.5	110	55
		1200	1700	.11	1.0	30	15
				.11	1.0	30	15
170	25	1115	1550	.23	2.0	65	30
				.28	2.5	80	35
		1170	1650	.40	3.5	110	55
				.40	3.5	110	55
210	30	1090	1500	.17	1.5	45	25
				.11	1.0	30	15
		1145	1600	.11	1.0	30	15
				.11	1.0	30	15
		1200	1700	.17	1.5	45	25
				.17	1.5	45	25

Source: NASA Technical Memorandum 79232, July 1979, 26 pages

TABLE IV. - ROOM TEMPERATURE NOTCHED IZOD IM-
PACT STRENGTH OF 36 V/O SiC/TITANIUM COMPOS-
ITES AND UNREINFORCED UNALLOYED TITANIUM
AS-FABRICATED AND AFTER AIR EXPOSURE AT
920 K (1200° F)

(a) Composites

Matrix alloy	Condition	Impact energy		Impact strength	
		J	in-lb	kJ/m^2	$ft-lb/in^2$
A-70	As fabricated[a]	1.36	12.0	75	35
	↓	1.36	12.0	75	35
		1.13	10.0	65	30
		1.41	12.5	80	35
	Exposed 100 hr	1.36	12.0	75	35
	↓	1.41	12.5	80	35
		1.19	10.5	65	30
		1.24	11.0	70	35
	Exposed 1000 hr	1.36	12.0	75	35
	↓	1.53	13.5	85	40
		1.07	9.5	60	30
		1.07	9.5	60	30
	As fabricated[b]	2.26	20.0	125	60
	As fabricated[b]	2.54	22.5	140	65
	Exposed 100 hr	1.80	16.0	100	45
	Exposed 100 hr	2.03	18.0	115	55
	Exposed 1000 hr	1.58	14.0	90	40
	Exposed 1000 hr	1.19	10.5	65	30
A-40	As fabricated[b]	4.52	40.0	250	120
	As fabricated[b]	3.56	31.5	200	95
	Exposed 100 hr	1.75	15.5	95	45
	Exposed 100 hr	1.70	15.0	95	45
	Exposed 1000 hr	.85	7.5	45	20
	Exposed 1000 hr	2.09	18.5	115	55

[a]Hot pressed at 1200 K (1700° F) and 210 MPa (30 ksi).
[b]Hot pressed at 1145 K (1600° F) and 140 MPa (20 ksi).

TABLE IV. - CONCLUDED.

(b) Hot pressed unreinforced titanium foil

Alloy	Condition	Impact energy		Impact strength	
		J	in-lb	kJ/m^2	$ft-lb/in^2$
A-70	As fabricated[a]	4.07	36.0	225	105
	As fabricated[a]	1.98	17.5	110	50
	Exposed 100 hr	2.03	18.0	110	50
	Exposed 1000 hr	2.71	24.0	150	70
	As fabricated[b]	4.41	39.0	245	115
	As fabricated[b]	4.41	39.0	245	115
	Exposed 100 hr	3.90	34.5	215	100
	Exposed 100 hr	4.12	36.5	230	110
	Exposed 1000 hr	4.52	40.0	250	120
	Exposed 1000 hr	4.52	40.0	250	120
A-40	As fabricated[b]	9.32	82.5	515	245
	As fabricated[b]	7.63	67.5	420	200
	Exposed 100 hr	8.59	76.0	475	225
	Exposed 100 hr	8.72	72.0	485	215
	Exposed 1000 hr	5.93	52.5	330	155
	Exposed 1000 hr	6.72	59.5	370	175

[a]Hot pressed at 1200 K (1700° F) and 210 MPa (30 ksi).
[b]Hot pressed at 1145 K (1600° F) and 140 MPa (20 ksi).

TABLE V. - ROOM TEMPERATURE NOTCHED IZOD IMPACT STRENGTH OF SiC/A-40 TITANIUM COMPOSITES CONTAINING VARIOUS AMOUNTS OF FIBER

Fiber content, vol %	Impact energy		Impact strength	
	J	in-lb	kJ/m^2	$ft-lb/in^2$
0	9.32	82.5	515	245
0	7.62	67.5	420	200
12	6.27	55.5	345	165
23	5.65	50.0	310	145
25	4.40	39.0	245	115
36	3.39	30.0	250	120
36	3.62	32.0	200	95
37	2.03	18.0	110	55
38	1.92	17.0	105	50
42	[a]1.53	13.5	85	40
74	[a]1.24	11.0	70	30
84	[a]1.53	13.5	85	40

[a]Shear failure.

TABLE VI. – ROOM TEMPERATURE NOTCHED IZOD IMPACT STRENGTH OF ANGLEPLY 36 V/O SiC/A-40 TITANIUM COMPOSITES

Ply angle, (θ) deg	Impact energy		Impact strength	
	J	in-lb	kJ/m^2	$ft-lb/in^2$
0	4.52	40.0	250	120
0	3.56	31.5	200	95
10	1.92	17.0	105	50
10	2.15	19.0	115	55
20	1.75	15.5	95	45
20	1.52	13.5	85	40
30	1.52	13.5	85	40
30	1.64	14.5	95	45

Source: NASA Technical Memorandum 79232, July 1979, 26 pages

TABLE VII. – ROOM TEMPERATURE AND ELEVATED TEMPERATURE TENSILE STRENGTH OF 36 V/O SiC/A-70 TITANIUM COMPOSITES, HOT PRESSED A-70 TITANIUM FOIL AND A-75 AND 6-4 TITANIUM BAR STOCK

(a) Composites

Test temperature		Tensile strength	
K	°F	MPa	ksi
RT	RT	650	94.5
645	700	550	80.0
810	1000	405	58.5
865	1100	265	38.5
920	1200	225	32.5
1035	1400	110	16.0

(b) Unreinforced Titanium

Alloy	Test temperature		Tensile strength	
	K	°F	MPa	ksi
A-75 Bar	RT	RT	770	112.0
	810	1000	190	28.0
	865	1100	145	21.0
	920	1200	60	8.5
	1035	1400	30	4.0
A-70 Foil (Hot pressed)	RT	RT	655	95.0
	810	1000	225	33.0
	865	1100	125	18.0
	920	1200	70	10.0
	1035	1400	30	4.5
6-4 Bar	RT	RT	1125	163.0
	420	300	880	128.0
	535	500	780	113.5
	675	750	710	103.0
	740	875	635	92.0
	810	1000	520	75.5
	865	1100	410	59.5
	920	1200	295	43.0
	980	1300	170	25.0
	1035	1400	100	14.5

TABLE VIII. - SONIC MODULUS OF ELASTICITY OF

36 V/O SiC/A-40 TITANIUM COMPOSITES AT ROOM

AND ELEVATED TEMPERATURE

Test temperature		Modulus of elasticity		Ratio: E_T/E_{RT}
K	°F	GPa	psi	
RT	RT	2.40	35.0×10^6	1.00
325	110	2.40	35.0	1.00
370	210	2.35	34.0	.97
420	295	2.25	33.0	.94
480	415	2.25	32.5	.93
535	500	2.20	32.0	.91
590	600	2.15	31.5	.90
645	700	2.10	30.5	.87
700	805	2.10	30.0	.86
755	900	2.05	29.5	.84
810	1000	2.00	29.0	.83
870	1110	1.95	28.5	.81

Source: NASA Technical Memorandum 79232, July 1979, 26 pages

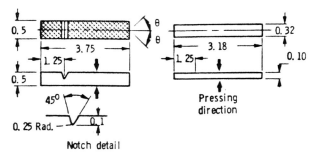

(a) - Notched miniature
Izod specimen.

(b) Unnotched thin-sheet
Izod specimen.

Figure 1. - Configuration and dimensions of notched, miniature Izod
impact specimens and unnotched, thin-sheet Izod impact specimens.
(All dimensions in centimeters.) Note: Ply angle θ is the angle
between the fiber and the longitudinal centerline of the specimen.

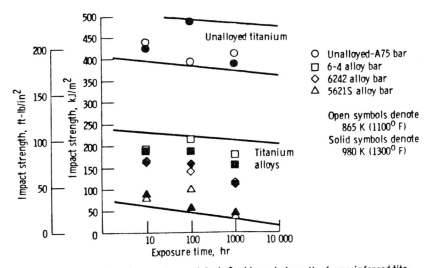

Figure 2. - Room temperature notched, Izod impact strength of unreinforced titanium and titanium alloys after air exposure at 865 K (1100° F) or 980 K (1300° F).

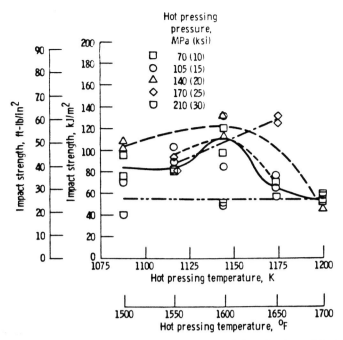

Figure 3. - Room-temperature Izod impact strength of unnotched thin-sheet 36 vol % SiC/A-70 Ti composites hot pressed at various temperatures and pressures.

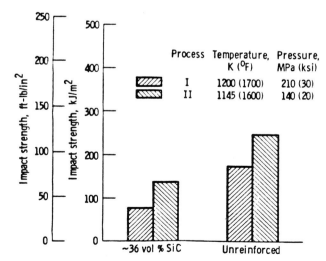

Figure 4. - Room temperature notched Izod impact strength of as-fabricated SiC/A-70 Ti composites and hot pressed unreinforced titanium foil made using different processing conditions.

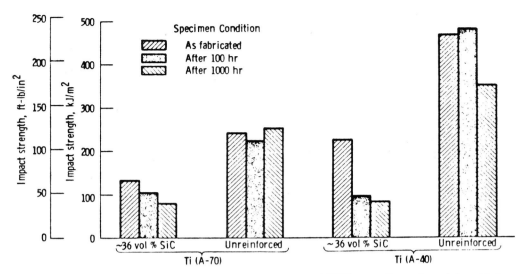

Figure 5. – Room temperature notched Izod impact strength of SiC/Ti composites and unreinforced hot-pressed titanium foil in the as-fabricated condition and after air exposure at 920 K (1200° F).

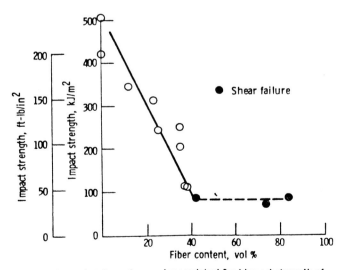

Figure 6. – Room temperature notched Izod impact strength of SiC/A-40 Ti composites containing various amounts of fiber.

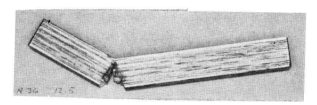

12.5 V/o

23 V/o

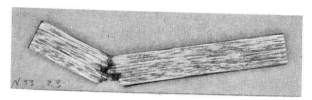

42 V/o

74 V/o

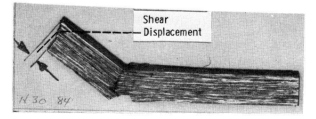

Shear Displacement

84 V/o

Figure 7. - Notched Izod impact specimens of SiC/A-40 Ti composites showing change from fibrous to shear displacement failure.

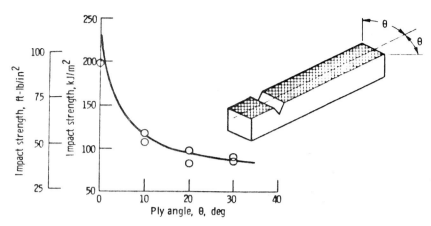

Figure 8. - Room temperature notched Izod impact strength of 36 vol % SiC/A-40 Ti composites containing alternate layers of fibers crossplied at various angles.

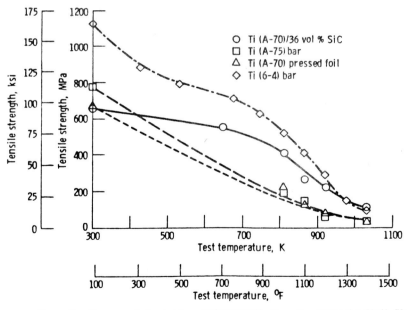

Figure 9. - Room- and elevated-temperature tensile strengths of 36 vol % SiC/A-70 Ti composites, hot pressed A-70 foil, and A-75 and titanium 6-4 bar stock.

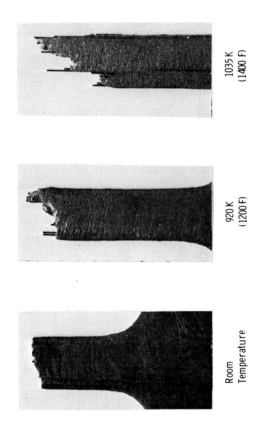

Room
Temperature

920 K
(1200 F)

1035 K
(1400 F)

Figure 10. – Tensile specimens of 36 $^V/o$ SiC/A-70 Ti composites showing fiber fracture and fiber pull-out failure.

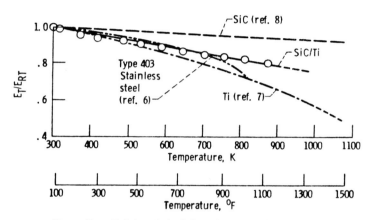

Figure 11. – Modulus of elasticity ratio for SiC fiber, 36 vol % SiC/A-40 Ti composite, unalloyed (A-75) titanium bar, and 403 stainless steel at various temperatures.

SECTION II
Joining

Joining of Metal-Matrix Fiber-Reinforced Composite Materials

by G. E. Metzger

Committee Foreword

That materials, judiciously combined, will perform differently and often more efficiently than the materials by themselves, has been an obvious and well-known fact for a long time. Not well known, however, is that this simple composites concept is an extremely useful and, in some aspects, a revolutionary way of thinking about the development and application of materials. Only lately, with the emergence of a unified and interdisciplinary approach to materials, have we begun to realize the full significance and the huge potential of composite materials.

With the field of composites just now emerging as a full-fledged materials technology, what directions the field will take in the future are difficult to predict. Nevertheless, some broad trends are evident. Dr. Metzger has provided, in this interpretive report, an initial review of the current problems and some solutions.

The greatest promise of composite materials lies in the concept itself, which can free us from the property bounds set by single, monolithic materials. It can also provide us with a means to multiply the variety of our presently existing materials. Moreover, the composites concept is teaching us to think of materials in terms of property systems, and thus is giving us a working approach with which to design, develop and apply materials and their joining methods to meet specific engineering needs.

M. M. Schwartz
Member, Interpretive Reports Committee

Abstract. The metal-matrix fiber-reinforced composites, their fabrication methods, properties, and some of the applications are briefly described. Virtually all composite applications have been for demonstration or test structures, and most of these have been in the aerospace industry. The available literature on composite joining has been summarized and evaluated, mainly on joining as a secondary fabrication process; not for primary fabrication of the base material.

The joining processes discussed include diffusion welding, fusion welding, resistance welding, brazing, soldering, mechanical fasteners, and adhesive bonding. Materials joined include Al/B, Al/graphite, Al/stainless steel, Al/Be, Ti/B, Ti/W, and Ti/graphite composites, although most work by far has been with Al/B materials.

The greatest effort has been expended on brazing, resistance welding, and mechanical fasteners, with considerable work also on soldering. The tensile strength joint efficiency of joints by these processes shows a marked decrease as the base material tensile strength increases. For brazing and resistance welding, the joint efficiency is about 60% for low strength composites of 600 MPa tensile strength and about 30% for the highest strength composites of 1400 MPa. The corresponding joint efficiencies for mechanical fasteners are about 40 and 20%.

Efficient joining of high-strength composites by diffusion welding, with the exception of titanium-matrix and ductile-fiber composites, or by adhesive bonding, appears to be of low probability. The great difficulty of fusion welding composites has resulted in a minimum of work by these processes, but their potential is believed to be good.

1. Introduction

The term "composite materials" encompasses a wide variety of materials from one of the earliest made by man; adobe brick formed from a mixture of mud and straw, to one of the latest; intermetallic rods or platelets grown in a melt of eutectic composition. Others, sometimes referred to as composites, include felted materials, laminates, clad materials, and dispersion-hardened materials.

In recent years, however, the greatest interest has been found in matrix materials reinforced with nonmetallic fibers, either continuous or discontinuous, because of their potentially very high strength and elastic modulus. Research and development in these fiber composites began in earnest about 1960 and reached its peak in perhaps 1971. Extensive investigation has been conducted on both resin-matrix and metal-matrix composites, with some work being done on ceramic matrixes. The present declining interest in metal-matrix fiber reinforced materials is due to a number of factors, but of major importance are the high cost and the difficult fabricability of these materials. Not the least of the fabrication problems is joining.

The subject of this report has been limited to metal-matrix fiber reinforced materials. Other materials, e.g., the dispersion-hardened metals and the resin-matrix composites have not been included because of their different strengthening mechanism, or because they are less appropriate to the functions of the Welding Research Council.

The purpose of this report is to evaluate and summarize, and place in the proper perspective, all of the available information on the joining of the metal-

G. E. Metzger is with the Air Force Materials Laboratory, Wright-Patterson AFB, O.

Publication of this report was sponsored by the **Interpretive Reports Committee** of the **Welding Research Council.**

matrix fiber composites. One approach to the production of structures from composites is to start with material containing a single layer of reinforcing fibers with thin matrix material on either side (usually referred to as tapes, although these may be much wider than what is commonly thought of as tapes) assemble sufficient tapes of the proper size and shape in a die and, by diffusion welding or brazing produce a completed structure, with the exception of minor finishing work, in one operation. This approach has been applied most often to the fabrication of aircraft gas-turbine engine blades. The fabrication of structures of composite materials, with tapes as the starting material, is somewhat analogous to the fabrication of conventional materials by casting, with melting stock as the starting material, or by forging, with a billet as the starting material. The manufacture of composite material and the production of structures from composite tape are considered to be primary fabrication processes, and the joining aspects of these processes are not a subject of this report.

Another approach to the fabrication of structures from composites is by the forming and joining of composite sheet, plate, bar, etc. Just as it is sometimes desirable or essential to fabricate structures of conventional materials by joining, methods for the application of composite materials as an assembly of simple forms and smaller members must be available, if these materials are to be used to their full potential.

A review and analysis of the joining experience with this latter approach to the fabrication of composite structures is the principal subject of this report.

1.1 Terminology, Common Conditions, and Units

To minimize misinterpretation of meaning and avoid repetitive lengthy description, some of the notation and description of terms used in this report is presented in this section.

The standard welding terminology recommended by the American Welding Society will be followed as closely as possible. In some cases it will be necessary to depart somewhat from the exact definition.

Nonmetallic refers to material such as boron, carbon, silicon carbide, and aluminum oxide, not to organic materials.

Unless otherwise stated, the conditions given in this paragraph are understood. The term "composite" refers specifically to metal-matrix fiber-reinforced materials. When the fiber orientation is essential to the understanding of a discussion, all fibers are oriented in one direction only, referred to as the longitudinal direction. The transverse direction is perpendicular to the fiber direction. The fiber direction is parallel to the principal stress axis in mechanical property specimens loaded in tension. The method of primary fabrication of the composite, in most cases, is by solid-state hot pressing of bare fibers between thin foils of matrix metal.

Joint efficiency, which is a measure of the load-carrying ability of a joint, normally refers to tensile strength and is defined by the following equation:

$$\text{Joint efficiency} = \frac{P \cdot 100}{\sigma \cdot A}$$

whereby P is the maximum load of the joint, σ is the average tensile strength of the base material and A is the original cross-section area of the base material. By extension, the same method can be used for other joint mechanical properties such as endurance limit or rupture strength.

The notation for composite will be to first give the matrix material, followed by the fiber material in volume percent (v/0). (This is the reverse of the notation followed in most reports on composites.) An alloy is identified by either its common designation or by an abbreviated form of its chemical composition in weight percent (w/0). The base metal is given first, followed by the alloying elements to the nearest whole percent, except where the exact percentage is essential to the alloy description. For example, Ti6Al4V/40SiC signifies a matrix of a titanium-base alloy with 6 w/0 Al and 4 w/0 V containing 40 v/0 silicon carbide fiber.

All units in this report are from the International System of Units (SI). For the convenience of the reader, a description of the SI Units and conversion factors for the units used in this report follows:

SI Prefixes

Exponential Expression	Prefix	Symbol
10^9	giga	G
10^6	mega	M
10^3	kilo	k
10^{-2}	centi	c
10^{-3}	milli	m

SI Units and Conversions

Property	Unit	Symbol	To Convert From	To	Multiply By
linear measurements	meter	m	m	inch	39.4
force	newton	N	N	pound force	0.225
pressure and stress	newton per meter squared	N/m²*			
	pascal	Pa*	Pa	pound force/inch²	0.000145

*N/m² = Pa.
Convenient, approximate conversions: 1 mm = 40 mils; 1 inch = 25 mm; 1 psi = 7 kPa; 1 ksi = 7 MPa.

2. Materials

To provide sufficient background for an understanding of the joining aspects of the metal-matrix fiber composites, this section must be somewhat longer than is usual for a report on joining. Some of the material components, the primary fabrication methods, and the properties, are unconventional and are not well known to many materials specialists.

Materials which have been reduced to a very small diameter possess tremendous strengths, e.g., copper or glass at a diameter of 1.27×10^{-2} mm has a tensile strength of 7000 MPa. Fibers of nonmetallic materials generally possess higher strength than those of metals, accompanied by a high modulus of elasticity and low density. However, fibers alone are not useful structural materials and have low ductility; and in the case of the nonmetallic fibers, are brittle.

By combining the tensile strength and the elastic modulus of fibers with the ductility of metals such as aluminum and titanium, a composite may be obtained with a higher strength/density or modulus/density ratio than any known alloy. High temperature properties and fatigue strength are also improved by the proper choice of matrix and fiber.

The metal-matrix fiber composite may be defined as a material consisting of two different substances, in which a fibrous substance is embedded within a continuous matrix metal so that each substance retains its identity and properties and the fibrous substance can be removed from the matrix intact.

A photomicrograph of a typical composite cross-section, Al/50B, is shown in Fig. 1, with an ideal boron filament distribution within the aluminum alloy matrix. A tungsten filament at the center of each boron filament is also visible.

2.1 Matrix/Fiber Combinations

A selection of the most important matrix/fiber combinations, based on the quantity produced, their potential for superior structural properties, or the amount of effort expended on their development, is given in Table 1. Many more combinations have been the subject of experimental or theoretical investigations, but most are considered to be of minor significance at this time.

The first two materials, boron and carbon, are polycrystalline and are normally used as continuous fibers, i.e., the individual fibers extend from one end of the composite to the other. Boron fiber is actually two materials, since it is manufactured by vapor deposition of boron on a 0.013 mm diameter tungsten filament to form a fiber of 0.1 to 0.13 mm diameter. A thin coating of silicon carbide (SiC) is sometimes applied to the boron fiber, which is then known as borsic fiber. Carbon fibers may be either amorphous, crystalline (graphite), or partially crystalline. Graphite is also produced as "whiskers."

Silicon carbide and Al_2O_3 (sapphire) have been produced as "whiskers," i.e., single crystals formed from the vapor phase with a diameter of about 0.013

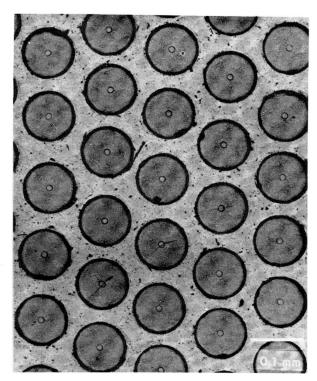

Fig. 1—Photomicrograph of Al/B composite cross-section, 150X

mm and a length considerably less than 25 mm. Both of these materials, SiC and Al_2O_3, are also manufactured as continuous fibers (filaments); the former material being manufactured by the same process as for boron fibers.

Ni_3Cb is an example of the reinforcement material in the eutectic type composites, whereby rod- or lamella-shaped intermetallic compounds are produced by the controlled unidirectional solidification of eutectic or near-eutectic alloys. These composites appear to be promising for high-temperature turbine engine application.

The metal fibers, also referred to as wires or filaments, do not appear to have as high a potential for structural composites as do the preceding nonmetallic fibers, but continue to be useful in studies concerned with composite concepts.

Each of the various types of fibers may be aligned in one (unidirectional) or more directions within the

Table 1—Matrix/Fiber Combinations

Matrix Alloy Base	Fiber Materials							
	Boron	Carbon	SiC	Al_2O_3	Ni_3-Cb	Steel	Be	W
Al	X	X	X	X		X	X	
Ti	X		X	X			X	
Ni		X	X	X	X			X

Table 2—Primary Fabrication Methods for Composites

Solid-State Hot-Pressing
Pressing and sintering of powdered matrix and fibers
Pressing of fibers coated with matrix material
Pressing of bare fibers between thin foils of matrix metal

Liquid
Infiltration of liquid matrix metal between fibers
Pressing of powdered matrix and fibers above solidus temperature of the matrix
Unidirectional solidification of a eutectic alloy

Deposition
Electrodeposition of matrix around fibers
Plasma spraying of matrix around fibers
Vapor deposition of matrix around fibers

Miscellaneous
Consolidation of matrix around fibers with explosive welding techniques
Co-extrusion of matrix and fibers
Consolidation of metal foil around fibers by resistance seam welding techniques

composite, or may have a random orientation, although there are very few examples of the latter type. Fibers may be continuous or discontinuous, i.e., the fiber length is much less than the length of the composite. The discontinuous fibers, as either whiskers or chopped filaments, have not been successfully applied to the manufacture of good-quality composites and are no longer considered to be of importance.

2.2 Primary Fabrication Methods

Primary fabrication methods are used to combine the fibers and matrix for the production of simple shapes such as sheet, plate, and beams. Occasionally, these methods result in a completed structure, without the need for secondary fabrication except for minor finishing operations. An example is the manufacture of jet engine blades by a solid-state hot-pressing operation.

The various methods, which have been used to manufacture composites, are given in Table 2, which is a modification of a table given in Ref. 2 of the bibliography. In some cases, more than one method is used, e.g., hot pressing of the materials of the third group is sometimes performed after deposition, to increase the matrix density. The manufacture of composites by explosive welding techniques and co-extrusion has been confined to ductile, i.e., metal, fibers.

The solid-state hot-pressing methods have found, by far, the greatest application; in particular, the pressing of filaments between thin foils of matrix metal.

2.3 Secondary Fabrication

The anisotropic properties of composites and their stiff reinforcements severely limit their formability. This is particularly true of the nonmetallic fibers, which are brittle and will tolerate no plastic deformation. Elevated temperature forming has produced the

best results, but time and temperature must be carefully controlled to prevent degradation of the material properties due to excessive matrix-fiber chemical reactions with the nonmetallic fibers, or annealing of cold-worked metallic fibers.

Since thin, narrow composites containing a single layer of fiber can be formed with relative ease, these "tapes" are being used as the starting material for more complex shapes or structures. After forming into the approximate desired configuration, the tapes are joined by brazing or diffusion welding. The final shape is produced during the joining process.

Damage of brittle fibers and excessive tool wear are among the major problems in the machining of composites. The erosive methods, such as electrodischarge and ultrasonic, appear to offer the most promise for success.

A detailed discussion of composite joining is presented in subsequent paragraphs.

2.4 Properties

The strength properties of the composite are achieved in a different manner than the precipitation-hardened or the dispersion-strengthened materials. The fine particles of the latter two materials cause strengthening by dislocation blocking, but the fibers of the composite carry the major portion of the applied load and are not generally considered as barriers to dislocation motion.

In composites with filaments, the matrix serves primarily to bind the fibers together in a bundle; very little load is transferred between the matrix and the fibers, except for the point of load introduction. However, with discontinuous fibers, the matrix must not only bind the fibers together but must also transfer the applied stress between fibers by shear stresses at the fiber-matrix interface. The fiber must be of a critical length, or greater, in order for the full strengthening effect of the reinforcements to be realized.

Many composite combinations have failed to achieve the excellent potential promised by the very high strength and; in the case of nonmetallics, high elastic modulus and low density of the reinforcement fibers. One of the major reasons has been low strength at the matrix-fiber interface; with other problems including fiber defects, fiber misalignment, physical or reaction damage to fibers during primary fabrication, and low strength at the matrix-matrix interface. Some combinations have not been produced with high enough fiber content to attain high composite strengths.

It is significant that with the combination on which the greatest effort has been expended, i.e., Al/B, composites have been produced with 50 v/0 fiber that utilize almost the full strength of the boron fiber. It may be inferred from this, that other combinations would prove equally successful, if sufficient development were accomplished.

Some of the mechanical properties of a selection of

Table 3—Tensile Properties and Density of Composites

Matrix	Fiber v/o	Fiber Material	Longitudinal T. S., MPa	Longitudinal El., %	Longitudinal E, GPa	Transverse T. S., MPa	Density, g/cm³
			Unidirectional Typical Values				
6061 Al	50	B	1450	0.6	230	120	2.7
2024 Al	50	B	1300	. . .	235	100	—
Ti6Al4V	50	B	1300	0.7	260	450	3.6
Ti6A14V	50	SiC	1050	0.4	240	400	3.9

composites are given in Table 3. The low transverse strength of the Al/B composites can be improved by placing some of the boron filaments at 45 or 90° to the longitudinal direction, or these cross-ply filaments may be of other materials such as stainless steel. Titanium alloy in foil form has also been used to increase transverse strength. However, all of these techniques result in lower strength in the longitudinal direction. Heat treatment of 6061 aluminum matrix to the T6 condition increases the transverse strength of the composites to 150–200 MPa, but is not always desirable because of adverse effects on other mechanical properties.

2.5 Applications

It is questionable whether the application stage in the development of the metal-matrix composites has been reached, or at best, it is only in its infancy. Virtually all of the applications consist of proposed uses or the production of demonstration or test articles.

As would be expected for a material with a high strength-to-weight ratio, most of the composite applications proposed are in the aerospace industry, e.g., gas turbine engines, aircraft structures, and space vehicles.

Most of the applications-oriented composite work involved the Al/B combination. Both General Electric[1] and Pratt & Whitney Aircraft Company[2] have fabricated and tested aircraft jet-engine blades of this material. Compressor blades made by GE of 2024 Al reinforced with boron filament and stainless steel wire mesh were reported to be equal to or better than a monolithic stainless steel blade in most respects. One of the fan blades, fabricated from 6061 Al/B by PWA, is shown in Fig. 2. These blades are 33% lighter than titanium alloy blades and would result in higher engine performance.

Other items of Al/B composite include: (1) the major portion of the bulkhead, illustrated in Fig. 3, constructed by Convair Division, General Dynamics[3] as one part of a full-size aft fuselage component of the F-111 aircraft; (2) landing gear components fabricated by Bendix Corp.[4]; (3) a shear beam and compression panel made by Convair Division[5] and a compression panel fabricated by McDonnell Douglas Corp.,[6] as structures applicable to space vehicles; (4) a subscale aircraft wingbox panel (Fig. 4) manufactured by brazing at Boeing Company[7]; (5) an aircraft wing section assembled by mechanical fasteners at Grumman Corp.[8]; and (6) a missile payload adapter (Fig. 5) manufactured by Convair Division.[9] The payload adapter, which has a diameter of 1.52 m and a height of 1.07 m, weighs 40% less than the original aluminum design. Joining of the structure was ac-

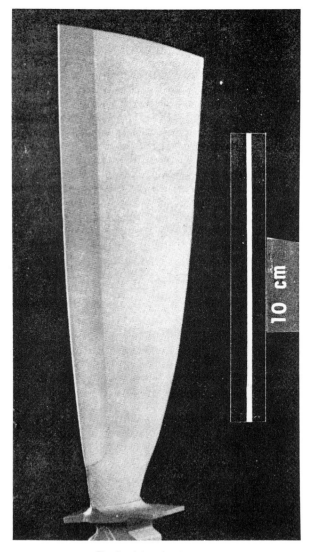

Fig. 2—Jet engine fan blade

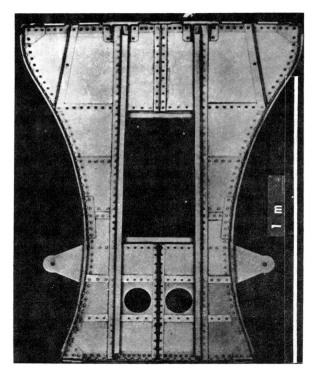

Fig. 3—Aircraft fuselage bulkhead

Fig. 4—Aircraft wingbox panel model

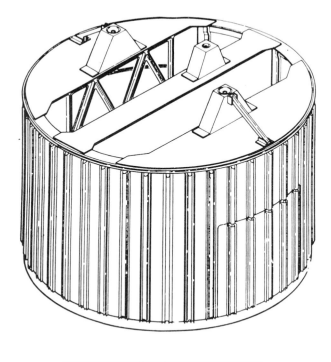

1 m

Fig. 5—Missile payload adapter

complished by the same methods used primarily for the bulkhead, shear beam, and compression panel fabricated by Convair Division, i.e., mechanical fasteners and spot welding.

North American Rockwell Corp. concluded from an applications study,[10] that Al/B composites warrant consideration for application to several space structures including pressurized cabins, shrouds, structures subject to motor-plume heating, solar-panel stiffeners, rocket engine skirt extensions, and missile interstage structures. Tubes manufactured of Al/B composite, with titanium alloy end fittings joined by diffusion welding, will be used on the space shuttle.[11]

Various aircraft jet-engine blades were fabricated of composites other than Al/B. The Allison Division, General Motors Corp. used Ti6Al4V/Be[12] for compressor blades and Ti6Al4V/B[13] for fan blades, while United Aircraft Research Laboratories manufactured turbine-blade shapes from a ternary eutectic alloy to yield a composite of Ni_3Al-Ni_3Cb.[14] None of these blades were subjected to operational tests, but it was concluded that the production of blades, with the materials and methods investigated, was feasible and would offer advantages over more conventional materials.

3. Solid-State Welding

3.1 Diffusion Welding of Aluminum/Boron Composites

The few attempts which have been made to diffusion weld the Al/B composites as a secondary fabrication process have demonstrated poor joint efficiency.

The diffusion welding of 6061 aluminum strips to 6061 Al/50B sheet and of double-lap shear joints (Fig. 6a) between the same two materials, heat treated to the T6 condition after welding, produced many unsatisfactory welds.[15] The welding conditions are not described. Both static and fatigue test results are reported as shear strength, but no shear area is given. Therefore, the results are meaningless as far as joint efficiency is concerned.

Diffusion welding of 7039 Al(Al4Zn3Mg)/20B composite sheet with silver-plated sheet faces was accomplished at 482° C and 28 MPa.[10] Butt joints, with a doubler of the same material on each side (Fig. 6b) and an overlap distance of about 7.6 mm on each side of the butt location, failed in shear at a tensile stress in the base material of about 240 MPa, resulting in a joint efficiency of about 50%. The poor results were attributed to composite material of low interlaminar shear strength.

The reports of two additional efforts on the diffusion welding of the Al/B composites, both concerned with the fabrication of aircraft gas-turbine engine blades, are not suitable for the evaluation of weld quality.[16, 17]

At first glance the failure of diffusion welding for joining the Al/B composites is puzzling, since this is one of the most prevalent processes used for primary fabrication of the material. At one time this process was considered to be the most promising for secondary fabrication.[18] However, it has been recently demonstrated that the diffusion welding of aluminum, even at temperatures near the solidus temperature of the alloy, requires appreciable plastic deformation at the faying surfaces in order to attain a joint efficiency of 100%.[19] There are a number of special techniques which will decrease the required deformation; however, most have very limited value for the diffusion welding of composites.

When composites are manufactured by hot pressing of alternate layers of aluminum foil and boron fibers to achieve full compaction of the aluminum, the plastic deformation of the aluminum, when it flows around the boron fibers and into the interstices between fibers, permits diffusion welding of the aluminum at the faying surfaces between the boron fibers. The same explanation applies when tapes of plasma-sprayed aluminum on boron fibers are joined by hot pressing. The inherent porosity of plasma deposits, plus the probable lack of complete penetration of aluminum between the boron fibers results in sufficient deformation during hot pressing for diffusion welding to occur.

It is also probable that a weld between adjacent aluminum layers of less than 100% joint efficiency is adequate for good structural performance. This weld in most structures fabricated of Al/B composites does not experience either high shear or tensile stresses; however, low joint strength can lead to poor fatigue strength of the composite.

On the other hand, to produce a high strength diffusion welded joint between two pieces of fully compacted Al/B composite, containing the usual 40 to 50 v/o filament, is not feasible. This requires so much aluminum plastic deformation that mechanical damage to the brittle boron fibers results.

3.1.1 Aluminum/Boron Composite to Titanium Alloy. Single-lap (Fig. 6c) shear joints of Al/50B composite to Ti6Al6V2Sn were diffusion welded and tested in tensile shear, in support of a program to de-

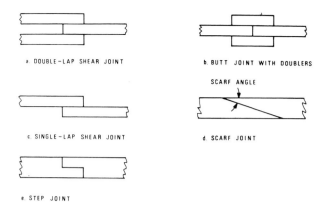

a. DOUBLE-LAP SHEAR JOINT

b. BUTT JOINT WITH DOUBLERS

SCARF ANGLE

c. SINGLE-LAP SHEAR JOINT

d. SCARF JOINT

e. STEP JOINT

Fig. 6—Longitudinal cross-sections of joint forms

velop an aircraft wing section.[8] The joints were welded at the same time as the composite material was fabricated by hot pressing at 34 MPa and 510–566° C for a minimum of 30 min. No heat treatment was performed either before or after welding.

The composite base material consisted of ten layers of boron filaments oriented with three outer layers on either side parallel to the sheet longitudinal direction and the center four layers at 45 deg to the longitudinal direction. The sheet thickness was 1.8 mm and the tensile strength 900 MPa at 20° C and 830 MPa at 177° C.

A series of 12 joints (25.4 mm wide and a nominal lap length of 7.6 mm) welded with an intermediate material of either Al8Si or Al12Si in foil form at the faying surface yielded an average shear strength of 48 MPa and a maximum of 62 MPa. Three joints with no intermediate material failed at an average shear strength of 44 MPa. A duplicate series of 12 specimens, tested at 177° C, indicated no appreciable change in the shear strength of the welds, with or without an intermediate material. All welds failed in shear at the joint. The lowest and the highest strength welds represent joint efficiencies, based on the Al/B composite, of 8 and 33%, respectively.

The diffusion welding process just described was applied to the development of a 6061 Al/50B to Ti6Al6V2Sn joint, with an intermediate material of Al12Si foil. A representative joint of this series is shown in Fig. 7. Aluminum/boron base materials of

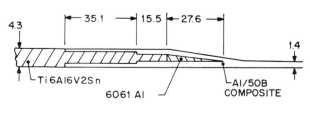

Fig. 7—Diffusion welded joint of Al/B composite to Ti6Al6V2Sn

various thicknesses (1.4–3.2 mm) and fiber arrangements were used, but all had some of the inner fiber layers oriented at 45 deg to the longitudinal direction. The tensile strengths of the Al/B material varied from 660 to 1020 MPa at 20° C and 610 to 950 MPa at 177° C.

The joining process, in some cases, involved a thermal cycle of 5 min at 585° C, which would result in a brazed joint. However, precisely when and how this operation was performed is not explained, and it is not always clear which specimens were subjected to this thermal cycle. Where it is possible to make a comparison, it appeared that there was no significant difference between those specimens which were diffusion welded only and those which were also subjected to the brazing cycle.

The tensile strength efficiency of the joints, which were tested at 20 and 177° C, varied from 74 to 100%, with an average value of 88%. Because of the large shear area, the shear stress on the joint at failure was very low; in the range of 11 to 16 MPa. All joints, with one exception, failed in the Al/B base material.

3.2 Diffusion Welding of Other Aluminum Matrix Composites

Eight scarf joints (Fig. 6d) of whisker-reinforced Al/10Al$_3$Ni composite, produced by unidirectional solidification of the Al-Al$_3$Ni eutectic alloy, were diffusion welded by pressing in a closed die in argon.[20] The base material had an average tensile strength of 280 MPa. Specimen thickness was 1.5, 1.8, or 2.8 mm. The specimen ends were machined to scarf angles varying from about 4 to 13 deg and prepared for welding by acid etching. Welding conditions were a pressure of 55, 69, or 138 MPa, a temperature of 500 or 525° C, and a time of ½ or 1 hr.

Despite conditions favorable to high joint efficiency, such as the low base material strength, the low v/0 of whiskers, and the low-angle scarf joint (which distributes the load over a large area without introducing bending load or stress concentration), the results were mediocre. The eight welded specimens failed at an average tensile stress in the base material corresponding to a joint efficiency of 63%, with a minimum and maximum value of 47 and 74%. Since six specimens failed in base material at an average tensile strength of 188 MPa, it appears that the base material was damaged during welding. A strong indication of this was the higher strength of joints welded at the lower temperatures and also at the lower pressures. In both cases the higher strength joints had greater scarf angles, i.e., smaller joint areas.

Step joints (Fig. 6e), with a lap length of 12 to 25 mm, welded as described for the scarf joints, failed at the weld with a joint efficiency of less than 20%.

A demonstration of the excellent results, that may be obtained by solid-state welding of high-strength aluminum matrix composites when suitable conditions are established, is presented in a report from the Massachusetts Institute of Technology.[21] The base material was a plate of 2024 Al/25 stainless steel with a tensile strength of 1200 MPa and a thickness of 6.3 mm. The steel filament was of Type 355 with a diameter of 0.23 mm and a tensile strength of 3450 MPa. Scarf joints with an angle of 6 deg were enclosed in a die for welding, probably with an air atmosphere. The faying surfaces were prepared for welding by metallographic polishing techniques.

The two joint members, when assembled for welding, were displaced in a manner which resulted in overlapping filaments in the completed weld, as shown schematically in Fig. 8. This also caused considerable plastic deformation at the faying surfaces during welding, even though the final dimensions would not differ greatly from the original.

The best results were obtained from a series of four specimens welded at 538° C; pressures of 84, 101, or 123 MPa; and 1 or 1.5 hr. The weld cross section was reduced by machining all four sides from 6.3 x 6.3 mm to 3.8 x 3.8 mm before tensile testing, which would remove any surface welding flaws. The average joint efficiency was 87%, with a maximum of 98% and a minimum of 74%. Lower welding temperatures, longer welding time, and a longer delay between faying surface polishing and pressure application yielded welds of lower strength.

Another technique for diffusion welding the 2024 Al/25 stainless steel composite, described above, in a thickness of 1.3 mm achieved a joint efficiency of up to 92%.[22] However, since the overlapping of each individual filament layer was required, this technique is not suitable for secondary fabrication.

3.3 Diffusion Welding of Titanium Matrix Composite

Published information is insufficient to permit a quantitative evaluation of diffusion welded titanium matrix composites. Since high quality welds in titanium alloys are produced, with a minimum of plastic

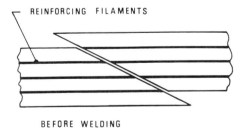

REINFORCING FILAMENTS

BEFORE WELDING

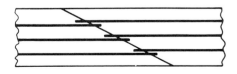

AFTER WELDING

Fig. 8—Scarf joint with overlapping fibers

deformation, by this process; diffusion welds of the composites should be feasible where matrix to matrix welding is required, e.g., a lap joint. However, with the Ti/B composites, it may also be expected that the time-temperature cycle normally required for diffusion welding of titanium, which is about ½ to 2 hr and 815–870° C, would result in decreased base material strength due to damage from matrix-fiber reaction.

An aircraft turbine-engine fan blade of Ti6Al4V was locally stiffened by diffusion welding an inlay of Ti6Al4V/50B composite in recesses in the blade surface.[23] The composite was in the form of fully compacted tape of about 0.20 mm thickness.

A rectangular recess was machined on opposite sides of the blade near the leading edge where the blade thickness varies from about 1.6 to 6.4 mm. A uniform thickness of 0.9 mm of the blade material remained at the center of the blade, i.e., separating the two recesses. After placing sufficient layers of tape, each cut to a predetermined size and shape, to fill the recesses, a 0.5 mm thick Ti6Al4V foil cover of a size greater than that of the recess is laser and electron beam welded around its perimeter to the blade surface on both sides of the blade. The boron fiber is

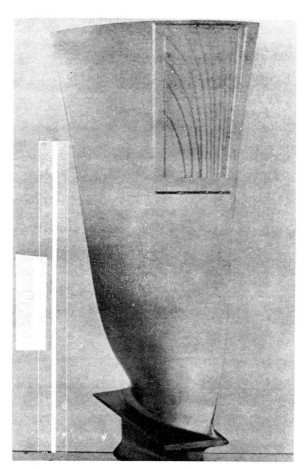

Fig. 9—Jet engine fan blade with diffusion welded insert

parallel to the length of the blade.

This assembly is next diffusion welded by isostatic helium gas pressure of 103 MPa at 815° C for 3 hr. One side of a blade, after diffusion welding and before removal of excess material to the original surface contour of the blade, is shown in Fig. 9. The cover foil shows the rectangular outline of the inlay and the curved outlines of the edges of the tape layers.

Metallographic examination showed excellent welding between tape layers, between tape and recess bottom and cover foil, and around the inlay perimeter between tape edges and recess walls. Various dynamic tests, including engine tests, of blades resulted in no failures at the composite inlay. However, maximum tensile and alternating stresses in the blade at the location of the composite inlay are much lower than the tensile strength or endurance limit of the Ti6Al4V base material. The advantage of the composite inlay, in this case, was a significant improvement in blade vibration stability due to the high elastic modulus of the composite.

4. Brazing and Soldering

4.1 Aluminum/Boron Composite

Double-lap shear specimens of a single strip of Al/50B brazed to 6061 Al were fabricated, in support of a program to develop a small, stiffened panel for an aircraft application.[7] The composite base material, 0.9 mm thick, was produced by hot pressing tape consisting of boron fiber in a matrix of plasma-sprayed 6061 Al on an Al8Si foil. The brazed specimens were 12.7 mm wide and the overlap distance was 6.4 mm. Faying surface preparation was by manual abrasion and wiping with methyl ethyl ketone. Brazing was done in vacuum with filler metal of 0.08 mm or 0.15 mm Al12Si foil and a mechanical pressure of 110 kPa perpendicular to the faying surfaces for a time of ½ hr at 588° C. Post-braze heat treatment is not clear.

Joint tensile-shear stress at failure from a total of 12 specimens varied from 31 to 48 MPa with an average of 40 MPa. Since the composite base material used for the braze specimen had a low tensile strength of about 620 MPa, due to a manufacturing deficiency, the higher strength brazed joints failed in the composite base material at 100% joint efficiency, but the average and minimum strength joints yielded efficiencies of 92 and 71%, respectively. By contrast, specimens of the same design and brazed in the same manner, but with a joint of 6061 Al to 6061 Al, exhibited shear strengths of about 90 to 110 MPa.

Better quality composite material, manufactured by the same process with about 55 v/0 boron fiber, did not suffer a loss in its average tensile strength of 1080 MPa, when subjected to the brazing cycle described above. However, when the structure shown in Fig. 4, consisting of Al/55B longitudinal stiffeners, titanium alloy edge members, and a flat Al/55B sheet, was fabricated by brazing, the strength of the Al/55B composite was reduced by more than 50%. This reduction is attributed to the long brazing cycle, i.e., about 5½ hr above 585° C, necessary for heating the structure with a massive holding fixture. During static load testing, the structure failed prematurely in the sheet and stiffeners.

The fabrication of aircraft landing-gear components with Al/B composite by brazing was largely unsuccessful.[4] The composite material and the brazing process were similar to that described directly above. Three trial components of complicated design, including brazed joints of Al/50B to itself, to Ti6Al4V and to H-11 steel were made. Two failed prematurely during load testing and the third was not of sufficient quality to warrant testing. A competing design with boron-fiber reinforced epoxy material was chosen for the manufacture of a full-size landing gear assembly.

Single-lap shear joints of 6061 Al/50B composite sheet of 1 mm thickness and an overlap distance of 13.4 mm were salt-bath dip brazed with Al12Si filler metal.[10] The specimens were preheated in air at 538° C for 20 min, followed by 2 min in the salt bath at 593° C. Sound joints, as determined by visual and radiographic inspection, failed at a tensile stress in the composite of 340 MPa for a joint efficiency of 33%. When the ends of the overlapped pieces were beveled at 15 deg to the plane of the sheet before brazing, an increase of joint efficiency to 38% was attained. All specimens broke by tensile failure in base material at one end of the lap.

A series of brazed specimens,[24] which illustrated the detrimental effect of stress concentration in laps and doublers on the tensile strength of Al/B composite joints, were fabricated from sheet material. One method for largely overcoming this disadvantage was developed, as explained in the following discussion. However, a lack of specific data for each joint specimen and other information deficiencies, essentially limits the readers' evaluation of the results to the conclusions of the authors. The base material was 6061 Al/50B with a thickness of either 0.6 or 1.0 mm and the average tensile strength varied from 930 to 1170 MPa. The width of all specimens was 8.5 mm, the length of overlap or doubler was 12.7 mm, and doublers were made of base material.

Faying surface preparation of all specimens was by either sand blasting or wire brushing immediately prior to brazing. The specimens, with filler metal foil of 0.025 or 0.05 mm thickness of either Al8Si or Al12Si, were clamped in a stainless steel fixture with a pressure, at room temperature, of about 240 MPa. Brazing was accomplished by heating in argon to 590° C for Al8Si, 580° C for Al12Si, with a holding time between 5 and 30 min. Other specimens were soldered by torch heating, using a Zn5Al5Cu alloy (melting temperature 380° C) as the filler metal. The best joints were obtained by abrasion pre-tinning of the faying surfaces with the filler metal, followed by fluxless soldering accompanied by agitation of the faying surfaces while in contact.

The tensile test results of basic brazed joints are

given as the first four items of Table 4. Deficiencies of the scarf joints were found to be fragmented boron fiber ends at the faying surfaces, caused by cutting and grinding, and poor wetting of boron by the filler metal. The addition of doublers to scarf joints increased the average failure tensile stress to 740 MPa. Scarf joints, with multiple steps rather than a plane surface, yielded low strength joints. A number of variations of the butt joint with doublers; including tapered doublers, pointed doublers, reduced doubler thickness, tri-layer doublers of decreasing length in a stepped arrangement, and a bi-layer doubler of aluminum and Al/B composite were brazed.

None of these variations demonstrated improved strength over the basic butt joint with doublers. A further variation, however, designed to counteract the stress concentration at the ends of the doubler, used a forked doubler as shown in Fig. 10, and did indeed increase the average tensile stress in the base material at failure to 910 MPa, which yields a joint efficiency of about 90%. It must be assumed that the filler metal was either Al8Si or Al12Si.

A comparison of all reported results for soldered joints (Zn5Al5Cu alloy) with those of brazed joints (Al8Si or Al12Si) shows an average strength only 10%

Table 4—Al/B Brazed and Soldered Joints

Joint Form	Filler Metal	Tensile[a] Stress, MPa	Joint[b] Efficiency
Single-lap shear Fig. 6c	Al8Si or Al12Si	510	49
	Zn5Al5Cu	590	57
Butt with doublers Fig. 6b	Al8Si or Al12Si	820	79
	Zn5Al5Cu	790	77
Scarf, 2 deg Fig. 6d	Al8Si or Al12Si	640	62
Scarf, 5 deg Fig. 6d	Al8Si or Al12Si	320	31
Butt with forked doublers Fig. 10	Al8Si or Al12Si	910	88

[a]Average tensile stress in base material at specimen failure.
[b]Approximate value.

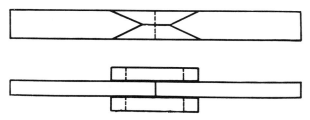

Fig. 10—Brazed butt joint with forked doublers

lower for the former. A shear strength of 48 MPa for joints soldered with the Zn5Al5Cu alloy is reported, although no supporting data is given.

Bending tests of brazed joints with Al8Si and Al12Si resulted in joint shear strengths of about 83 to 96 MPa.

A summary of the shear strength results from work, conducted as part of larger programs extending over a period of about three years, on soldered lap joints between Al/B composite sheet and 6061 aluminum sheet is presented in Table V.[5, 15, 25, 26] The composite base material of Groups 1, 2, and 3 was 6061 Al/47B sheet of 1 mm thickness with a tensile strength of 1207 MPa and of Groups 4, 5, and 6 was 6061 Al/50B sheet of 1.5 mm nominal thickness with an average tensile strength of 1430 MPa and an endurance limit of about 760 MPa. The composite of the latter three groups was solution treated and aged prior to soldering. All 6061 aluminum material was 3.1 mm thick and in the T6 heat treat condition prior to soldering.

All specimens were 12.7 mm wide and the faying surfaces were prepared by electroless plating with about 0.05 mm nickel. Joining was accomplished with the use of flux by manual oxyacetylene torch soldering. The melting temperature of the Cd5Ag, Cd17Zn, and Zn5Al solders are 335–390, 265, and 382° C, respectively. No clear explanation of the discrepancy between the shear strength of single-lap (Group 1) and double-lap joints (Groups 4 and 5) with Cd5Ag solder is given. Later work indicated that joints with Cd17Zn, which yielded the highest joint efficiency in Table 5, were brittle and cracked during cooling from the soldering temperature.[15, 26]

Elevated temperature shear strength of the single-lap soldered joints was also determined.[25] At 93° C the strength with Cd5Ag and Zn5Al increased to about 90 MPa, while at 204° C the strength of Cd5Ag-soldered joints was 47 MPa. Fatigue results ($R = +0.1$) for the double-lap joints, soldered with Ag5Cd alloy indicated an endurance limit joint efficiency of less than about 15%.[15]

Double-lap shear specimens of the same design and material as described above for Groups 4, 5, and 6 of Table 5 were also joined by dip brazing in a molten salt bath with Al8Si as filler metal.[15] The joint overlap varied from 6.1 to 8.4 mm and the specimens were heat treated to the T6 condition after brazing. Shear stress on the brazed joint at failure for 19 specimens varied from 28 to 83 MPa with an average of 48 MPa. Joint efficiency varied from 20 to 58% with an average of 34%. Fatigue results ($R = +0.1$) indicate that the endurance limit of the brazed joint would be less than 40% of the endurance limit of the composite material. Static tested joints failed by tension in the composite for the higher strength joints and by shear at the brazed joint for the lower strength joints, while fatigue failures were a combination of the two failure modes. Joints of thicker material (4.3 mm) failed at a very low tensile stress of 39 MPa and a shear strength of 73 MPa.

A resistance welding machine was used to braze single-lap joints of 1 mm thick 6061 Al/50B sheet.[25] The material was cross-ply Al/B with about one-half of the boron filament layers arranged at 90 deg to the others. One spot braze was made at the center of a 25 mm overlap of 25 mm wide specimens, with a 0.025 mm thick Al8Si filler metal foil preplaced at the faying surface. Three specimens brazed with a 25 by 25 mm square, flat electrode, resulting in a brazed area of about 3.2 cm², failed in tensile shear at an average load of 7.5 kN which was equivalent to a tensile stress in the base material of 290 MPa. Post-braze heat treatment of two specimens to the solution treated and aged condition resulted in an increase of the average failure load to 9.2 kN for a base material tensile stress of 356 MPa. The specimen with the lowest strength; base material tensile stress of 268 MPa, failed in shear at the brazed joint. The remaining four specimens failed in base material tension at one edge of the brazed spot.

A series of joints in 6061 Al/50B composite sheet material was brazed or soldered and tested, in support of a program to develop an aircraft wing section.[8] The base material is described in the second paragraph of Section 3.1.1. No heat treatment was performed either before or after brazing or soldering.

The brazed or soldered specimens were single-lap

Table 5—Al/B to Al Soldered Joints*

Joint Form	Group	Filler Metal	Lap, mm	Spec. Quant.	Shear Strength, MPa	Tensile[a] Stress, MPa	Joint[b] Eff., %	Refs.
Single-lap shear	1	Cd5Ag	3	3	81	240	20	24, 25
	2	Cd17Zn	3	3	74	220	18	24, 25
	3	Zn5Al	3	3	80	240	20	25
Double-lap shear	4	Cd5Ag	3	2	50	200	14	14
	5	Cd5Ag	6	5	54	430	30	14
	6	Cd17Zn	6	8	71	590	41	14

*Base materials described in text of report. All joints failed in tensile shear.
[a] Average tensile stress in Al/B member at joint failure.
[b] Based on Al/B composite.

joints of 25 mm wide strips and were tested in tensile shear. Faying surface preparation and brazing or soldering processes were not described. Joints brazed with Al8Si filler metal foil and a lap length of 6.6 mm at either 582 or 610° C, a mechanically applied brazing pressure of 69–103 kPa and a brazing time of 5 min exhibited, at a test temperature of 20° C, an average shear strength of 43 MPa. At a testing temperature of 177° C, joints brazed at 582° C had a shear strength of 29 MPa and those brazed at 610° C, a strength of 47 MPa. Joints soldered with Cd5Ag filler metal and a lap length of 16.5 mm at 432° C, a soldering pressure of 69–103 kPa, and a soldering time of 5 min failed at an average shear stress of 7 MPa. The tensile strength joint efficiency varied from 16 to 19% for brazed joints at 20° C, 11 to 22% for brazed joints at 177° C, 8 to 15% for soldered joints at 20° C, and 6 to 7% for soldered joints at 177° C.

4.1.1 Aluminum/Boron Composite to Titanium Alloys. The work on the aircraft wing section,[8] discussed in the immediately preceding paragraphs, also included brazed double-lap shear joints of 6061 Al/50B to Ti6Al6V2Sn. The composite base material had a thickness of 3.6 mm and a tensile strength of 1490, 1390, and 1060 MPa at temperatures of 20, 177, and 288° C, respectively. A single strip of composite was brazed to two strips of 4.8 mm thick titanium alloy with Al12Si filler metal at 585° C for 5 min. The nominal width of the specimens was 25.4 mm.

The results of the tensile tests are shown in Table 6. All failures occurred by shear at, or adjacent to, the joint. The strength for a lap length of 9.5 mm, tested at 20° C, does not appear to be representative; joint efficiency normally decreases with decreasing lap length and the strength is based on the results of a single specimen.

The work discussed in the first and second paragraphs of Section 4.1[7] also included brazed joints between Al/50B and titanium (although not specified, this was probably Ti6Al4V alloy). The composite was the low strength material previously described with a thickness of 3.2 mm. A single-lap shear joint with a width of 12.7 mm was used. A group of three specimens, brazed as previously described, with an overlap distance of 6.1 mm and a 0.15 mm filler metal foil of Al12Si, failed in tensile shear with maximum and average values of 68 and 56 MPa, respectively. A second group of three specimens, with the outer layer of aluminum at the faying surface of the composite member removed by grinding to expose the underlying boron fibers, yielded the same shear strength results as the first group. A third group of two specimens, with the same faying surface preparation as for the second group, an overlap of 9.6 mm, and a change of pressure from 110 to 70 kPa, failed at 52 and 44 MPa.

The decrease in strength of the latter two specimens is in accordance with the normal behavior of brazed lap joints. As the shear area increases, the shear strength decreases; but the load which can be transferred across the joint increases.

Stiffener specimens of annealed Ti6Al4V alloy and 25.4 cm long were reinforced with strips of Al/46B composite as shown in Fig. 11.[27] The composite material was manufacutred as described in the first paragraph of Section 4.1 and had an average flexure strength of 807 MPa and an elastic modulus of 221 GPa. The composite strips were joined to the stiffener by brazing in vacuum with an 0.08 mm filler metal foil of Al12Si at 610° C and 172 kPa pressure on the joint.

Compression testing of the reinforced specimens parallel to the 25.4 cm length indicated an improvement of more than 25% in buckling strength and maximum strength over the range from room temperature to 427° C, when compared to unreinforced Ti6Al4V alloy specimens on an equivalent-weight basis. Performance at room temperature was not affected by prior exposure at 315° C for 1000 hr in air or by 400 thermal cycles between −54 and 315° C.

4.2 Brazing Other Aluminum Matrix Composites

A feasibility study of the joining of a graphite fiber

Table 6—Al/B to Titanium Brazed Joints*

Test Temp., °C	Lap, mm	Spec. Quant.	Shear Strength, MPa	Tensile[a] Stress, MPa	Joint[b] Efficiency, %
20	9.5	1	77.1	414	27
20	12.7	3	39.0	276	19
20	25.4	3	34.6	496	33
177	9.5	3	41.4	207	15
177	12.7	3	27.2	193	14
177	25.4	3	30.4	427	31
288	6.2	3	19.6	70	7
288	12.7	3	28.3	200	19
288	25.4	3	27.1	386	36

*Double-lap shear joints between Al/B and Ti6Al6V2Sn sheet.
[a] Average tensile stress at specimen failure.
[b] Based on Al/B composite.

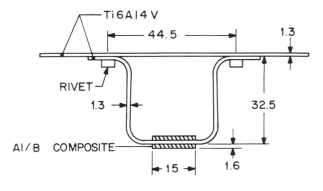

DIMENSIONS IN mm

Fig. 11—Cross-section of stiffener specimen reinforced with Al/B composite

reinforced aluminum alloy matrix composite, produced by the infiltration of graphite yarn with molten aluminum alloys, included brazing experiments.[28] Brazing was performed at 590° C without flux in an argon atmosphere with preplaced filler metal foil at the faying surfaces, which were the flat surfaces of the sheet. It is assumed that some mechanical pressure was applied across the joint during brazing.

Joints of 6061 Al to Al7Zn/C composite and to Al/C composite with Al12Si filler metal indicated excellent wetting and joints of Al7Zn/C to Al7Zn/C with a combination of 6061 Al and Al12Si as filler metal showed fair wetting, but attempts to braze composite to composite with Al12Si filler metal were not successful. The demonstrated requirement for the presence of Mg at the faying surfaces in order to accomplish brazing of the composite is consistent with vacuum brazing of aluminum alloys, where the presence of Mg is also believed to be required.

4.3 General

It has been established that Al/B composite suffers a loss in tensile strength, when exposed to the heating cycle of conventional aluminum brazing at 590–600° C. Although the available information[4, 6, 7, 29] does not permit an accurate prediction of the quantitative effect, it appears that a 10 to 15 min period in this temperature range results in a decrease in tensile strength of about 10 to 20%.

5. Resistance Welding

5.1 Aluminum/Boron Composite

Nearly all investigations of joining composites by resistance welding was performed on Al/B by General Dynamics during the fabrication of several experimental aircraft and spacecraft structures.[3, 5, 9, 15, 25, 30–35] However, despite the many publications on the results of this work, incomplete data, confusing presentation, and a multitude of various material and thickness combinations, make an accurate analysis of the results impossible.

The results of what is believed to be a selection of representative joints, which required assumptions regarding some of the data, is summarized in Table 7. In general, the spot spacing is about equal to what would be used with good resistance welding practice. Therefore, the calculated joint efficiencies are considered to be valid. Exceptions are the multiple spot welded joints, which have insufficient spot spacing. There does not appear to be much difference in joint strengths of composite welded to composite or to aluminum alloys. Failure of most joints was by tensile fracture of the composite material at the edge of a weld, while others failed by combination of tensile fracture at the edge of a weld and shear through the composite parallel to the loading direction.

Average joint efficiency for high-strength Al/B composite spot welded to Al/B composite or to aluminum alloys with a single weld, varied from 14 to 38%. Joints welded with two spots on a line parallel to the loading direction had joint efficiencies of 32 to 53%, but the latter value was obtained with low strength composite. A single multiple-spot welded joint failed at 63%, but others of similar design failed at 37%. A sensitivity of the spot welded joints to bending stresses is indicated by the very low ratio of cross-tension strength to tensile shear strength of about 0.1. This compares with a ratio of about 0.5 for spot welded carbon and low-alloy steels.

Joint No. 15 of Table 7 was also fatigue tested ($R = +0.1$).[15] The results indicate that the endurance limit of the spot welded joints was less than about 170 MPa, as compared to an endurance limit of about 760 MPa for the composite material. This yields an efficiency of about 23%.

A feasibility study of resistance welding of 6061 Al/50B composite sheet, 1 mm thick with an average tensile strength of about 1000 MPa, was made with insufficient material to develop the optimum weld schedule.[10] A seam welded single-lap shear joint exhibited a joint efficiency of 15%, while a similar joint with overlapping spot welds failed at 21%.

Single-lap resistance spot weld joints in 1100 Al/B of very low strength (180 MPa), with fibers at 45 deg to the longitudinal specimen axis (both base material and spot welded specimens) and sheet thicknesses of 0.8 and 1.5 mm, attained a joint efficiency of only about 50 to 70% in tensile shear tests.[6]

5.2 Discussion

Resistance welded joints of medium strength Al/B composites may be achieved with average efficiencies up to 40 to 45% (with the exception of a single weld at 63%), if a large overlap, very closely spaced spots, or a seam weld is used. All of these measures have disadvantages; a large overlap increases the joint weight, very closely spaced spots are not reliably consistent in strength, and seam welds often cause excessive distortion.

Single spot welds in medium strength Al/B composites vary in average efficiency from 14 to 38%. The average failure load of these welds from Table 7 is 6.5 kN and the average sheet thickness is 0.92 mm. This average failure load is less than the minimum failure load obtained in commercial practice with the same thickness of other high strength metals. For example, the minimum failure load for AISI 4340 steel is about 7 kN and for AISI 8630 steel, about 7.5 kN.[36]

The joint efficiency of spot welds in unidirectional composites loaded in the transverse direction is about 50 to 60%.

6. Fusion Welding

Few attempts to join composites by fusion welding have been made, because of the difficulty of locally fusing one substance of the composite without damage to the second, when using the very high temperature and highly concentrated heat sources of the fusion welding processes. However, the great difference in melting points of matrix and fiber in, e.g., the Al/B

Table 7—Al/B Resistance Welded Joints*

| | | Specimen Material | | | Specimen | | | | | At Failure | | |
Joint No.	Member	Base Material	t, mm	T.S., MPa	Type^a	W, mm	Lap, mm	Quantity	Load,^b kN	Stress in Comp., MPa	Joint^c Eff., %	Ref.
1	A	Al/50B	0.51	1170	I	19	...	5	2.79 / 2.67–2.92	294	25	9
	B	←——— Same ———→										29
2	A	Al/50B	0.51	1170	II	19	...	3	0.29 / 0.28–0.32	...	...	9
	B	←——— Same ———→										
3	A	Al/50B	0.89	1124	III	13	25	5^d	5.89 / 5.38–6.32	505	45	9
	B	←——— Same ———→										29
4	A	Al/50B	0.64	848	IV	19	...	4	5.51 / 5.07–6.00	453	53	9
	B	←——— Same ———→										
5	A	Al/50B^e	0.76	1240	I	25	...	7	3.35 / 2.49–3.93	174	14	30
	B	←——— Same ———→										
6	A	Al/50B	0.51	1170	I	19	...	6	2.15 / 1.27–2.49	222	19	30
	B	2024-T3Al	0.51	480								
7	A	Al/50B	0.51	1170	II	19	...	6	0.19 / 0.16–0.21	...	...	30
	B	2024-T3Al	0.51	480								
8	A	Al/50B	0.51	1170	IV	19	...	4	3.60 / 3.43–4.00	372	32	30
	B	2024-T3Al	0.51	480								
9	A	Al/47B	1.02	1207	I	25	25	3	11.78 / 9.79–13.94	462	38	30
	B	Al/43B^f	1.52	1170								
10	A	Al/47B	1.02	1207	IV	25	...	2	11.97 / 9.30–14.66	469	39	24
	B	Al/43B^f	1.52	1170								
11	A	Al/47B	1.02	1207	II	25	...	2	1.26 / 1.12–1.41	...	...	24
	B	Al/43B^f	1.52	1107								
12	A	Al/47B	1.02	1207	V	38	...	1	29.67	763	63	24
	B	Al/43B^f	1.52	1170								
13	A	Al/47B	1.02	1207	VI	46	19	2	17.90 / 16.30–18.90	384	32	24
	B	Al/43B^f	1.52	1170								
14	A	Al/43B	1.02	1207	VII	38	25	2	17.41 / 17.13–17.70	448	37	24
	B	Al/43B^f	1.52	1170								
15	A	6061-T6Al	3.18	310	VIII	13	25	6	10.16 / 8.50–10.93	526	36	14
	B	Al/50B	1.52	1434								
	C	6061-T6Al	3.18	310								
16	A	Al/50B	1.52	1490	I	25	38	3	15.80 / 12.90–18.68	409	27	5
	B	Al/50B	5.08	1490								

*All joints tested in tensile shear, except cross-tension joints.

^a Specimen type: I—Single spot weld. II—Single spot weld joint tested in cross tension. III—Seam weld cut into 13 mm wide strips for testing. IV—Two spot welds on longitudinal axis of specimen. V—Four spot welds in a square pattern. VI—Four spot welds on a common line perpendicular to longitudinal axis of specimen. VII—Four spot welds in a diamond-shape pattern. VIII—Single spot weld in double lap joint.

^b Numerator is average value; denominator is minimum and maximum value.

^c Based on Al/B composite.

^d Five test specimens were cut from one seam weld.

^e Boron filaments coated with vapor-deposited silicon carbide.

^f Material also contained 6 v/o stainless steel filaments in the two outer layers 90 deg to the boron fibers.

composites, would indicate that the development of fusion welding methods for such composites may be feasible.[18, 37]

An experimental demonstration that it is possible to successfully join composites by the gas tungsten arc (GTA) welding process has been reported.[38] Although the composite combinations titanium/tungsten and titanium/graphite, are not considered to be promising for structural application, and the fiber content was not sufficient for a high strength composite, the materials did serve well as models for the investigation of the fusion weldability of composites.

The matrix material was commercially pure titanium (Ti75A), the tungsten filament was 0.2 mm diameter, and the graphite fiber was 0.008 mm diameter. Composite sheet thicknesses were 0.9 to 1.4 mm.

The Ti/W composite varied in filament content from 3 to 19% with composite tensile strengths of 685 to about 760 MPa. Bead-on-sheet specimens welded parallel to the fiber direction and tested in the same direction, exhibited higher tensile strengths than comparable base material specimens for filament contents up to 19%. Tensile elongation was about the same for base material and welds.

Specimens of Ti/10W composite were prepared for butt joints with a square-groove weld transverse to the filament direction by chemical etching to remove the titanium matrix from the faying surfaces, which exposed about 0.25 mm lengths of W filaments protruding from each faying surface. After intermeshing of the protruding filaments, the welds were made. Transverse weld tensile specimens, i.e., with the longitudinal specimen axis parallel to the filament direction, failed in the base material at 100% joint efficiency. The same results were achieved with plane faying surfaces, when the weld thickness was increased by the addition of sufficient filler metal.

Fabrication defects in the Ti/C composite caused welding difficulties and no mechanical testing results are given.

Metallographic examination showed, for both the Ti/W and the Ti/C composites, that the thickness of the reaction zone at the matrix-filament interface could be reduced when welding with the lower energy of machine welding, rather than manual welding.

The GTA welding process was also employed for a limited study[39] of arc welding thermal effects on Al/B composite sheet material. Alternating current was used for the study.

Arc-spot welds in 6061 Al/40B composite (1.2 mm thick), with no filler metal, caused severe fiber damage and poor flow and wetting of the fibers by the molten aluminum matrix. However, when Al5Si filler metal (ER4043) was used, these problems were largely eliminated, although some weld porosity was present. Apparently, the additional molten metal offers some shielding of the boron fibers from the high temperature welding arc and allows the aluminum alloy to wet the fibers. The silicon addition may also play a role in the flow and wetting behavior.

Bead-on-sheet welds in 6061 Al/50B material (0.6 mm thick) were made with Al5Si filler metal by manual welding at an energy input of about 200 kJ/m and one pass on each side. Examination by optical and scanning electron microscopy and by electron microprobe analysis indicated uniform alloying of matrix and filler metal through the sheet thickness, excellent wetting of the boron fibers, and no apparent fiber damage.

From a report[28] of superficial experiments with the GTA welding of an Al7Zn/C composite sheet, it is difficult to determine precisely what was done. However, it appears that, under some conditions, manual welding with Al5Si filler metal results in flow of the weld metal and wetting of the graphite fibers to form a weld, accompanied by considerable aluminum carbide formed at the matrix-fiber interface. The use of machine welding, with lower energy, would be expected to decrease the thickness of the carbide phase, which decomposed in the presence of water, but would not eliminate it.

An early exploratory joining program[10] included GTA and electron beam (EB) welding experiments with Al/B sheet composites. The work was handicapped by the lower base material quality of this earlier period in the primary fabrication of composites. Their greater void content would result in the entrapment of various contaminants, which could not be removed. The contaminants could then cause difficulties such as porosity and poor flow during welding.

All GTA welding was performed manually using alternating current. Butt welds, transverse to the fiber direction, in 7039 Al(Al4Zn3Mg)/20B of 1.2 mm sheet thickness and 480 MPa tensile strength with Al5Mg filler metal (ER5183) did not achieve full penetration, even with one weld pass from each side, when a square-groove joint preparation was used. A 90 deg single-vee-groove joint preparation was much better in joint penetration and in preventing voids. With sufficient weld bead reinforcement, tensile testing transverse to the weld resulted in failure at the weld toe at 70% joint efficiency.

Much less success was attained when GTA butt welding a 1 mm 6061 Al/50B sheet (tensile strength about 1000 MPa) with a 120 deg single-vee-groove joint preparation. Poor weld metal flow, gross weld porosity, and inadequate penetration were serious problems. A tensile specimen with the weld reinforcement machined flush yielded a joint efficiency of 9%.

Butt welds made by the EB process with a square-groove joint transverse to the fiber direction, in the 7039 Al/20B material previously described, produced numerous weld voids. A joint efficiency of 49% was achieved. No information is given on the use of filler metal. Much improved weld behavior resulted when EB welding the 6061 Al/50B composite with Al5Si filler metal. A butt weld with a square-groove joint, made with one weld pass from each side, failed at a joint efficiency of 19%. This increased to 27% with a 20 deg single-vee-groove joint, welded with two passes from the same side.

6.1 Discussion

Although the basic work described in the preceding section has proven that composites are weldable by fusion processes, much further development remains to be done before these processes could find significant practical application for the efficient joining of high-strength composites.

The filler metal composition may be critical in the fusion welding of Al/B composites. It is not stated specifically what filler metal was used in the earlier work on the GTA welding of 6061 Al/50B,[10] but it may be assumed that it was the same as used for 7039 Al/20B in the same investigation, that is, Al5Mg. If this is the case, it may explain the poor results with regard to the weld behavior (penetration, wetting, flow), when compared with the favorable results of later work[39] on 6061 Al/50B composite using Al5Si filler metal. Furthermore, the EB welds with Al5Si filler metal, produced during the first study,[10] were of satisfactory weld behavior. The evidence tentatively indicates that a filler metal containing silicon may be essential to the fusion welding of Al/B composites.

7. Mechanical Joining

The tensile shear test results from a selection of typical mechanically fastened joints reported in the literature are summarized in Table 8. The formulas used for stress calculations are given in Table 9, the notation for joint dimensions in Fig. 12, and joint failure types in Fig. 13. The dimension "e" of Fig. 12 is referred to as edge distance. The joint lap length for joints 12a and 12b is equal to $2e$; for joint 12c, it is

Table 8—Al/B and Al/Be Mechanically Fastened Joints*

Joint No.	Specimen Material Base Material[a]	Other[b]	t, mm	T.S., MPa	Type[c]	Joint Fastener	Dimensions, mm[d] D	w	e	x	p	Quan.	Load, P kN	Test Results Stress, MPa[e] σ_B	σ_S	σ_TN	σ_T	Failure[f] Type	Joint Eff., %	Refs.
1	7002 Al/24B	...	1.0	620	II	rivets	4.0	15.9	22.2	...	...	2	4.77	1192	108	401	300	B and T	48	39,40
2	7002 Al/24B	...	1.0	620	II	2 rivets	4.0	31.8	25.4	...	15.9	1	9.63	1204	95	405	303	B and T	49	39,40
3	7002 Al/37Be	...	0.8	520	II	rivets	4.0	15.9	20.1	...	...	3	3.26	1019	101	342	256	B and T	41	39,40
4	7002 Al/37Be	...	0.8	520	II	2 rivets	4.0	31.8	20.8	...	15.9	2	6.13	958	92	322	241	B and T	39	39,40
5	6061 Al/47B CP	same	2.8	855g	I	2 bolts	4.8	25.4	14.3	28.6	...	3	19.48g	725	122	338	274	S and T	32	8
6	6061 Al/47B CP	same	2.8	855g	I	2 bolts	6.4	38.1	19.0	38.1	...	3	31.58g	881	148	356	296	S and T	35	8
7	6061 Al/47B CP	same	2.8	855g	I	2 bolts	7.9	50.8	23.8	47.5	...	3	49.28g	1114	185	410	346	S and T	40	8
8	1100 Al/40B	...	1.2	700	II	pin	4.7	33.2	16.6	...	...	3	2.53	448	64	74	64	S	11	41
9	1100 Al/40B	...	1.2	450h	II	pin	3.2	22.2	11.1	...	...	3	1.00h	260	38	44	38	S	8	41
10	1100 Al/40B 90°	...	1.2	70	II	pin	3.2	15.9	7.9	...	...	3	0.94	244	50	62	49	T	70	41
11	6061 Al/53B CP	...	2.4	400	II	pin	17.5	30.3	21.6	...	...	3	8.31	198	80	270	114	T	28	4
12	6061 Al/45B	steel	6.4	1510	I	bolt	8.0	31.6	15.8	...	...	3	21.97	429	109	145	109	S	7	42
13	6061 Al/50B	6061 Al	1.5	1430	I	rivet	6.4	12.7	12.7	...	...	10	6.4	667	168	677	336	H	23	14
14	6061 Al/50B	6061 Al	1.5	1430	II	bolt	6.4	12.7	12.7	...	...	8	8.3	865	218	878	436	H	30	14
15	6061 Al/50 CP	Ti6Al4V	4.4	250	I	2 swagged collar	6.4	28.0	25.4	25.4	...	2	26.2	465	59	275	213	T	85	5
16	6061 Al/50B CP	Ti6Al4V	4.4	250	II	swagged collar	6.4	28.0	25.4	...	...	2	27.4	973	122	288	222	T	89	5

* Stress, failure type, and joint efficiency refers to composite. Additional information is given in the text.

[a] The notation "90°", indicates material which was tested perpendicular to the longitudinal direction. If some of the fiber layers are not parallel to the longitudinal direction, this is noted by "CP," i.e., cross ply.

[b] Material to which the composite was joined.

[c] I, single-lap shear; H-double-lap shear.

[d] See Fig. 12.

[e] σ_B = bearing stress; σ_S = shear stress; σ_{TN} = net tension stress; σ_T = tension stress. See Table IX for stress calculations.

[f] See Fig. 13.

[g] Tested at 177° C.

[h] Tested at 260° C.

Table 9—Stress Calculations*

	Bearing Stress, σ_B	Shear Stress, σ_S	Net Tension Stress, σ_{TN}	Tension Stress, σ_T
Single fastener	Dt	$2et$	$t(w\text{-}D)$	tw
Double fastener, Fig. 12b	$2Dt$	$4et$	$t(w\text{-}2D)$	tw
Double fastener, Fig. 12c	$2Dt$	$4et^a$	$t(w\text{-}D)$	tw

* Basic formula is: $\sigma = P/A$, with area (A) values from table body. See Fig. 12 for letter definitions.

^{a} If joint is between composite and a second, different material, the formula for area is: $2(e + x)t$.

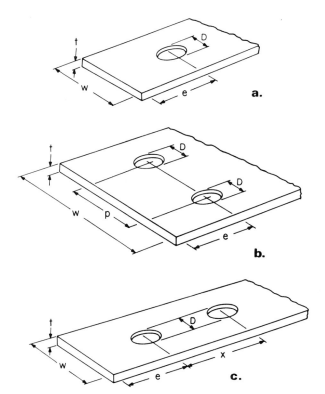

Fig. 12—Dimension notation for mechanically fastened joints

equal to $2e + x$. To avoid unnecessary complications in data presentation, both in the tables and in the text, averages and simplifications have been used, where this could be done without significant distortion of the results. The stress calculation formulas and joint failure types are not intended to be applicable to all fastener joints, although they are generally consistent with current fastener technology. The geometric arrangement of joints with two fasteners in Table 8 is indicated by the joint dimensions, with reference to Fig. 12.

The joint efficiencies reported in Table 8 and elsewhere in this section are consistent with the definition given in Section 1.1. That is, the calculation of the specimen area includes the full specimen width; the diameter of the fastener hole is not deducted, as is normally done in fastener technology.

Similar joints to those of numbers 1–4, with lesser edge distances, failed in composite shear at efficiencies of 16 to 36%.[40, 41] Joints of the same design as numbers 1–4, but with 7002-T6 aluminum alloy base material (tensile strength 420 MPa), were also tested, with a resulting efficiency of about 80%.

Swaged collar fasteners, rather than bolts, used for joints similar to numbers 5–7 yielded approximately the same joint efficiency.[8] One joint, tested at room temperature, had the same efficiency as a similar joint tested at 177° C.

Composite was tested in joints 8–10 by pinning a sheet strip in a clevis.[42] Joints of similar design with cross-ply material exhibited a tensile stress in the composite at failure almost twice that of the unidirectional material, but the joint efficiency is unknown, since no information on the tensile strength could be found.

Some improvement in the efficiency of joint 12, but also at a sacrifice in base material tensile strength, was achieved by the inclusion of stainless steel wire mesh in the composite or the orientation of some boron layers at 45 deg to the tension axis.[42]

Fatigue testing of joint 13 at $R = +0.1$, indicated an endurance limit of about 170 MPa.[15] With an endurance limit for the base material of about 760 MPa, the fatigue efficiency of the riveted joint was

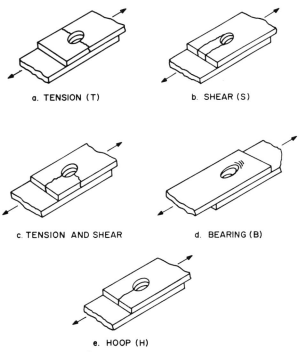

a. TENSION (T) b. SHEAR (S)

c. TENSION AND SHEAR d. BEARING (B)

e. HOOP (H)

Fig. 13—Failure types for mechanically fastened joints

about 23%. Joint 14 was also fatigue tested; but, since most of the fatigue strengths were higher than the strength of the bolted joint tested in static loading, the results are considered invalid.

The efficiency of joint 15 decreased to an anomalous 64% at a testing temperature of 93° C and to 70% at 316° C, while joint 16 decreased to 85 and 69%, respectively, at the same temperatures.[5]

Two additional publications,[9, 25] which contain reports of work on mechanically fastened joints in Al/B composite, are not included in Table 8. The many different combinations of materials, conditions, and joint designs, obscured or missing essential data, and a minimum of logical order in the data does not permit the results to be efficiently summarized in tabular form. However, an approximation of the more pertinent results follows.

The material 6061 Al/50B, with a tensile strength of 1200 MPa and a thickness of 0.5 to 0.6 mm, was tested as single-lap shear joints with two rivets, arranged as shown in Fig. 12c.[9] As the specimen width was increased from 18 to 36 mm, the rivet diameter was increased from 2.4 to 4.0 mm, and the lap length from 30 to 75 mm. The joint efficiency of these specimens varied from 12 to 22%, with failure occurring in the composite. Similar joints, but with doublers adhesively bonded to the faying surfaces of the base material, yielded an increase in joint efficiency from 26 to 29%. The doublers were mostly of 0.4 mm aluminum alloy, but also included stainless steel foil and Al/B composite. At a testing temperature of 150° C, the joints without doublers failed at an efficiency of 11%, and those with doublers at 22%, based on a tensile strength of 1100 MPa for the base material at 150° C.

The rivet diameter of the joints with doublers was limited to 3.2 mm and joint failure was by rivet shear. However, it is improbable that a significant increase in joint efficiency could have been attained with rivets of greater diameter or strength. Similar joints with doublers and 4.0 mm bolts, failed in the composite at 27% joint efficiency.

Single-rivet joints attained a 7% efficiency with no doubler and 20% with doublers of Al/B composite.

The composite was also tested at 90 deg to the filament direction (about 105 MPa tensile strength), using the joint with two fasteners. Joint efficiency was about 30 to 40% with no doubler, and 100% with doublers of 0.4 to 0.6 mm thick aluminum alloy.

Cross-ply 6061 Al/45B composite, with a tensile strength of 550 MPa and thicknesses of 0.6 and 0.9 mm, was tested as the two-fastener joints; and with approximately the same range of joint dimensions, previously described for the 6061 Al/50B unidirectional material.[9] Riveted joints with no doubler yielded efficiencies of 21 to 38% and riveted and bolted joints with 0.4 to 0.6 mm aluminum alloy doublers, 46 to 51%. At a testing temperature of 150° C the joint efficiencies, for joints with and without doublers, was about the same as at room temperature, based on a base material strength of 410 MPa at 150° C.

An extensive series of riveted single-lap joints were produced in both 6061 Al/50B composite, with a tensile strength of 1140 MPa, and 6061 Al/47B cross-ply composite, with a tensile strength of 820 MPa.[25] Rivet diameters for all of the specimens varied from 3.2 to 4.7 mm and the failure type of virtually all joints was other than rivet shear.

At specimen widths of 6 to 15 mm and lap lengths of 28 to 81 mm, single-rivet joints in the 6061 Al/50B material exhibited efficiencies of 17 to 36%. Double-rivet joints, as shown in Fig. 12c, with about double the width and lap length of the single-rivet joints, failed at joint efficiencies of 22 to 39%.

The joint efficiency of both single- and double-rivet specimens with cross-ply material was about 1.15 to 1.20 times that of joints in the unidirectional material. Joint dimensions of the cross-ply specimens was about the same for the width and about one-half for the overlap, as those of the unidirectional material.

Heat treatment of the 6061 aluminum matrix had little or no effect on the efficiency of joints in either material.

As would be expected, joints with four rivets, arranged in a square pattern, yielded about the same efficiency as double-rivet specimens.

Adhesive bonded joints, in addition to riveting, with about one-half the lap length of comparable specimens without adhesive showed a joint efficiency of one-half or less for both composite materials, compared to joints with no adhesive, when polyurethane adhesive was used. An epoxy-phenolic adhesive improved the joint efficiency, but it was still less than for comparable joints in unidirectional material with no adhesive and no better than comparable joints in cross-ply material with no adhesive. The effect of increased lap length on joint efficiency was not determined.

7.1 Discussion

The low strength of mechanically fastened joints in composites, illustrated in the foregoing section of experimental results, has been recognized by the designers of a recently constructed aircraft wing section.[44] The results of a study to develop reliable design strengths for various composite materials, based on lamination theory supported by testing and empirical extrapolation, included bolted joints in Al/B composite.

The base material was 6061 Al/47B with a tensile strength of 1480 MPa. All test specimens were single-lap shear joints with a single rivet, tested in tension. The range of test parameters were a D/t ratio of 1.7 to 2.8, a w/D ratio of 5.2 to 6.4, and an e/D ratio of 3.0. Although the material thickness was not given, the program for which the data was developed,[8] included Al/B composites of about 1 to 3 mm thickness.

By combining the average design net tension strength of about 480 MPa with the design bearing strength, which is dependent upon the D/t ratio, a maximum joint efficiency of about 20% is attainable. To realize this efficiency with a design shear strength of about 40 MPa, a lap length of about 90 mm would be required in a joint with a 12 mm joint width per fastener.

Various techniques to improve the low joint efficiency in the longitudinal direction of mechanically fastened joints have been applied; such as the orientation of some fiber layers at 45 deg to the longitudinal direction, the use of doublers, and an increase of the number of rivet rows. However, these techniques cause a reduction in the material strength or increase the joint weight; in a joint which already is at a weight disadvantage, when compared to other joining methods.

Despite the low joint efficiency of mechanical fasteners, this joining process has found the widest application in the fabrication of composite structures. This is probably due to a higher reliability, resulting from less difficult process control, than other joining processes for composites, and to the ease of joint disassembly.

8. Adhesive Bonding

Double-lap shear joints in 1100 Al/40B composite sheet (tensile strength about 700 MPa) were bonded with an epoxy-phenolic adhesive using specimens of 19.5 mm width and a lap length of 12.7 mm.[42] Joints of 0.9 mm thick material failed in the base material at about 60% joint efficiency; those of 0.6 mm thick material yielded about 100% joint efficiency, also failing in the base material. The joint efficiency for joints in the thinner composite at a testing temperature of 260° C was about 95%, with failure occurring by shear in the adhesive. The tensile strength of the base material at 260° C was about 450 MPa.

Similar results were experienced by another investigator[10] with an adhesive (elastomeric modified epoxy) bonded butt joint with doublers. The base material was 6061 Al/50B sheet with a thickness of 1 mm and a tensile strength of about 1000 MPa. The joint failed in base material at an efficiency of about 70%.

Filler strips in ring form of 6061 aluminum were adhesive bonded[9] to each end of the cylindrical section, fabricated of Al/B composite, shown in Fig. 5. Many of the Al/B structural shapes used as stiffeners for the aircraft fuselage bulkhead shown in Fig. 3, were adhesive bonded, in combination with bolts, to the Al/B panel.[3] Quantitative data on the structural behavior of the adhesive bonded joints is not available.

The lack of interest in adhesive bonded joints for composite materials, indicated by the paucity of reported results, is probably consistent with the potential for producing efficient joints, from the aspect of both weight and strength, in high strength composites by this process.

9. Summary

Metals reinforced with fibers of high strength and high elastic modulus have the potential for yielding structural materials with greater strength-to-weight and stiffness-to-weight ratios, combined with acceptable toughness, than the present metal alloys. The composite which has reached the highest stage of development, aluminum alloy matrix with continuous boron fibers (Al/B), has demonstrated a tensile strength of 1450 MPa and an elastic modulus of 230 GPa at a density of 2.7 g/cm^3. Other composites of various combinations of the metal matrixes Al, Ti, and Ni with reinforcing fibers of boron, carbon, SiC, Al_2O_3, Be, etc., have also been produced.

Difficulties in manufacturing and fabrication and high cost have prevented the metal-matrix composites from attaining the application stage as structural materials. Virtually all of the uses consist of proposed applications or the production of demonstration or test articles. Composites have been manufactured by many methods, but the most widely used method is the solid-state hot-pressing of multiple layers of bare fibers between foils of the matrix metal.

Since most development effort has been with the Al/B composites, and very few other composites have been developed sufficiently for joining studies, most of the literature on composite joining is concerned with the Al/B composite. However, much of the joining technology developed should be applicable to other composites.

Four major problems in the joining of the composites are evident: (1) Many fiber materials, e.g., boron, graphite, and Al_2O_3, are not wetted by many filler metals which are compatible with the matrix material. (2) Unlike conventional materials, the composites depend on a continuity of fibers across any given plane transverse to the major stress axis for their strength. The interruption in this continuity, inherent in a joint, is difficult to compensate for with any known joining method. (3) Stress concentrations, such as in a lap shear joint or a bolt hole, results in decreased composite strength. (4) Fusion of one component of the composite, i.e., the fiber, cannot be tolerated without almost complete loss of the composite mechanical properties. Lesser difficulties include the degradation of composite strength when exposed to brazing conditions, the matrix of the major composite, Al/B, cannot be diffusion welded at conditions compatible with the composite, and bolted joints and spot-welded joints subject the composite to an unfavorable stress condition, i.e., shear parallel to the fibers.

The plastic deformation required for the solid-state pressure welding of aluminum is more than can be sustained, without fiber damage, by the composites containing brittle fibers, e.g., Al/B. On the other hand, an Al/stainless steel composite of 1200 MPa tensile strength has been welded by this process with high joint efficiency. The diffusion welding of Al/B composite to titanium alloy, although producing joints of low shear strength, has resulted in high joint efficiency with a joint of large shear area and complicated design.

Most development work on composite joining has been with the brazing, resistance welding, and mechanical fastening processes. Considerable effort has also been devoted to soldering. The process which yields the lowest joint efficiency, mechanical fastening, is also the process which has found the widest application in the assembly of composite structures; probably due to a higher reliability.

Even when consideration is limited to the joining of the Al/B composites, the wide variety of base materials and joint designs and the large scatter in results makes it difficult to arrive at generalized quantitative conclusions. However, a trend to lower joint efficiency with increasing composite strength is evident. It is estimated that the average joint efficiency for very low tensile strength composites of 600 MPa is about 60% for brazing and resistance welding and 40% for mechanical fasteners, and the joint efficiencies for 1400 MPa tensile strength composites is about 30% for brazing and resistance welding and 20% for mechanical fasteners. From the more limited results for soldered joints, it is estimated that the joint efficiency is about the same as for mechanically fastened joints.

The joint strengths are generally higher than these averages; at an increased weight penalty, when joint doublers or greater lap lengths are used.

Adhesive bonded joints of high weight (double-lap shear and butt joint with doublers) in low strength, thin Al/B composite exhibited a joint efficiency of 70 to 100%. No results of more significant tests have been reported.

Insufficient investigation has been accomplished to allow more than a preliminary estimate of the feasibility of fusion welding the composites. Early experiments with GTA and EB welding, which did not include significant development work, produced butt welds of 50 to 70% joint efficiency in very low strength Al/B composite (480 MPa tensile strength) and 9 to 27% joint efficiency in low strength Al/B composite (1000 MPa tensile strength). Later work with the GTA welding of Al/50B composite yielded promising results, based on metallographic examination only. Gas tungsten-arc welds attained a joint efficiency of 100% with Ti/10W composite of about 700 MPa tensile strength.

In view of the good potential of the fusion welding processes for producing efficient joints in high strength composites, the limited success achieved with a minimum of experimental work, appears to justify further investigation.

References

1. Steinhagen, C. A. and Stanley, M. W., "Boron/Aluminum Compressor Blades," General Electric Co., Cincinnati, Ohio. Rpt AFML-TR-73-285, 1973.

2. Boll, K. G., "Advanced Composite Engine Development Program," Pratt & Whitney Aircraft Co., United Aircraft Corp., East Hartford, Conn. Rpt AFML-TR-72-108, Parts I and II, 1972.

3. Swazey, E. H. and Wennhold, W. F., "Advanced Composite Technology Fuselage Program," Convair Aerospace, Div. General Dynamics Corp., San Diego, Calif. Rpt AFML-TR-71-41, Vol. III, 1972.

4. Anon. "Filament Composite Material Landing Gear Program," Bendix Corp., South Bend, Ind. Rpt AFFDL-TR-72-78, 1972.

5. Miller, M. F., Christian, J. L., Wennhold, W. F., and Spier, E. E., "Design, Manufacture, Development, Test, and Evaluation of Boron/Aluminum Structural Components for Space Shuttle," Convair Aerospace Div., General Dynamics Corp., San Diego, Calif. Rpt No. GDCA-DBG73-006, Contract NAS8-27738, 1973.

6. Niemann, J. T. and Garrett, R. A., "Eutectic Bonding of Boron Aluminum Structural Components," Weld. Jnl., 53, 175s–184s and 351s–360s (1974).

7. Cheatham, R. G., Bulloch, D. F., Dobyns, A. L., et al., "Development of Fabrication Techniques for Borsic-Aluminum Aircraft Structures," Boeing Company, Seattle, Wash. Rpt AFML-TR-72-25, 1972.

8. Donohue, P. J., et al., "Composite Box Beam Optimization," Grumman Aerospace Corp., Bethpage, N. Y. Rpt AFML-TR-74-105, 1974.

9. Schaefer, W. H., Christian, J. L., et al., "Evaluation of the Structural Behavior of Filament Reinforced Metal Matrix Composites," Convair Aerospace Division, General Dynamics Corp., San Diego, Calif. Rpt AFML-TR-69-36, 1969.

10. Happe, R. A. and Yeast, A. J., "Metal Matrix Composites Annual Report," Space Div., North American Rockwell Corp., Los Angeles, Calif. Rpt SD 68-971, 1968.

11. Weisinger, M. D., Forest, J. D., and Miller, M., "Feasibility Demonstration Program for the Application of Boron/Aluminum to Space Shuttle," Convair Aerospace Div., General Dynamics Corp., San Diego, Calif. Rpt No. CASD-NAS-74-017, 1974.

12. Goodwin, V. L. and Herman, M., "Beryllium Wire-Metal Matrix Composites Program," Allison Division, General Motors Corp., Indianapolis, Ind. Final Rpt EDR 6518, Contract N00019-69-C-0234, 1969.

13. Stevens, C. and Hanink, D. K., "Titanium Composite Fan Blades," Allison Div., General Motors Corp., Indianapolis, Ind. Rpt AFML-TR-70-180, 1970.

14. Thompson, E. R., George, F. D., and Kraft, E. H., "Investigation to Develop a High Strength Eutectic Alloy With Controlled Microstructure," United Aircraft Research Laboratories, United Aircraft Corp., East Hartford, Conn. Final Rpt J910868-4, Contract N00019-70-C-0052, 1970.

15. Hersh, M. S., "Fatigue of Boron/Aluminum Composites," Convair Aerospace Div., General Dynamics Corp., San Diego, Calif. Rpt No. ZZL-72-006, Contract NAS8-27437, 1972.

16. Anon., "Research and Development of an Advanced Composite Technology Base and Component Fabrication Program for Gas Turbine Compressor and Fan Blades," Allison Div., General Motors Corp., Indianapolis, Ind. Rpt AFML-TR-68-258, pp. 343–347, 1969.

17. Toth, I. J., "Comparison of Manufacturing Processes for Filamentary Reinforced Aluminum and Titanium Alloys," Presented at Meeting of National Materials Advisory Board, Washington, D. C., 9 May 1973.

18. Metzger, G. E., "Welding of Metal-Matrix Fiber-Reinforced Materials," Wright-Patterson AFB, Ohio. Rpt AFML-TR-68-101, 1968.

19. Metzger, G. E., "The Brazing and Forge Welding of Aluminum Alloys," Paper presented at American Welding Society Annual Meeting, Chicago, Ill., 1973.

20. Salkind, M. J., "Diffusion Bonding of Whisker-Reinforced Aluminum," Trans. Met. Soc. AIME, 242, 2518–2520 (1968).

21. Olster, E. F. and Jones, R. C., "Joining of a Metal Matrix Composite," Massachusetts Institute of Technology, Cambridge, Mass. Rpt R69-25, 1969.

22. Sumner, E. V. and Cole, D. Q., "Development of Ultrahigh Strength Low Density Aluminum Plate Composites," Harvey Aluminum Co., Torrance, Calif. Rpt HA-2166, Contract NAS-11508, 1965.

23. Sippel, G. R. and Swain, K., "Turbine Engine Fan Blade Stability Using Metal Matrix Composites," Allison Div., General Motors Corp., Indianapolis, Ind. Rpt AFAPL-TR-73-125, 1973.

24. Breinan, E. M. and Kreider, K. G., "Braze Bonding and Joining of Aluminum Boron Composites," Met. Eng. Quart., 9 (4), 5–15 (1969).

25. Miller, M. F., Schaefer, W. H., Weisinger, M. D., et al., "Development of Improved Metal-Matrix Fabrication Techniques for Aircraft Structure," Convair Aerospace Div., General Dynamics Corp., San Diego, Calif. Rpt AFML-TR-71-181, 1971.

26. Robertson, A. R., Miller, M. F., and Maikish, C. R., "Soldering and Brazing of Advanced Metal-Matrix Structures," Weld. Jnl., 38, 446s–453s (1973).

27. Herring, H. W., Carri, R. L., and Webster, R. C., "Compressive Behavior of Titanium Alloy Skin-Stiffener Specimens Selectively Reinforced with Boron-Aluminum Composite," Langley Research Center, Hampton, Va. Rpt NASA-TN-D-6548, 1971.

28. Goddard, D. M., Pepper, R. T., Upp, J. W., and Kendall, E. G., "Feasibility of Brazing and Welding Aluminum-Graphite Composite," Weld. Jnl., 51, 178s–182s (1972).

29. Klein, M. J. and Metcalfe, A. G., "Effects of Interfaces in Metal Matrix Composites on Mechanical Properties," Solar Div., International Harvester Co., San Diego, Calif. Rpt AFML-TR-71-189, 1971.

30. Hersh, M. S., "Resistance Welding of Metal Matrix Composite," Weld. Jnl., 47, 404s–409s (1968).

31. Hersh, M. S., "Resistance Spot Welding of Metal Matrix Composite," Weld. Jnl., 49, 254s–258s (1970).

32. Hersh, M. S., "Correlation Between Boron/Aluminum Sheet Quality and Resistance Weld Quality and Strength," Weld. Jnl., 50, 515s–521s (1971).

33. Hersh, M. S., "The Versatility of Resistance Welding Machines for Joining Aluminum/Boron Composites," Weld. Jnl., 51, 626–632 (1972).

34. Hersh, M. S., "Resistance Diffusion Bonding Boron/Aluminum Composite to Titanium," Weld. Jnl., 52, 370s–376s (1973).

35. Anon., "Design and Fabrication of a Boron Aluminum Composite Wing Box Test Specimen," Convair Aerospace Div., General Dynamics Corp., San Diego, Calif. Rpt No. CASD-NSC73-005, Contract N6ss69-72C-0414, 1973.

36. "Recommended Practices for Resistance Welding," Miami, Fla.,

American Welding Society, AWS C1.1, 1966.

37. Hauser, D., "Methods of Joining Composites to Produce Structures," Battelle Memorial Inst., Columbus, Ohio. Summary Report, 1966.

38. Kennedy, J. R., "Fusion Welding of Titanium-Tungsten and Titanium-Graphite Composites," Grumman Aerospace Corp., Bethpage, N. Y., Rpt RM-519, 1971.

39. Kennedy, J. R., "Microstructural Observations of Arc Welded Boron-Aluminum Composites," *Weld. Jnl.*, **52**, 120s–124s (1973).

40. McDanels, D. L., "Riveting of Filament-Reinforced Aluminum Composite," North American Aviation, Los Angeles, Calif. Rpt NA-66-1220, 1966.

41. Toy, A., Atteridge, D. G., and Sinizer, D. I., "Development and Evaluation of the Diffusion Bonding Process as a Method to Produce Fibrous Reinforced Metal Matrix Composite Materials," North American Aviation, Los Angeles, Calif. Rpt AFML-TR-66-350, 1966.

42. Brown, N. M., et al., "Boron Aluminum Composite Structure Program," McDonnell Douglas Corp., St. Louis, Mo. Rpt No. EO424, Contract NAS8-26295, 1970.

43. Matoi, T. T., "Exploratory Development and Testing of Large Cross-Sectional Area Hardware of Boron/Aluminum," Rockwell International Corp., Los Angeles, Calif. Rpt AFML-TR-73-306, 1973.

44. Cairo, R. P. and Torcyzner R. D., "Graphite/Epoxy, Boron-Graphite/Epoxy Hybrid and Boron/Aluminum Design Allowables," Grumman Aerospace Corp., Bethpage, N. Y. Rpt AFML-TR-72-232, 1972.

Bibliography

1. Rauch, H. W., Sutton, W. H., and McCreight, L. R., "Survey of Ceramic Fibers and Fibrous Composite Materials," Space Sciences Lab., General Electric Co., King of Prussia, Pa. Rpt AFML-TR-66-365, 1966.

2. Jackson, C. M. and Wagner, H. J., "Fiber-Reinforced Metal-Matrix Composites: Government-Sponsored Research, 1964–1966," Defense Metals Information Center, Battelle Memorial Inst., Columbus, Ohio. DMIC Rpt 241, 1967.

3. Hanby, K. R., "Fiber-Reinforced Metal-Matrix Composite—1967," Defense Metals Information Center, Battelle Memorial Inst., Columbus, Ohio. DMIC Rpt S-21, 1968.

4. Hanby, K. R., "Fiber-Reinforced Metal-Matrix Composite—1968," Defense Metals Information Center, Battelle Memorial Inst., Columbus, Ohio. DMIC Rpt. S-27, 1969.

5. Hanby, K. R., "Fiber-Reinforced Metal-Matrix Composites, 1969–1970," Defense Metals Information Center, Battelle Memorial Inst., Columbus, Ohio. DMIC Rpt S-33, 1971.

6. Fleck, J. N., "Bibliography on Fibers and Composite Material, 1969–1972," Metals and Ceramics Information Center, Battelle Memorial Inst., Columbus, Ohio. Rpt MCIC-72-09, 1972.

7. "Structural Fabrication Guide for Advanced Composites," Lockheed-Georgia Co., Lockheed Aircraft Corp., Marietta, Ga. Contract F33615-72-C-1215, 1974.

Braze Bonding and Joining of Aluminum Boron Composites

E. M. Breinan and K. G. Kreider

COMPOSITE MATERIALS fabricated from silicon carbide coated boron (Borsic) filaments and aluminum alloy matrix materials have demonstrated excellent strength properties and low density. These properties derive from the high ultimate tensile strength ($>$ 400,000 psi average), high elastic modulus (58×10^6 psi) and low density (2.6 g/cc) of boron filament (1). The aluminum alloy matrix provides ductility and tough-

The authors are associated with United Aircraft Research Laboratories, East Hartford, Conn. This paper was presented at the 1969 Westec Conference, Los Angeles.

ness. Unidirectional 50 vol % Borsic-6061 aluminum alloy matrix composites with axial tensile strengths in excess of 160,000 psi and elastic moduli of 3.3×10^7 psi have been produced (2).

The diffusion bonding step which has been used in the conventional fabrication of these composites typically requires pressures in the range of 5000 psi at elevated temperature (3). Fabrication of large parts could thus require quite large presses, and difficulties may be encountered in maintaining uniform high pressures in complicated or closed shapes. It was consequently deemed desirable to develop

a means of fabricating composites with little or no pressure, but which achieved properties in the composites similar to those fabricated by diffusion bonding at high pressures. A fabrication technique utilizing braze bonding for consolidation of the composites was developed and composites fabricated by this technique were evaluated as one part of this study.

The second part of the investigation was concerned with the evaluation of braze bonding as a means of joining sections of Borsic aluminum composite. The high strength and filamentary nature of the composite causes obvious problems in joining. Conventional bolt or rivet methods of joining are not ideally suitable for uniformly transferring loads from the filaments in one piece to the filaments in the adjacent section, or even to an adjacent section of an isotropic, noncomposite material. Areas of conventional seam or butt welds contain no filaments and thus do not possess adequate strength to transfer the high loads. Only a transfer of loads by shear in overlapped sections seems practical, and brazing apparently offers both higher shear strength and higher temperature capability than currently available epoxy glues (4,5). If the advantages of the higher temperature capability of the metal matrix composites are to be utilized, the joining material must possess similar strength. Several brazed joint configurations were designed, fabricated, and evaluated in the second portion of this investigation.

Experimental Procedures

Composite Fabrication by Braze Bonding

The plasma-spray monolayer tape process was used to prepare tapes incorporating a 0.001-in. thick Alcoa 713 brazing foil (composition 7.5% Si, 0.8% Fe, 0.2% Zn, 0.1% Mg, bal Al). The monolayer tape was made in the standard manner described by Kreider, et al (3) by plasma spraying the 6061 alloy over a filament-wound layer of 0.004-in. diam. Borsic on the 1-mil brazing foil. The tapes were then cut to size and laminated for braze bonding, which was accomplished in vacuum or argon without flux. The tapes were cleaned with an acetone rinse only, since attempts to deoxidize them with acid and with Na(OH) were unsuccessful because the deoxidizer could not be completely removed from the porous plasma sprayed matrix.

A braze cycle consisting of melting the 713 alloy at 590-600 C for 15 min was used to bond the tapes together and to eliminate the porosity in the plasma-sprayed 6061. This consolidation was accomplished by a combination of infiltration, liquid phase sintering, and incipient melting of the 6061 which has a solidus temperature of 582 C, and is enhanced by pressure applied to the composite during the heat treatment. The temperature was carefully controlled to avoid excessive melting of the 6061 alloy. Pressures ranging from 15 psi to 200 psi were applied.

Composite Fabrication by Diffusion Bonding

A series of composites was fabricated by diffusion bonding of plasma-sprayed monolayer tapes which consisted of Borsic filament, wound at 185 turns per inch, sprayed with 6061 aluminum alloy matrix (3). The tapes were hot pressed in vacuum for 1 hr at 550 C and 5500 psi to form 1 in. × 5 in. composite panels which contained 50 vol % filament. These panels were cut into strips and used to fabricate brazed joint specimens.

A transverse cross section of a Borsic-aluminum composite fabricated from monolayer tapes is shown in Fig. 1. The average ultimate tensile strength of these composites in the filament axis direction was 135,000 to 170,000 psi, with transverse tensile strength of 12,000 to 15,000 psi and Poissons ratio of 0.22 (2). These properties were taken as the properties of the parent material used in the joining studies.

Joint Fabrication

All experimental joints were made between strips of 50 vol % Borsic-aluminum diffusion-bonded composites 0.300 in. wide and several inches long, with the filaments in the axial direction. Fig. 2 illustrates the simple lap joint, the single-strap reinforced butt joint, the double strap reinforced butt joint, and the scarf joint. Load transfer in these four joints is primarily by shear in the joining material between the overlapped parts. Several other variations of these basic joint types were also fabricated.

Brazed joints were made using Alcoa 713 and 718 aluminum-silicon alloy brazing foils with thicknesses of 0.001 and 0.002 in. Surfaces to be bonded were prepared either by sandblasting or wire brushing immediately prior to fabrication of the joints. The brazing foil was then sandwiched between these surfaces and the sandwich was clamped in a stainless steel alignment fixture with a pressure of about 35,000 psi at room temperature. No flux was used, as it would have

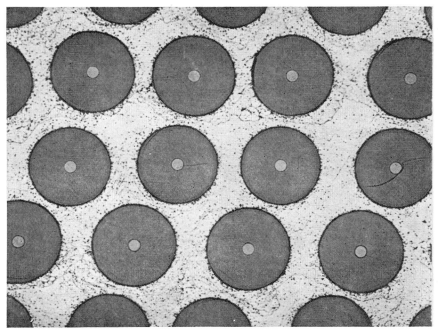

Fig. 1. Borsic-aluminum composite microstructure (composite fabricated from plasma-sprayed monolayer tapes).

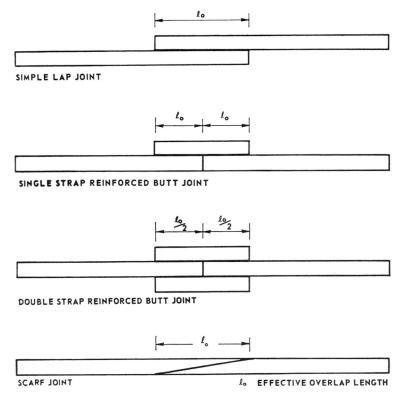

SIMPLE LAP JOINT

SINGLE STRAP REINFORCED BUTT JOINT

DOUBLE STRAP REINFORCED BUTT JOINT

SCARF JOINT l_o EFFECTIVE OVERLAP LENGTH

Fig. 2.   Four basic types of joint configurations investigated.

cuts of 0.001 in. per pass and finishing cuts of 0.0002 in. per pass. Low angles (5° and 2°) were used in order to provide a maximum overlapping of fiber ends. The scarf joints were brazed by the same technique as was used for the butt joints.

Fabrication of Composites with Interrupted Layers

In order to study the effects of non-reinforced lap joints in composites, a six-layer composite 8 in. long and 6 in. wide was fabricated from plasma-sprayed Borsic-aluminum tapes by the same diffusion bonding process which was used to fabricate the specimen material for the joining studies (see materials). This composite was bonded with a series of transverse gaps in the layers which ran across the entire width, Fig. 3. This composite was expected to perform in a manner similar to a composite containing a properly fabricated layered joint, and was evaluated as a model for such a joint in the same way as the joint specimens.

Tensile Testing

All joints were evaluated by tensile testing at room temperature. The standard gage length used was the joined length plus 0.5 in. on each end and the specimens were tested without a reduced section.

The tensile test specimens were mounted in a fixture and aligned in friction grips using a 10× microscope. Unbonded foil doublers (0.001-in. thick aluminum) were used to insure firm gripping with a minimum of damage to the filaments. The test specimens were transferred to the testing machine in a rigid fixture and mounted in the self-aligning loading

been difficult to prevent the entrapment of flux in joints with the large, flat geometries that were employed. The brazing cycle consisted of heating in argon to 590 C for 713 alloy (580 C for 718 alloy) and holding for between 5 and 30 min, then cooling to ambient temperature.

Low-temperature brazing of joints with geometries similar to those of the conventionally brazed joints was accomplished using Ney 380 aluminum-copper-zinc alloy. Surfaces to be joined were cleaned by either sandblasting or wire brushing and then wetted with the alloy. The melting temperature of the alloy is 380 C. Since it was not possible to achieve sufficient temperature uniformity or control when a torch was used to heat the specimen directly, the joined parts were held on a copper or steel plate which was heated with a torch at a location remote from the specimen.

A necessary step in the low-temperature joining process was the scraping of the aluminum under the coating of molten alloy in order to break up

the coating of aluminum oxide and provide a good bond. After coating with the braze alloy, the surfaces to be joined were placed together and heated to remelt the alloy. It was necessary to agitate the two surfaces together in order to get a good bond. When this was not done, poor joints were obtained. The use of a flux was also tried, but better results were obtained using fluxless brazing, provided the proper brazing technique was used.

When scarf joint configurations were fabricated, specimens were mounted on an angled ramp and surface ground to the desired angle. A diamond wheel was used, with rough

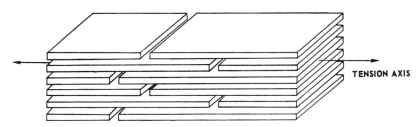

TENSION AXIS

Fig. 3.   Schematic drawing of a 6-layer diffusion-bonded composite containing interrupted layers.

Table 1. Tensile Tests of Brazed Composites

Specimen No.	Borsic fiber, %	Young's modulus, 10^6 psi	Ultimate tensile strength, 10^3 psi	Strain at fracture, 10^{-3} in./in.	Pressure during brazing psi
506 B3-1	48	27.1	117	4.3	15
-2	49	27.9	125	4.5	15
-3	50	29.2	140	4.8	15
-4	52	30.3	145	4.9	15
506 B2-1	49		121		15
-2	49		127		15
503 A4-1	54		148		25
-2	54		143		25
-3	51		144		25
-4	54		149		25
-5	54	36.3	141	4.5	25
-6	54	34.7	143	4.6	25
-7	54	34.5	157	5.5	25
-8	55	37.3	154	4.7	25
503 J -1	52	31.1	147	4.6	50
-2	52	32.6	137	4.2	50
506 B2-1	52		160		100
503 A5-1	48	26.3	124		100
-2	52	30.0	126		100
-3	52		140		100
-4	52	32.5	128		100
503 G -1	50		8.5*		100
-2	50		7.5*		100
-3	50		8.6*		100
364 J -1	41		15.1*		200
-2			15.4*		200

* Tested at 90° to fiber axis.

train. Testing was performed using a Tinius Olsen four screw testing machine with torsion bar LVDT load cell at a strain rate of 0.01 in./min.

The braze-bonded composites were similarly tested, with the exception that specimens used contained a 1-in. long reduced gage section. The gripped section in these tests was 0.240 in. wide, while the gage section was 0.200 in. wide. Specimens of braze-bonded composites which were tested with the fibers at 90° to the loading axis, however, did not have reduced gage sections.

Both strain gages and a clip-on differential transformer extensometer were used to measure elastic strain and modulus in these specimens. The use of an extensometer was hindered by the tendency for slippage if not tightly attached to the specimens, or the possibility of inducing damage at the knife edges if attached too tightly. The use of strain gages provided an increase in sensitivity and were found to be superior to the extensometer.

Results and Discussion

Evaluation of Braze-Bonded Composites

The results of tensile tests of braze-bonded composites are summarized in Table 1. Tensile tests on eight specimens (503 A4) brazed at 25 psi for 10 min demonstrated tensile strengths ranging from 141,000 psi to 157,000 psi, and elastic modulus values averaging 35×10^6 psi. These were unidirectional composites with up to 55 vol % filament. Additional panels were brazed at pressures up to 200 psi, and specimens brazed at 100 psi exhibited axial tensile strengths of up to 160,000 psi and transverse tensile strengths of 7500 psi to 8500 psi with 50 vol % filament. The use of higher pressures promotes full and even consolidation of the composite; however, high pressure can cause expulsion of matrix material, especially if the optimum brazing temperature is exceeded. The composites pressed at 200 psi were found to exhibit transverse tensile strengths of more than

15,000 psi; however, these composites also had a somewhat lower filament content than those pressed at the lower pressures.

Figures 4a and 4b are micrographs of a braze-bonded composite. Fig-

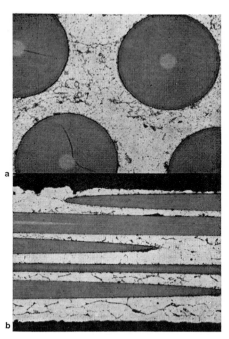

Fig. 4. Microstructures of braze-bonded composites: (a) transverse, (b) longitudinal.

ure 4a, a transverse cross section, clearly illustrates the nearly homogeneous matrix with the silicon precipitate. Figure 4b, a longitudinal cross section, shows a gross precipitate, which is silicon, located at the grain boundaries. This morphology may explain why the strength at 90° to the fibers was lower (8000 psi) than that normally achieved in diffusion bonded composites.

Evaluation of Joints

All composite joints were evaluated by room-temperature tensile testing. It was originally intended to evaluate the necessary length of overlap (l_o in Fig. 2) as a function of composite thickness in order to determine what lengths were necessary for the joints so that the brazed layers would not fail in shear. It was found, however, that the braze exhibited quite high room-temperature shear strengths when estimated from the behavior of the actual joints ($>$ 12,000 psi for 713 and 718, and $>$ 7000 psi for 380) and so for the composites investigated, which varied in thickness from 0.022 to 0.040 in., even the joints with the shortest overlap lengths (l_o = 0.375 in.) did not fail in shear when the joints were properly brazed. It was decided that there would be little to gain by further reducing joint length below 0.375 in., since the difficulty in fabricating the joints increased substantially when small pieces were involved.

In a double-strap-reinforced butt joint with l_o equal to 0.375 in., for example, the actual overlap of the reinforcing strap on the parent piece ($l_o/2$) is only 0.188 in. Since cutting of Borsic-aluminum composites with a diamond wheel or electrical discharge machining causes considerable filament breakage near the cut ends, it was decided not to attempt to reduce joint overlap to the point where shear failures occurred. Instead, l_o was standardized at 0.5 in. for both 5-layer (0.022 in.) and 8-layer (0.040 in.) thick composites and efforts were concentrated on obtaining a joining method and design with as near the parent material strength as possible. When thicker composite panels are joined it will be necessary to increase the shear overlap area in order to transmit the higher loads which these panels will be capable of carrying.

The shear strengths achievable by brazing composites with 713 and 718 alloys were more thoroughly evaluated by standard short-beam shear tests (6) using specimens which were fabricated with a brazed joint along the central plane of maximum shear stress (7). Tensile failures, rather than shear failures, were obtained in tests with span-to-depth ratios ranging from 5 to 2, and smaller span-to-depth ratios could not be used due to the fact that the specimens tended to crush, rather than fail in tension or shear. The results of short-beam shear tests of 713 and 718 brazed joints are summarized in Table 2. Since the specimens failed in tension at the outer surface, the actual shear strengths must be somewhat higher

Table 2. Short-Beam Shear Tests of Brazed Joints

Braze Alloy	Span/depth ratio	Calculated shear stress on joint, psi	Type of failure
713	5	11,750	Tensile
718	5	12,700	Tensile
713	4	13,550	Tensile
718	4	13,100	Tensile
713	3	14,350	Tensile
718	3	14,050	Tensile
713	2	17,100	Tensile and crushing
718	2	15,700	Tensile and crushing

than those calculated. The values given in Table 2 are corrected for the effect of width/depth ratio according to the analytical work of Sattar (8). The correction applied has the effect of increasing the calculated shear stress values by approximately 15%.

Three modes of failure were found to occur in properly bonded joint specimens (Fig. 5). Tensile failure of the composite at the end of the joint (mode-a) and tensile failure of the single reinforcement strap (mode-b) were found in all three types of lap and butt joints and involved no shear of brazed layers. Mode-c failures, which involved shear failure of one bond occurred in some double-strap-reinforced joints and are probably indicative of a poor braze which caused this bond to fail prior to the other three similar bonds. Since these failures exhibited strengths nearly as high as the other modes of failure, they are included as properly brazed joints. It is probable that the shear portion of the failure occurred first, since the shear strain capability of the brazing alloys was rather low, especially in the 380 alloy which was

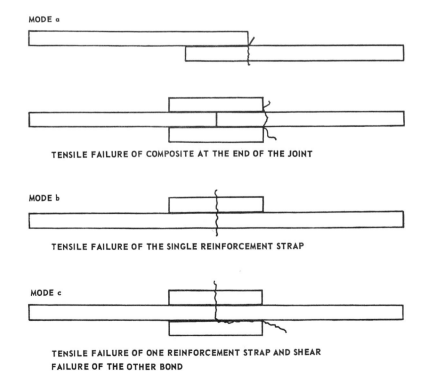

MODE a

TENSILE FAILURE OF COMPOSITE AT THE END OF THE JOINT

MODE b

TENSILE FAILURE OF THE SINGLE REINFORCEMENT STRAP

MODE c

TENSILE FAILURE OF ONE REINFORCEMENT STRAP AND SHEAR FAILURE OF THE OTHER BOND

Fig. 5. Modes of failure encountered in properly bonded lap and butt joints.

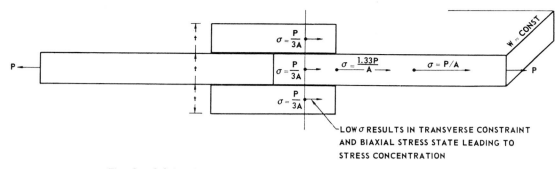

σ = P/3A

σ = P/3A   σ = 1.33P/A   σ = P/A

σ = P/3A

P   P

W = CONST

LOW σ RESULTS IN TRANSVERSE CONSTRAINT
AND BIAXIAL STRESS STATE LEADING TO
STRESS CONCENTRATION

Fig. 6. Origin of stress concentration at end of a reinforced joint.

used in a large majority of the specimens in which mode-c failures occurred (380 also has somewhat lower shear strength). After the shear step occurs, the remaining portion of the sample is essentially a single-strap-reinforced butt joint and fails by mode-b.

The location of the fractures in Fig. 5 indicates the presence of a stress concentration due to restriction of the Poisson contraction in the lateral direction due to the lower load in the reinforced section of the joint. This is illustrated in Fig. 6. The failure naturally occurs adjacent to the reinforced section in the most highly stressed region of the specimen. Analysis of a similar situation involving the influence of end constraint (gripping) on tension and compres-

sion testing of composites has been carried out by Schile (3). Assuming complete restraint at the ends, he has calculated stress concentrations as high as 1.5 adjacent to the grips in a tensile test.

The results of tensile tests of lap joints and reinforced butt joints are summarized in Fig. 7. It can be seen that the double-strap butt joints exhibited the highest strengths, probably because of the more nearly pure tensile loading on this symmetric joint configuration. Even the highest strengths, however, were only approximately 75% of the average parent material tensile strength, and this was attributed to the presence of an effective stress concentration of about 1.33.

The effect of the stress concentration was verified by fabricating speci-

mens in which two composite reinforcing straps either 1 layer (0.005 in.) or 5 layers (0.025 in.) thick were brazed to a specimen of parent material in which there was no joint. One specimen was also similarly fabricated but with 10-mil aluminum straps, as it was expected that the aluminum would be nearly as effective in reinforcing in the transverse direction as the composites. The results of tensile tests on these specimens were that the brazing on of a single reinforcement layer on each side reduced composite strength to 147,000 psi and a mode-a failure occurred at the end of the reinforcement. Ten-mil aluminum reinforcements decreased the strength to 134,000 psi, and 5-layer composite reinforcements decreased the strength to 105,000 psi, which is substantially lower than the strength of a double-strap butt joint similarly fabricated.

The decrease in strength with increased amounts of bonded reinforcement and the persistence of a mode-a failure at the end of the reinforcement proves the significant detrimental effect of the stress concentration on composite strength. It is even understandable that the specimen with 5-layer reinforcement and no joint is weaker than the similar specimen containing a joint. As shown in Fig. 6, the load in the reinforcing straps increases to P/2A at the plane of the joint, but when no joint is present it remains at the lower level of P/3A, and hence yields a more severe stress concentration. Since the estimated stress concentration in the joined specimens is approximately 1.33, which indicates that these specimens do not act fully restrained, the lack

JOINT TYPE

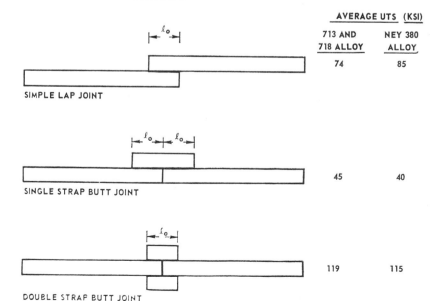

| | AVERAGE UTS (KSI) | |
JOINT TYPE	713 AND 718 ALLOY	NEY 380 ALLOY
SIMPLE LAP JOINT	74	85
SINGLE STRAP BUTT JOINT	45	40
DOUBLE STRAP BUTT JOINT	119	115

Fig. 7 Results of tensile tests of lap and butt joints.

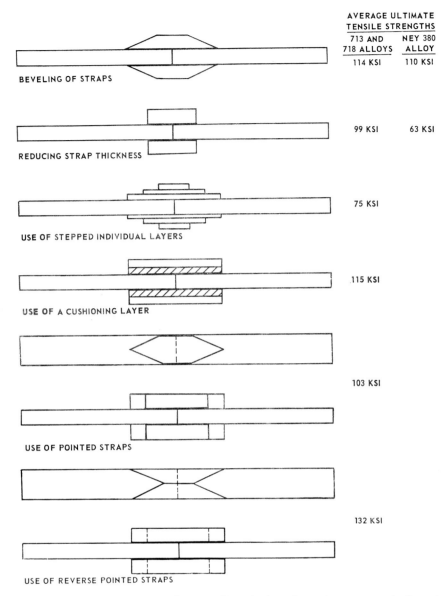

	AVERAGE ULTIMATE TENSILE STRENGTHS	
	713 AND 718 ALLOYS	NEY 380 ALLOY
BEVELING OF STRAPS	114 KSI	110 KSI
REDUCING STRAP THICKNESS	99 KSI	63 KSI
USE OF STEPPED INDIVIDUAL LAYERS	75 KSI	
USE OF A CUSHIONING LAYER	115 KSI	
USE OF POINTED STRAPS	103 KSI	
USE OF REVERSE POINTED STRAPS	132 KSI	

Fig. 8. Summary of results of some attempts to reduce stress concentrations in double-strap butt joints.

of a joint can still further increase the stress concentration toward the fully restrained value of 1.5.

The results of a series of attempts to reduce the stress concentration by changing the joint design are summarized in Fig. 8. Lower strengths were often obtained due to greater difficulty in fabricating the more complex modifications of the basic double-strap-reinforced butt joint. The specimens failed primarily by mode-a and occasionally by mode-b. Failures of several modified joints are shown in Fig. 9. Lack of improvement of strength in the majority of the specimens can be traced to the detailed nature of the stress concentration as was given by Schile (3).

The axial and transverse stress distributions are shown in Fig. 10. It can be seen from the transverse stress distribution that the maximum stress concentration occurs at the outer edges of the specimen and so the first five attempts to reduce the stress concentration which are illustrated in Fig. 8 did not prove to be effective.

The sixth modification, use of reverse pointed straps, was designed specifically to compensate for the stress distribution given in Fig. 10 by having the bonded-on reinforcement gradually narrow toward the edge of the specimen in order to gradually distribute the restriction of the reinforcements along the length of the specimen. Tapering the reinforcement away from the edge was important, since the stress concentration is maximum at the edges. This configuration definitely increased the strength of the joined specimens to values as high as 90% of the average parent material strength. It can be seen in Fig. 9 that only for the reverse pointed strap configuration did the fracture path deviate from the location of the end of the reinforcing strap. It is expected that in wider joints even higher percentages of the parent material strength can be achieved due to the decreased importance of edge effects.

Longitudinal and transverse microstructures of brazed lap joints are shown in Fig. 11. It can be seen that considerable grain growth occurs in the aluminum matrix of the composite as a result of the brazing thermal cycle. The silicon in the braze alloys generally precipitates at the grain boundaries, although the silicon is distributed over a sufficiently large area, so that good bonding is achieved. It is expected that the presence of the silicon at the grain boundaries is the major limiting factor for ductility in the brazed layers. Basically, both 713 and 718 brazed specimens exhibited similar microstructures except for the larger amount of silicon precipitates in the 718 brazed material. Since similar shear strengths were obtained for both brazing alloys, it is recommended that, for better ductility, the lower Si braze (713) be used. Actually, joints brazed with 713 exhibited slightly higher strengths. Both 0.001 and 0.002-in. brazing foils were used, and since the 0.001-in. foils proved adequate, the use of thicker foils is not recommended.

Scarf joints were also investigated in an effort to fabricate unreinforced

UNMODIFIED JOINT (MODE – a)

ALUMINUM CUSHION LAYER
(MODE – a)

BEVELED REINFORCING STRAPS
(MODE – a)

POINTED REINFORCING STRAPS
(MODE – a)

REVERSE POINTED REINFORCING
STRAPS

Fig. 9.   Failures of modified reinforced butt joints.

LATERAL CONTRACTION IS PREVENTED AT THE ENDS

OUTER FIBER
STRESS –1.5 τ_0

NOMINAL
STRESS

AXIAL DISTRIBUTION OF τ_{11} AT EDGE OF SPECIMEN

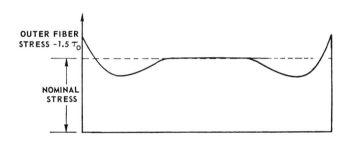

τ_{11} τ_{22} τ_{12}

TRANSVERSE STRESS DISTRIBUTIONS

Fig. 10.   Restrained tension of a laminated strip (due to Schile).

joints with adequate load transfer area. The results of investigations of both unreinforced and reinforced scarf joints are summarized in Fig. 12.

It can be seen that strengths in the unreinforced scarf joints were far below the parent material strength. Scarf joints commonly exhibited a combination of tensile failure and some shear failure in the brazed scarf plane (Fig. 13).

Simulated joint specimens fabricated with built-in gaps in the layers as shown in Fig. 3 exhibited strengths of $\frac{N_i}{N_t} \times$ parent material strength where N_i = number of intact layers and N_t = total number of layers. The fractures were located in transverse planes which included the original gaps. In an ideal scarf joint of sufficiently low angle, each transverse plane may include no more than one

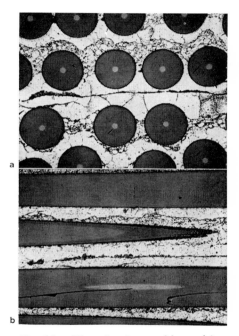

Fig. 11.   Microstructures of brazed lap joints: (a) transverse, (b) longitudinal.

broken or gapped layer and gaps are spaced more than the critical load transfer length apart to allow reloading of the fiber ends between the gaps. Ideally then, the expected strength of an unreinforced scarf joint would be $\frac{N_t - 1}{N_t} \times$ parent material strength.

It can be seen from the microstructures of brazed scarf joints, such as those shown in Fig. 14, that the filaments were actually quite severely fragmented close to the plane of the joint. This breakage occurred during grinding of the angle, even though

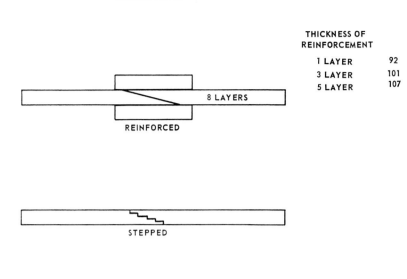

Fig. 12. Summary of scarf joint results.

Fig. 13. Reinforced scarf joint failures.

very fine cuts were taken. The exposed angled surface area was composed of 50% boron and 50% aluminum. Such surfaces can be seen in Fig. 13. The photographs in Fig. 13 show that the adherence of the braze alloy to the exposed boron was not good since the boron is again exposed after failure, indicating that the failure occurred at the interface and not within the braze. The difficulties in making strong scarf joints were thus attributed to the above two factors; fragmentation of fibers and poor bonding to fibers, and also to the inability to maintain even pressure on the joints with the apparatus used. Attempts to make reinforced scarf joints, usually with the reinforcing straps thinner than the pieces being joined are also summarized in Fig. 13. Though some increases in strength were realized, this approach was rapidly trending toward the double-strap butt joint with little advantage being gained from the presence of the scarf.

The possible use of stepped scarf joints is also included in Fig. 12. Here, individual layers could theoretically be terminated at different locations and strengths of $\dfrac{N_t - 1}{N_t} \times$

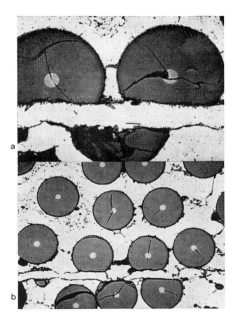

Fig. 14. Microstructures of brazed scarf joints: (a) 500×, (b) 200×.

parent material strength were again expected if the joint could be properly fabricated. The use of single reinforcing layers might then bring the strength up to that of the parent material. In the relatively small number of attempts to fabricate these joints, difficulty was encountered in trying to machine close fitting stepped pieces, without damaging the fibers, and no strong step joints were achieved.

In order to demonstrate the feasibility of fabricating composite beams by braze bonding, a series of beams was fabricated using welded aluminum honeycomb core material and Borsic-713 aluminum brazing alloy top and bottom sheets. An example is shown in Fig. 15. These beams were tested in three-point bending. Failure occurred by crushing and buckling of the honeycomb cores, but the composite top and bottom sheets remained bonded to the honeycomb. When similar specimens were fabricated using epoxy bonding, failure occurred by peeling or delamination, and then buckling of the top and bottom sheets.

Conclusions

1. It was determined that plasma sprayed monolayer Borsic tapes backed with 1-mil 713 alloy brazing foil could be successfully consolidated into composites which had essentially the same strength as diffusion bonded composites. The braze bonding step should be carried out in vacuum and with dies to apply a moderate pressure (about 50 psi) to obtain best results.

2. Microstructures of braze-bonded specimens exhibited silicon precipitates in the grain boundaries which are thought to limit transverse tensile strength.

3. Brazed lap and butt joints in composites which exhibited shear strengths greater than 12,000 psi using 713 and 718 alloys and 7000 psi for 380 alloys were fabricated. Due to the high shear strength values achieved, it was not necessary to attempt to reduce joint overlap distances in this study. A standardized overlap distance of 0.5 in. was thus used for composites ranging from 0.022 to 0.040 in. thick.

4. Joint shear strengths in excess of 17,100 psi were achieved in short-

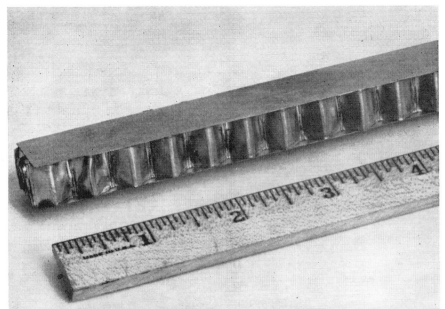

Fig. 15. Borsic-aluminum composite sheets brazed to welded aluminum honeycomb core.

beam shear tests of specimens brazed with 713 alloy foil. Joints brazed with 718 alloy exhibited shear strengths greater than 15,700 psi.

5. Lap and butt joint specimens failed in tension due to stress concentrations which were caused by restriction of the Poisson contraction in the lateral direction by bonded-on reinforcements. The severity of the stress concentration was estimated to be approximately 1.33. The detrimental effect of the stress concentration was confirmed by testing specimens which contained no joint, and thus acted even more fully restrained.

6. Double-strap butt joints exhibited the highest strengths in tensile tests of reinforced lap and butt joints, probably due to the more nearly pure tensile loading on this symmetric joint configuration. The highest strengths achieved were approximately 75% of the average parent material strength.

7. A modification of the double-strap butt joint which gradually distributes the restriction of the reinforcement was developed. Joint strengths of up to 90% of the average parent material strength were achieved.

8. Microstructures of brazed joints exhibited silicon precipitates in the grain boundaries similar to the braze bonded composites. It was concluded that the lower silicon brazing alloy (713) was preferable, and also that a foil thickness of 0.001 in. was adequate for good joining.

9. Tests of a model for a scarf joint which consisted of a composite with built in gaps in the layers exhibited tensile strengths of $\frac{N_i}{N_t} \times$ parent material strength where $N_i =$ number of intact layers and $N_t =$ total number of layers. Based on these results, it was predicted that an ideal scarf joint could achieve a strength of $\frac{N_t - 1}{N_t} \times$ parent material strength.

10. The strengths achieved in both reinforced and unreinforced scarf joints were below those predicted for ideal joints. The reasons for the low strengths were found to be fragmentation of fibers near the joint, poor bonding of the braze to exposed boron, and inability to maintain even pressure on the joints during fabrication. It was concluded that reinforced butt joints were potentially as strong or stronger, more reliable, and far easier to fabricate than scarf joints.

11. The feasibility of fabricating composite beams by braze joining of Borsic-aluminum sheet to welded aluminum honeycomb core material was demonstrated. Failure in three-point bend tests occurred by crushing and buckling of the core, but the composite plates remained bonded to the honeycomb.

ACKNOWLEDGMENT

The authors wish to acknowledge the technical advice and assistance of Dr. R. D. Schile and Mr. L. Pennington, and the metallographic studies of Mr. W. J. Hermann.

REFERENCES

1. A. H. Lasday and C. P. Talley, Boron Filament for Structural Composites, Paper D-1, Advanced Fibrous Reinforced Composites, SAMPE, Vol. 10 (1966).
2. K. Kreider and M. Marciano, Mechanical Properties of Boron Aluminum Composites, Transactions AIME (in press).
3. K. G. Kreider, R. D. Schile, E. M. Breinan and M. Marciano, Plasma Sprayed Metal Matrix Fiber Reinforced Composites, Technical Report AFML-TR-68-119, United Aircraft Research Laboratories (July 1968).
4. C. Clement Anderson, Adhesives, Industrial and Engineering Chemistry, Vol. 60 (Aug. 1968).
5. C. H. Wick, All About Adhesive Bonding, Machinery (Nov. 1967).
6. 1968 Book of ASTM Standards, Part No. 26, Standard D 2344, American Society for Testing and Materials, Philadelphia, Pa. (1968).
7. F. B. Seely and J. O. Smith, Resistance of Materials, 4th Ed., John Wiley and Sons, New York (1956).
8. S. Sattar, The Effect of Geometry on the Mode of Failure in Short Beam Shear of Composites (to be published in the Proceedings of the ASTM Conference on Composite Materials Testing and Design, held in New Orleans, La., Feb. 1969).

Aluminum-Boron Composites for Aerospace Structures

Advantages of this structural material for aerospace components include light weight, a high strength-to-weight ratio, stiffness, postbuckling strength, excellent fabricability, and ease of joining by a variety of techniques.

By J. L. CHRISTIAN, J. D. FOREST, and M. D. WEISINGER

THE COMPOSITE adapter assembly shown in Fig. 1 was static tested in March 1969 at the Convair Structural Test Facility in San Diego. The most severe combined flight loads of shell bending moment, axial compression, normal shear, and shell crushing pressure were applied simultaneously to failure (in increments of 10% of design limits). The test setup and load magnitudes were essentially the same as that for static testing the original all-aluminum production version of this adapter.

At about 40% of limit load, noticeable skin wrinkling began as a result of combined shear and pressure acting between longitudinal stiffeners. This had been predicted in the design analysis and did not constitute a failure of the structure. Design ultimate loads (150% of limit loads) were reached without loss of stability or excessive yielding. When ultimate loads were removed, some permanent deformation of the skins remained in the form of a flattening of the skin contour (originally a 30 in. radius) between stiffeners. The observed deformation was not severe and would not have compromised the success of a flight.

All loads were reapplied in progressive increments until failure at 133% of design ultimate. The failure was a localized instability collapse of three stiffeners near the base of the composite shell (Fig. 2). Massive

Mr. Christian is staff scientist, Mr. Forest is senior design engineer, and Mr. Weisinger is research group engineer, Convair Div., General Dynamics Corp., San Diego.

Fig. 1 — Adapter assembly is made of Al-B composite sheet material.

Fig. 2 — Under test loading, experimental adapter assembly failed at 133% of design ultimate by stiffener collapse near base of unit.

fracturing did not occur even though aluminum-boron is a brittle material in contrast with conventional aluminum and steel alloys. The mode of failure, in fact, was quite common to that of aluminum structures of similar configuration.

The test successfully demonstrated the high weight reduction payoff (40%) of metal matrix composites in a structure. It also proved out the design concept and unique fabrication methods. At the beginning of this two-year effort, only the most rudimentary data on the basic composite material were available, and there was no knowledge of the fabricability or structural behavior of aluminum-boron. Perhaps the major factor contributing to the success of the program was the co-operation between the disciplines of materials, processing, analysis, and design engineering.

Prior to the construction and final test of the adapter assembly, a series of structural tests was conducted on critical elements. The goal was to define the structural behavior of aluminum-boron, verify analytical methods for predicting strength and stability, and prove the adequacy of adapter component designs.

A crossplied layup was selected for the skin because of failures of the unidirectional material in acoustical panel and pressure buckling tests. The crossply configuration successfully passed both the acoustic and pressure environment and showed significant postbuckling shear strength in a tension field beam web test. Column, beam, and crippling tests on the hat section longitudinal stiffener indicated that its design was adequate with critical failures occurring within a few per cent of the design load. Elemental buckling investigations and joint tests verified the analysis and provided empirical design data.

Materials for the Shell

A number of shell concepts and materials were studied to determine the lightest construction for the adapter. Included were designs in aluminum, fiber glass, boron-epoxy, and Al-B. As shown in the table below, the minimum weight was

provided by an Al-B buckled skin shell, which was chosen for the adapter.

Shell Concepts	Weight
Original aluminum monocoque	129.9 lb
Aluminum skin stringer shell	111.4
Fiber-glass sandwich shell	120.0
Epoxy-boron sandwich shell	111.5
Epoxy-boron monocoque	99.1
Aluminum-boron monocoque	90.2
Aluminum-boron skin stringer shell	73.3

The buckled skin type of shell is common in aluminum designs but had not been considered for advanced composite materials. The shell is composed of rigid longitudinal stiffeners (stringers) and circumferential frame stiffeners which carry all axial forces resulting from applied and internal loads. Attached to this rigid framework is a thin skin which is allowed to relax under load. The applied shear loads are carried by the skin membrane action following a snap-through to reverse curvature. The snap-through and shear wrinkling of the skin impose significant transverse loads and strains on the composite matrix.

The most advanced composites include resin matrix materials which have low strain and strength allowables compared with those of metals. Resins commonly used with advanced composites may fracture under either snap-through or shear wrinkling, and consequently have not been specified for a buckled skin design. The significant postbuckling strength of aluminum-boron is one important advantage of this material over the resin-base composites.

Another major advantage of Al-B is the variety of joining techniques that are available. Resin-based composites can be joined only by adhesive bonding or mechanical bolts. Aluminum-boron can be joined by those two methods and brazing, resistance welding, or diffusion bonding. The last three processes give joint efficiencies superior to adhesive bonding or mechanical fastening.

The final composite design (Fig. 3) has strength and stiffness equivalent to the original aluminum adapter but it is 40% lighter.

Fabricating the Adapter Section

Although a variety of joining

methods were evaluated, spot welding was chosen because rapid heating does not affect filament properties. Cutting and machining tests revealed that conventional tool materials were inadequate, especially for milling and drilling. New types of diamond-coated tools were developed for these operations. Other machining operations such as shearing and abrasive cutoff worked well in final processing. Because of concern for the formability of the unidirectional composite layup (bend axis parallel to filament direction), special high-temperature heating and cooling dies were developed that allowed a reasonably sharp radius to be formed.

Aluminum-boron was selected for skins, door, bulkhead, skin splices and stringers. Subcomponents were placed in a fixture, drilled, and fastened with rivets and bolts made of aluminum and steel.

Stringers were made from unidirectional Al-B composite sheets with selectively placed filaments. Filaments were deleted in areas requiring sharp radius bends. Four bends were brake formed at room temperature along the bend lines where the filaments were deleted. The cap section, or center portion, was hot formed to eliminate matrix cracking. After forming, the stringer was slowly cooled in a wooden fixture to prevent distortion. Final trimming of the stringer was done with a diamond-plated milling cutter (Fig. 4).

Four crossplied skin segments were trimmed on a shear prior to assembly. Cutouts in the skins were made, using diamond-coated routing cutters. After splicing two skin segments together, the stringers were attached to the skins by spot welding (Fig. 5) and by riveting.

Strength of individual spot welds depends on the quality of the composite sheet, as well as welding schedule and nugget diameter. However, tensile-shear strengths in excess of 600 lb force per weld spot are readily achieved. A critical operation was drilling the holes for mechanical fasteners. A special diamond coating on a regular drill bit was required to minimize tool wear. The tool was

Fig. 3 — Components of adapter assembly include bulkhead, access door, skin and stringers, plus miscellaneous hardware. Cylindrical section is 41.5 in. high by 59.5 in. in diameter. The direction of the boron filaments is indicated for all parts.

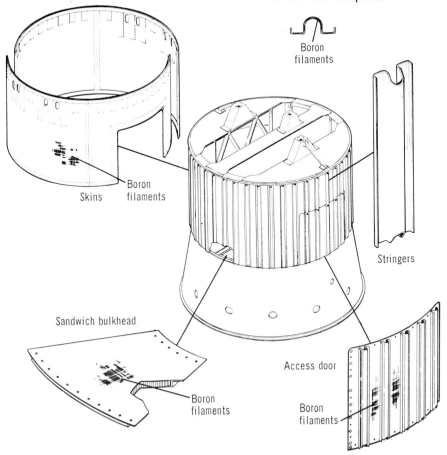

Boron filaments

Skins | Boron filaments

Stringers

Sandwich bulkhead

Boron filaments

Access door

Boron filaments

graphic examination.

Visual examination can detect surface flaws and evaluate flatness, primarily to correct layup, tooling, pressing, or other process parameters to assure quality material. Nondestructive tests included X-radiography and ultrasonic C-scans to determine filament location and orientation, detect misplaced or broken filaments, and evaluate soundness of diffusion bonding. Tensile, compression, and shear tests evaluated mechanical properties, provided design data, and verified conformance with specifications.

Developing the Composite

During the past four years, Convair has evaluated 30 different material systems, including boron, silicon carbide, graphite, metallic wires, whiskers, and coated filaments in combination with aluminum, titanium, nickel, or superalloy matrices. Aluminum-boron evolved as one of the more promising metal matrix systems.

Al-B can be produced in many ways: diffusion bonding alternate layers of aluminum foil and boron filaments, casting or molten metal infiltration, plasma spraying, powder metallurgy, or by electroplating. Diffusion bonding and plasma spraying are the more popular methods. The availability of Al-B sheet has broadened from "postage-stamp" size in early 1966 to 4 by 4 ft pieces today. Al-B material is made by Hamilton-Standard, Div. of United Aircraft Corp.; Harvey Aluminum Co.; and Marquardt Corp. Several other companies are producing limited quantities and may be considered as future suppliers.

The cost of the material (50 vol.% filament, 0.040 in. thick sheets) is $700 per pound, including the cost of boron filaments (25% of the total). Material suppliers estimate future costs of these sheets at $100 to $200 per pound by 1972, based on lower filament costs and quantity production.

There has been a considerable improvement in tensile strength of Al-B during the past four years (Fig. 6). Increased strength is due to improvements in boron filament prop-

Fig. 4 — Milling of Al-B stringer requires a diamond-coated cutter.

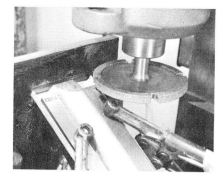

Fig. 5 — Stringers are attached to the skin by resistance spot welding.

dipped in a diamond dust-nickel slurry and coated by electroless nickel plating. Proper support for the part as well as optimum feeds and speeds are necessary to prevent damage to the composite.

Qualification Testing

Qualification tests by the producer

(Harvey Aluminum) and user (Convair) were performed on each of the Al-B composite panels. Quality control tests included visual and nondestructive examinations, volume per cent of filament, thickness, mechanical properties, and metallo-

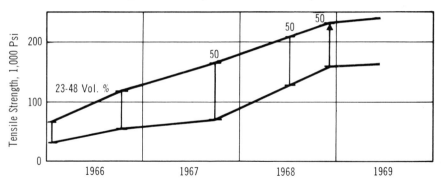

Fig. 6—Strength of Al-B composite sheets has grown steadily in the last three years.

Table I — Effect of Filament Orientation and Volume Per Cent on Mechanical Properties of Al-B Sheets

	Unidirectional			0-90° Crossply	± 30° Crossply
	25*	37	50	45	50
Tensile strength, 1,000 psi — Longitudinal	78.8	125	167	75.5	76.3
Transverse	14.5	13.2	12.1	68.3	17.2
Elastic modulus, 10⁶ psi — Longitudinal	18.6	25.6	32.6	19.5	21.4
Transverse	14.1	16.0	20.6	20.1	21.5
Shear strength, 1,000 psi	12.2	12.4	13.0	14.9	—

* Volume per cent of filament.

erties, processing, and refinements in testing. Typical tensile properties of 50 vol.% Al-B sheet material are 175,000 psi with a spread of ±25,000 psi. The recent preparation of a Convair material specification represents a significant advancement because it calls for a guaranteed ultimate tensile strength of 160,000 psi.

Mechanical properties (tension, compression, shear, creep, axial and flexural fatigue, notched tension, crack propagation, and thermal cycling) have been determined to provide design and engineering data. Tests were at room and elevated (300, 500, 700 F) temperatures. Typical tensile and shear data are summarized in Table I.

We evaluated corrosion resistance of bare and coated Al-B composites in several environments, including salt spray and industrial sea-coast atmospheres and under various conditions of stress. Results indicate that standard aluminum coatings, applied by dipping, spraying, or brushing, provide adequate protection.

Future Looks Promising

The successful completion of this program has demonstrated that Al-B is well suited for lightweight aerospace structures. However, the future use of this advanced composite material depends on obtaining more experience and test data. We need more information on fabrication, material properties, structural testing, and flight experience. Assuming continued success, it is believed Al-B will be ready for large-scale production by 1972.

Convair Div. of General Dynamics Corp. performed this study under contract from Materials Laboratory Research and Technology Div., Air Force Systems Command, Wright-Patterson AFB, Ohio.

Eutectic Bonding of Boron-Aluminum Structural Components

Studies show that copper can be used as a eutectic former with aluminum to provide a reliable, low pressure method for fabricating boron-aluminum composites

EVALUATION OF CRITICAL PROCESSING PARAMETERS

BY J. T. NIEMANN AND R. A. GARRETT

ABSTRACT. This study was undertaken as part of a major program to develop a low pressure technique for fabricating boron-aluminum structural components. The method selected consists of joining monolayer foils together by a brazing process which relies on the diffusion of a thin surface layer of copper into the aluminum matrix to form a liquid phase when heated above the copper-aluminum eutectic temperature.

Laboratory investigations conducted to establish a suitable bonding cycle are described. First, the effect of holding time within a selected bonding temperature range on filament degradation was determined. Then, a copper coating thickness was selected to be compatible with a bonding cycle that would minimize the loss of filament strength. These studies showed that sound, strong joints could be produced by limiting the coating thickness to 20 microin. and restricting the bonding tempera-

The authors are associated with McDonnell Douglas Astronautics Co. - East, St. Louis, Mo. J. T. NIEMANN is Senior Group Engineer, Materials and Processes Dept. R. A. GARRETT is Program Manager, Advanced Composites Programs.

Paper was presented at the AWS 54th Annual Meeting held in Chicago during April 2-6, 1973.

ture to the range between 1030 F and 1060 F with the time not to exceed 15 min at the lower temperature limit or 7 min at the upper limit. Techniques for including titanium interleaves in boron-aluminum laminates are described and typical properties of eutectic bonded B/Al laminates are presented.

Introduction

During the past several years, the McDonnell Douglas Astronautics Co.-East has been evaluating metal-matrix composite systems for application to advanced spacecraft and missiles. These efforts have concentrated on the design, fabrication and testing of boron-aluminum structures. One of the first problems faced was selecting a joining method that could be used to produce structural components from monolayer foils of the composite. This led to the application of a brazing process that relies on the interdiffusion of dissimilar metals to form a liquid phase when heated above their eutectic temperature. This type of brazing has been used in a variety of specialized joining applications, (Refs. 1,2) and is known by various names such as eutectic brazing and diffusion brazing. Joining of boron-aluminum by the process described in this paper depends on the diffusion of a thin surface layer of copper into the aluminum matrix to form a eutectic liquid when heated above 1018 F.* This particular development has been termed eutectic bonding.

Several steps were involved in the evolution of eutectic bonding from a laboratory development to a production process. The more critical processing parameters were examined first to establish a bonding thermal cycle and to define copper thickness requirements. Then, production processing methods for chemical cleaning, copper coating, bonding, machining and joining were evaluated and selected. When this was accomplished, mechanical property tests were conducted to establish design allowables. Finally, large, complex structural components were designed, fabricated and tested. Major investigations conducted during this period will be discussed in this paper. Part I will describe the metallurgical studies conducted to select a thermal cycle and copper coating thickness for joining boron-aluminum to itself and to titanium and will present typical mechanical properties. Part II will be concerned with process development and component fabrication.

*All units are presented in the English system; factors for converting to the International System are presented in Table 1.

Selection of a Joining Process

The high strength, high elastic modulus and low density of boron-aluminum make it an attractive candidate material for weight-critical aerospace structures. It is much superior to conventional materials from the standpoint of specific strength and specific modulus. For example, boron-aluminum with the boron filaments oriented longitudinally combines the strength and stiffness of high strength steel with the density of aluminum. A less desirable feature is its high degree of anisotropy. In the transverse direction, the mechanical properties are determined by the aluminum matrix and consequently they are much lower than the longitudinal properties which are determined. by the very high strength boron filaments.

Because of the potential offered by boron-aluminum, there has been a concerted effort to develop methods of fabricating structural components from this material. The objective of these studies has been to utilize the good longitudinal mechanical properties of the composite while compensating for the low transverse properties through techniques such as cross-plying and adding local reinforcement.

Two general approaches have been followed to achieve this objective. Diffusion welding was developed first and proved to be an economical method of producing flat plates and sheets and simple structural shapes which could be formed after consolidation by the diffusion process. The major drawback to this process is its limited versatility. Greater design flexibility is offered by a second technique in which laminated structures are built up from monolayer foils of the composite. These individual foils are first formed to the desired shape and then bonded together. This approach facilitates cross-plying, adding local reinforcement through interleaving with another material and varying the thickness of detail parts for maximum weight savings. A typical laminated structure incorporating these features is shown in Figure 1.

In order to apply the built-up laminate approach, a method of joining individual foils together had to be selected. Brazing appeared to be an ideal choice when combined with vacuum bagging and an externally applied hydrostatic pressure to provide intimate contact between adjacent monolayers and conformance to the desired shape. However, the conventional silicon-containing aluminum base brazing filler metals require heating to temperatures in excess of 1080 F. Even very short time exposures at this temperature result in a fiber-matrix interaction which drastically lowers composite strength. This interaction can be minimized by using boron filaments which have been coated with silicon carbide, but this solution is accompanied by increased cost and lower mechanical properties. To avoid these penalties, a study was undertaken to develop a lower temperature joining process.

The need for a lower temperature brazing process led to the investigation of eutectic bonding as a possible method of fabricating B/Al structures. Although many elements form eutectics with aluminum, copper was selected as the most promising for B/Al fabrication because of its low cost, availability and ease of deposition by electroplating and physical vapor deposition. Furthermore, it is a common element in aluminum alloys and did not present compatibility problems with the 1100 aluminum matrix. In fact, substantial solid-solution strengthening of the matrix by the copper addition was anticipated.

Mechanics of Eutectic Bonding

The eutectic bonding process resembles brazing in that joining is effected through a liquid phase which forms at a temperature below the melting points of the base metals involved. The major difference lies in the manner in which the liquid is formed. Brazing utilizes a low-melting filler metal while eutectic bonding relies on the diffusion of copper into the aluminum matrix to form a liquid at the joint interface when heated above the aluminum-

Table 1 — Conversion Factors for International System of Units [a]

To convert from	To	Multiply by
Farenheit	kelvin	$K = 5/9 \ (F + 459.67)$
inch	meter	$2.54(10)^{-2}$
pound/inch³	kilogram/meter³	$2.76799(10)^4$
pound/inch²	newton/meter²	$6.89476(10)^3$
torr	newton/meter	$1.3332(10)^2$

(a) From Ref. 3

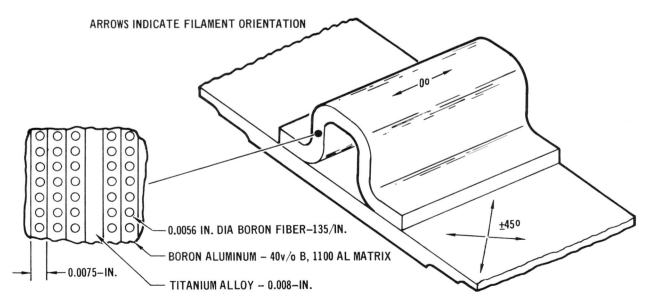

ARROWS INDICATE FILAMENT ORIENTATION

0°

±45°

0.0056 IN. DIA BORON FIBER—135/IN.

BORON ALUMINUM — 40v/o B, 1100 AL MATRIX

TITANIUM ALLOY — 0.008—IN.

0.0075-IN.

Fig. 1 — Typical structural application for boron-aluminum

copper eutectic temperature of 1018 F.

The metallurgical reactions which lead to the formation of a liquid phase and determine the final microstructure can be described in terms of Al-Cu phase relationships and diffusion theory. Basically, these two metals will interdiffuse when heated while in close contact, and a concentration gradient varying from 100% Al to 100% Cu will exist across their interface.

The metallurgical structure within this gradient will be determined by the temperature and phase equilibrium relationships (Fig. 2). Below the 1018 F eutectic temperature, the Al-Cu system contains two primary solid solutions and a number of intermediate phases. Heating within this temperature range will result in the formation of the two solid solutions and between them distinct bands of each intermediate phase. The width of these bands will be determined by the time at temperature and relative diffusivities of the two metals. Near the eutectic temperature the diffusivity of copper in nearly pure aluminum is about 2000 times greater than that of aluminum in nearly pure copper (Ref. 5). Therefore, the direction of diffusion will be predominantly from the copper into the aluminum. As a result, the aluminum-rich solid solution zone will be much wider than the copper-rich zone.

A eutectic liquid results from the interdiffusion of aluminum and copper and the resulting formation of the aluminum-rich solid solution. Therefore, the mechanics of eutectic bonding, described below, are based on the formation of this solid solution; existence of the other phases is disregarded.

As shown in Fig. 3, the process begins with the copper-coated composite monolayer foils sandwiched together (the copper content of the 1100 aluminum matrix is essentially zero). As the monolayers are heated, diffusion begins. At any time in the heating cycle, the amount of copper in solid solution will vary from a maximum near the original interface to zero a short distance away. The depth of diffusion is determined by the rate of heating and temperature while the copper content of the solid solution is determined by the solid solubility limit of copper in aluminum. This limit varies with temperature, ranging from less than 0.5% at room temperature to a maximum of 5.65% at 1018 F. During the heating stage, the concentration of Cu in solid solution is determined by the instantaneous temperature and associated solubility limit while the depth of copper penetration continually increases with time.

When the temperature reaches 1018 F, melting begins in the aluminum which contains 5.65% Cu. The first liquid formed at 1018 F is the eutectic composition, Al-33.2% Cu. Further heating to the bonding temperature, nominally 1050 F, produces additional melting in the aluminum-rich solid solution to a depth where the copper content is equal to about 4% — the solid solubility limit at the nominal 1050 F bonding temperature. Normally at this temperature, the liquid phase of an Al-4% Cu alloy would contain about 25% Cu. However, the liquid can contain about 40% Cu so any remaining copper coating is dissolved in the unsat-

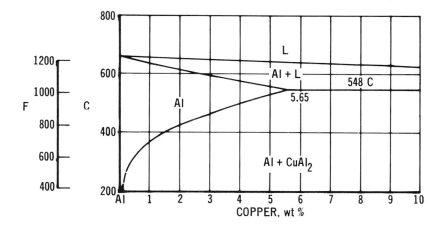

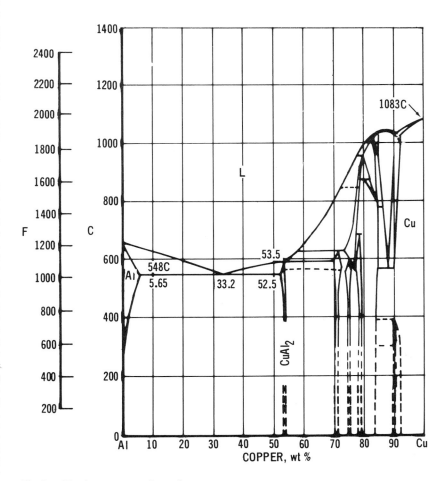

Fig. 2 — Aluminum-copper phase diagram

urated liquid or depleted through the continuation of the diffusion-melting process.

The metallurgical reactions which occur during solidification determine the microstructure and properties of the joint and are dependent upon the time the assembly is held above the eutectic temperature. During this time, copper continues to diffuse into the matrix so that the copper content of the liquid is continually decreasing.

When the process is applied to boron-aluminum, the time is sufficiently long so that all the liquid is depleted and solidification occurs isothermally at the bonding temperature. Upon cooling, the final microstructure will consist of fine particles of $CuAl_2$ intermetallic compound randomly distributed in an aluminum matrix containing a small amount of copper in solid solution. These particles are precipitated from a solid solution as the temperature decreases and the solubility limit is exceeded. As shown in Fig. 3, the final copper distribution will vary on a microscale from a maximim value of less than 5.65% at the original interface to a minimum value some distance away.

A less desirable microstructure can be formed if the joint is cooled prematurely and the liquid undergoes the eutectic decomposition to form a solid solution and the intermetallic compound. Then, the solid solution will blend into the previously solidified material. This will leave the intermetallic compound isolated at the joint centerline either as isolated particles distributed along the interface or, under the most extreme conditions, as a continuous (or near continuous) stringer along the bond line. This condition is avoided during eutectic bonding by carefully balancing the copper thickness and time/temperature profile to ensure isothermal solidification.

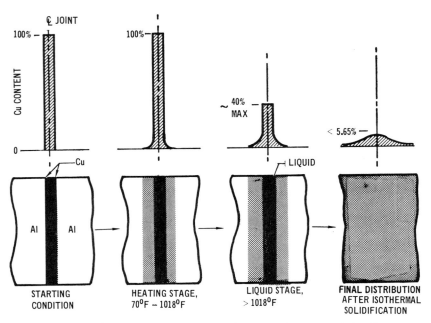

Fig. 3 — Variation in copper concentration and distribution during eutectic bonding

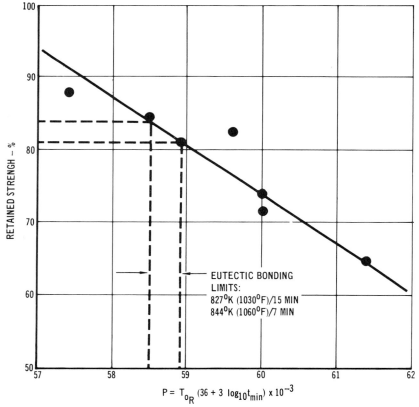

Fig. 4 — Effect of elevated temperature exposure on strength of boron-aluminum monolayer foil

Definition of Bonding Cycle Parameters

Metallurgical reactions associated with eutectic bonding suggested the most critical features of the process were related to the thermal cycle and the copper thickness. It was apparent that successful bonding of boron-aluminum would require careful balancing of these two variables to provide high strength joints free of brittle intermetallic compound networks through complete diffusion of all the copper coating. At the same time, extended high temperature exposure that would result in excessive fiber degradation had to be avoided. Two studies were undertaken to achieve this balance. The first was made to establish the effects of time-temperature exposures likely to be encountered in eutectic bonding on filament strength. Based on these results, limits were established for production control of bonding parameters and a second study then made to establish compatible coating thickness limits.

Bonding Cycle Study No. 1 — Fiber Degradation

It has been well documented that high temperature exposure of B/Al composite can result in lowered strength properties and that the magnitude of the loss is directly proportional to both time and temperature. Generally, degradation has been attributed to the formation of the compound AlB_2. Reaction zones at the fiber/matrix interfaces have been observed at 5000X magnification on samples subjected to severe exposure conditions such as 1000 F for 500 h. However, measurable strength degradation will occur at 1000 F in as short a time as one hour (Refs. 6,7).

Although the problem of fiber degradation has been studied by several investigators, their test results were not directly applicable to the eutectic process. Time/temperature conditions were not comparable nor were data available on the effect of the copper added to the matrix through diffusion. Therefore, a two part study was undertaken to determine the effect of eutectic bonding on fiber/matrix interaction. First, the effects of thermal exposures which simulated eutectic bonding on composite strength were evaluated. Then, tests were conducted to determine if copper influenced the reaction between boron and aluminum.

The first series of tests was made on uncoated samples cut from a single monolayer foil. This foil was cut into seven strips, 1-1/8 in. wide × 21 in. long. Five specimens from each group of nine were degreased, thermally exposed and tested. The remainder were tested in the as-received condition to establish baseline properties. This procedure was followed to minimize data scatter, which normally is high in composite materials, by ensuring that all specimens within a group shared common filaments.

Samples were heated in a vacuum furnace at times and temperatures ranging from 1020 F for 7 min to 1160 F for 7 min. The specimens were heated to the desired temperature at a rate of 20 F per min, and held for a predetermined time under a vacuum of approximately 1×10^{-5} torr. After the exposure time had elapsed, the specimens were fast cooled to 900 F by back-filling the furnace with argon, and then allowed to slow cool to 250 F before removal from the furnace.

Both the exposed and unexposed specimens were tensile tested to failure at room temperature using a cross-head travel rate of 0.05 in./min. Pneumatically tightened grips, which contained linings of hard rubber, were used to minimize specimen damage. Also, the specimens were aligned carefully to avoid introducing bending stresses.

The tensile test results were expressed both as monolayer strength and filament bundle strength. The latter value is derived by assuming that the matrix contribution to longitudal strength is not significant from the following expression:

$$\sigma_B = \frac{F}{A_f \, W \, C_f}$$

where

σ_B = filament bundle strength
F = failing load of coupon
A_f = cross-sectional area of a single filament
W = width of coupon
C_f = filament count (number of filaments per unit width of composite)

Bundle strength is often the preferred method for evaluating monolayer and laminate test results because it eliminates specimen-to-specimen variation in filament volume content and inaccurate thickness measurements of the thin monolayer attributable to surface irregularities. In this study, the fibers in each specimen were measured over a 1-in. width to obtain an accurate filament count and calculation of bundle strength.

Analysis of the tensile test results listed in Table 2 showed that boron fiber degradation occurred throughout the time/temperature range evaluated; the degree of degradation was governed by the severity of the exposure. For example, at 1060 F a holding period of 7 min resulted in a strength loss of about 15%; extending the holding time to 30 min increased the loss to 35%. A similar effect occurred when the temperature was increased. Holding at 1030 F for 7 min resulted in a 12% strength loss while exposure to 1100 F for the same amount of time produced a 29% loss.

The final fiber degradation tests were made to determine if the copper added to the 1100 aluminum alloy matrix as a result of eutectic bonding would influence fiber/matrix interaction. In these tests, copper coated and uncoated B/Al samples were exposed simultaneously to selected thermal cycles, tensile tested, and compared with as-received specimen results. The previously described procedures for sampling and testing were followed.

A total of 30 coated and 30 uncoated samples was thermally cycled and tested. One half of each type were held at 1060 F for 15 min which represented an extreme condition and the others at 1030 F for 15 min to represent a eutectic bonding cycle. These tests showed about 5% increase in the amount of degradation measured in coated samples held at 1060 F for 15 min. An increase of only 2% was noted for the simulated eutectic bonding cycle at 1030 F; this amount was considered tolerable.

To better define fiber degradation in terms of time-temperature relationships, the filament bundle strength data reported in Table 2 were expressed in terms of a Larson-Miller parameter. This representation is shown in Fig 4. Then maximum and minimum conditions were selected for eutectic bonding based on anticipated production limitations. It was

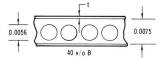

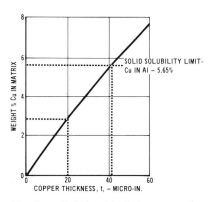

Fig. 5 — Relationship between coating thickness and copper content of matrix

Table 2 — Room Temperature Longitudinal Tensile Strength of Thermally Exposed and Unexposed Boron-Aluminum

Exposure condition	Average monolayer ultimate strength, ksi	Average filament bundle strength, ksi	Degradation, %
Unexposed controls	185	441	—
1020 F - 30 min	150	364	17.6
1030 F - 7 min	167	389	11.9
1030 F - 15 min	159	358	19.0
1060 F - 7 min	157	374	15.5
1060 F - 15 min	133	317	28.3
1060 F - 35 min	123	289	34.6
1100 F - 7 min	135	316	28.6

A. 900°F – NO HOLDING TIME

B. 1,000°F – NO HOLDING TIME

C. 1,000°F – HOLD 60 MIN

Fig. 6 — Effect of time and temperature on diffusion of copper coating during eutectic bonding heating cycle

ALL SAMPLES COATED WITH
16 MICRO–IN. COPPER

COUNTS/SEC →

C

B

A

0.002   0.004   0.006

APPROXIMATE DISTANCE FROM INTERFACE – IN.
(REPRESENTATION OF MICROPROBE LINE SCANS)

a target maximum value to further minimize coating thickness and thereby ensure that the other objectives would be met.

Before proceeding with eutectic bonding using 20 microin. thick copper coatings, an analysis was made to determine if enough of the coating would survive the diffusion process during heating to provide a liquid phase. For this purpose it was again assumed that the aluminum-rich solution region was of prime importance and that all the copper which crossed the interface went into solid solution. This total amount was defined by the expression:

$$x\rho_c = 1.1284 \sqrt{Dt} \quad (C_s - C_o) \quad \text{(Ref. 5)}$$

where x = thickness of coating lost through diffusion, cm.

ρ_c = density of copper = .324 lb/in³

D = diffusion coefficient of Cu in Al, cm²/sec.

t = time, sec.

C_s = copper concentration maintained at surface; for this analysis $C_s = C\rho_a$ where C is solid solubility limit of Cu in Al and ρ_a is density of alloy, $\sim$.1 lb/in³

C_o = initial copper concentration in Al =0

Substituting and changing units to have t in hours and x in inches:

$$x = 8.25 \, C \sqrt{Dt}$$

The factors that determine the rate of coating loss are the diffusion coefficient, time, and solid solubility limit of Cu in Al. Both the diffusion coefficient and solid solubility increase with temperature. Because of the interrelationships the rate of coating loss during heating at a uniform rate cannot be calculated readily. However, the derived expression was used to calculate the rate of diffusion under constant temperature conditions. Temperatures near the eutectic bonding temperature were considered because of the rapid change in diffusion rate expected in this range. Between 900 F and 1000 F, the diffusion coefficient increases by an order of magnitude (Ref. 4) and the solubility limit increases about 50% (Fig. 2).

Calculations showed that holding at 850 F would deplete a 7 microin. coating in about 5 h, but at 1000 F only about 4 min would deplete the same. This rapid change with temperature was of concern because for large production parts as much as 10 min or more might elapse in heating between 900 F and the eutectic tem-

assumed that bonding conditions might range from 15 min at 1030 F to 7 min at 1060 F. With these extremes, it was predicted that eutectic bonding could result in a maximum strength loss ranging from about 16% to 19%. While this amount was considered appreciable, it was still well below the 35% or more that would occur in a brazing operation which normally requires heating near 1100 F.

Bonding Cycle Study No. 2 — Copper Coating Thickness

The thickness of the copper coating applied to boron-aluminum monolayer was selected to accomplish the following objectives:

(a) complete consumption of the copper layer.

(b) isothermal solidification at the bonding temperature to avoid both eutectic decomposition and concentration of CuAl₂ along the bond line.

(c) total amount of copper available for dispersion through the aluminum matrix must not exceed the maximum solid solubility of copper in aluminum, (5.6%), to ensure (a) and (b) above.

These goals could be accomplished only by using very thin copper coatings.

Figure 5 shows the relationship between coating thickness and matrix copper content. The upper limit for coating thickness, assuming no loss of liquid during bonding, is about 42 microin. to stay within the solubility limit. A coating thickness of 20 microin. per side was selected as

perature of 1018 F. Therefore, a series of tests was considered necessary to determine how much copper is depleted during heating to the eutectic temperature, and if the 20 microin. target value would be sufficient to survive heating and provide a liquid for bonding.

The effect of the heating cycle was evaluated by heating 0.006 in. thick 1100 aluminum samples, which had been coated by physical vapor deposition on one side with 16 microin. of copper, to several temperatures below the eutectic temperature. The diffusion specimens were heated to 900 F at a rate of 20 F/min, and then at about 7 F to the maximum temperature. When the preselected temperature was reached, the specimens were immediately cooled to room temperature and then analyzed by electron microprobe line scans and raster images to determine how much of the copper had been depleted. One sample, used as a control, was held at 1000 F for 60 min to deplete the coating entirely.

Results of representative microprobe analyses are shown in Fig. 6. The x-ray rasters and line of scans show that the 16 microin. thick copper coating was completely diffused in the 1000 F/60 min treatment as expected. Some diffusion occurred as a result of heating to 900 F and 1000 F, with the higher temperature resulting in significantly more diffusion. The line scans were used to estimate the amount of copper depleted from the coating. This was accomplished by measuring the areas under the concentration curves and determining the ratio of the area of the heating cycle in question to that of the completely diffused sample (curve C, Fig. 6). This ratio multiplied by 16 microin. — the original thickness of the completely diffused coating — provided the amount of coating lost during heating. The coating loss was calculated to be:

Test Temp.	Depleted Zone
900 F	1.98 microin.
1000 F	8.05 microin.

These data for 0.006 in. aluminum foil indicated that about 8 microin. of copper would be lost in heating to 1000 F and it was estimated that probably another 2 or 3 microin. would be lost in reaching the eutectic temperature. A smaller loss would be encountered with a composite monolayer because the matrix would be equivalent to a thinner (~.0045 in.) aluminum foil which would be coated on two sides rather than just one. Consequently, diffusion completely through the thickness would be expected. This would lower the concentration gradient and reduce the rate of diffusion. On this basis, a coating thickness range from 17.5 to 22.5

microin. (20 microin. nominal) was selected for eutectic bonding B/Al. This amount would provide a liquid film between about 100 and 160 microin. thick even with the loss of 10 microin. of copper per surface during heating. This amount was considered adequate since bonding would be done under pressure to provide good contact between adjacent monolayer foils.

In order to verify the adequacy of the selected coating thickness range, 1 × 1 in. laminates were prepared from B/Al monolayer foils coated with 14-20 microin. of copper and heated within the eutectic bonding temperature range of 1030 F and held from seven to fifteen minutes at temperature. The as-bonded samples were characterized by droplets of solidified Cu/Al alloy along the edges

where liquid had been squeezed out of the joint during bonding. Electron microprobe analyses were made to determine if $CuAl_2$ was present at any of the bond lines. These analyses showed that the bonds essentially were free of intermetallic compound except in the form of discrete, isolated particles. Chemical analyses showed the average copper content of the aluminum matrix was about 2%.

The typical microstructure shown in Fig. 7 was a result of isothermal solidification at the bonding temperature when all the copper was taken into solution in the solid aluminum. On cooling, most of the copper was precipitated as fine, randomly distributed particles of $CuAl_2$. Some agglomeration of the precipitates occurred which can be seen on the x-ray raster image. Microprobe line scans

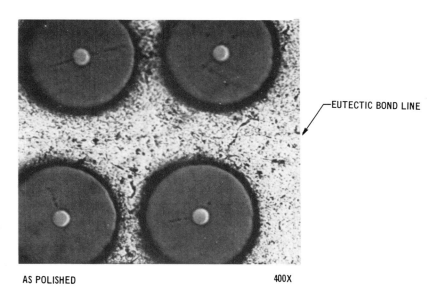

EUTECTIC BOND LINE

AS POLISHED   400X

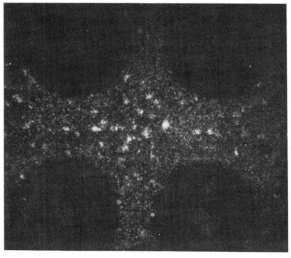

400x

Fig. 7 — Eutectic bonded joint in boron-aluminum laminate (reduced 17%)

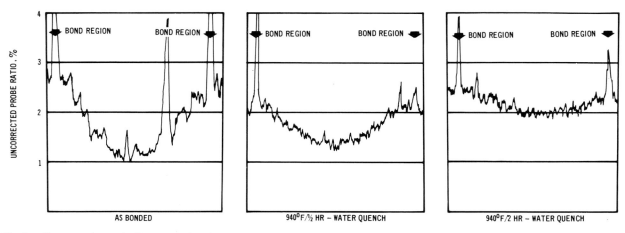

Fig. 8 — Electron microprobe line scans showing effect of postbonding heat treatment on the distribution of copper in the matrix of eutectic bonded boron-aluminum composite laminate

showed a copper concentration gradient existed from bond lines to the center of the monolayer foils.

A study was undertaken to determine if a more homogenous structure might be obtained by heat treatment. This possibility was evaluated by subjecting 4 ply laminates to postbonding heat treatments and then conducting microprobe line scans to determine copper distribution. The microprobe scans in Fig. 8 show the pronounced gradient between bond lines in the as-bonded condition with numerous sharp peaks indicating localized areas of copper enrichment. A solutioning treatment at 940 F for 30 min appeared to narrow the gradient and reduce the tendency for the compound to agglomerate. Increasing the time at temperature from 30 min to 2 h produced a further improvement in copper distribution.

These studies showed that the copper distribution within the aluminum matrix could be improved by post-bonding heat-treatment. However, localized areas of copper segregation were not eliminated completely under any of the conditions evaluated. Even in the as-bonded condition the lack of homogeneity did not appear severe enough to be detrimental to composite properties. Therefore, post bonding heat-treatments were not considered essential to the eutectic bonding process.

As a final check on the suitability of the 20 micron. coating thickness, the strength of eutectic bonded joints was evaluated by short-beam interlaminar shear tests. Details of the specimen design are shown in Fig. 9. With this type specimen, interlaminar shear strength is determined from the expression:

$$\gamma = 3/4 \ (P/bh)$$

where:

P = applied load at center of beam

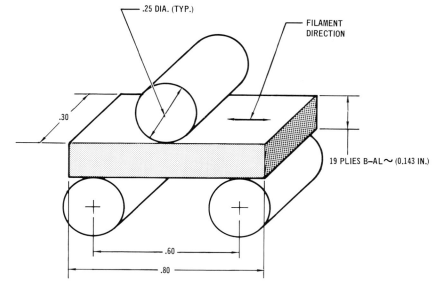

Fig. 9 — Configuration and test setup for short-beam interlaminar shear specimens

Table 3 — Typical Properties of Eutectic Bonded Boron-Aluminum [a]

Type of loading	Ultimate strength, ksi at		ultimate strain, microin./in. at		initial modulus 10^6 psi at	
	R.T.	600 F	R.T.	600 F	R.T.	600 F
Longitudinal tension						
coupon	160	155	6500	5800	29.8	28.1
beam	194	—	7200	—	28.7	—
Longitudinal compression						
beam	343[b]	—	10500	—	36.8	—
Transverse tension						
coupon	16.7	3.9	4200	6500	19.5	14.0
Transverse compression						
beam	37.5	9.6	2400	10700	20.0	17.4
In-plane shear						
rail shear	10	3.7	16000	13000	10.0	7.8

(a) All tests conducted on 42-45 v/o B/Al
(b) Failure occurred in honeycomb core—no failure in B/Al face sheet

b = beam width
h = beam depth

However, the actual stress distribution within a short-beam specimen may differ significantly from that predicted by the above expression. Therefore the test was used solely to determine if the bond line would fail by shear. All of the specimens tested failed by flexural tension rather than by shear. At the failure loads, the calculated shear stress on the bond lines exceeded an average value of 10,000 psi. These tests provided further evidence that eutectic bonds made with 20 microin. of copper were strong and of good quality.

Co-Bonding of Titanium and Boron-Aluminum

One of the unique advantages of eutectic bonding proved to be its ability to incorporate local reinforcement in the form of titanium interleaves during the normal processing of boron-aluminum laminates. Interleaving significantly improves matrix dependent properties, particularly at elevated temperature, in the following areas:
- Mechanical joint bearing strength
- Ultimate transverse strain capability
- Shear strength of cross-plied (±45 deg) laminates
- Crippling strength at elevated temperature
- In-plane shear strength of undirectional laminates

When structural analysis indicated the advantages to be gained from local reinforcement, tests were initiated to determine if titanium interleaves could be incorporated into boron-aluminum laminates during eutectic bonding. Initial tests were made on a multi-ply laminate consisting of copper coated boron-aluminum and bare Ti-6Al-2Sn-4Zr-2Mo alloy. The lay-up contained three plies of boron-aluminum interleaves with two plies of 0.012 in. thick Ti alloy. Metallographic examination was used to evaluate the joint and showed the bond to be continuous and of high quality with an interaction zone less than 40 microin. thick (Fig. 10). Microprobe analyses were made to determine the depth of copper diffusion into the titanium. No evidence of copper was detected beyond the interaction zone.

Because the initial results were encouraging, the evaluation was extended to determine the interlaminar shear strength between the Ti alloy and boron-aluminum. For these tests, specimens of the type shown in Fig. 9 were cut from a multi-ply laminate and tested in three point bending.

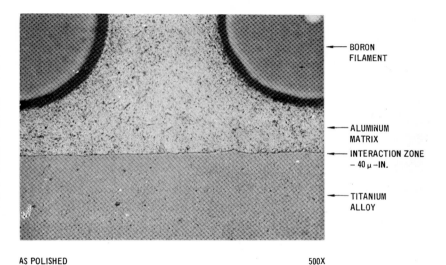

BORON FILAMENT

ALUMINUM MATRIX

INTERACTION ZONE – 40 μ –IN.

TITANIUM ALLOY

AS POLISHED   500X

Fig. 10 — Titanium joined to boron-aluminum by Al-Cu liquid formed during eutectic bonding cycle (reduced 24%)

The basic laminate contained nineteen plies, of which three were Ti-6Al-2Sn-Zr-2Mo alloy. These Ti plies were located in the center of the laminate and were separated from each other by a single boron-aluminum ply.

Five interlaminar shear specimens were tested at room temperature. Failure loads ranged from 1190 to 1210 lb and the average was equivalent to a nominal shear strength in excess of 16,000 psi at the titanium alloy-boron-aluminum interfaces. However, all the specimens failed in tension and the failures originated in the outer boron-aluminum plies at stress levels in excess of 160,000 psi. On the basis of tese tests, it was concluded that titanium interleaves could be included in a boron-aluminum laminate during a normal eutectic bonding cycle.

Subsequent tests showed that adding one titanium ply, 0.008 in. thick, for each four boron-aluminum plies increased the transverse tensile strength by a factor of three while increasing the strain-to-failure about three times to about 1.2%. Additional tests were made to demonstrate the improvement in bearing strength that could be attained by interleaving. Test specimens were 1.5 in. wide × 9 in. long with 0.312 in. diam fastener holes centrally located 0.75 in. from each end. These samples contained 28 plies of boron-aluminum (0.0075 in./ply) and 8 plies of Ti-6Al-4V alloy (.012 in./ply). These specimens failed at an average load of 14,000 lb in a combined tension/shear mode. A pure boron-aluminum specimen of equivalent weight would withstand only about 4,000 lb before failure. These and other test results indicate that the inclusion of titanium interleaves is a major advancement in boron-aluminum.

Determination of Mechanical Properties

During the development of the eutectic bonding process, more than 500 mechanical property tests were conducted. Initially, bonded tensile coupons were tested to verify that eutectic bonding could be used to produce laminates with the high properties generally offered by B/Al. Once the critical processing parameters were defined and process feasibility demonstrated, a comprehensive program was undertaken to establish design allowable mechanical properties. Both transverse and longitudinal properties were determined at room temperature and 600 F. Properties of interest include ultimate tensile strength, strain to failure, modulus of elasticity, shear strength, crippling strength. A full description of this phase of the investigation including the test techniques, specimen designs, and test variables is beyond the scope of this paper. However, some of the test results are summarized in Table 3. These data demonstrate that the high strength, high modulus, low density potential of B/Al can be attained in laminated structures fabricated by eutectic bonding.

Summary

The results of the initial investigation showed that boron-aluminum laminated structures can be fabricated by coating the monolayer foils with a copper layer about 20 microin. and heating in the temperature range from 1030 F and 1060 F. Unacceptable degradation can be avoided by restricting the bonding time to less than fifteen minutes at the low end of the range and seven minutes at the high end.

Eutectic bond lines characteristically are free of brittle compounds that would lower joint strength. If present, these compounds exist as discrete, widely spaced particles. The very high interlaminar shear strength of eutectic bonded joints and the fact that joint failures have seldom been observed attest to their high over-all quality. Further evidence is given by the good mechanical properties measured on eutectic bonded B/Al laminates.

Eutectic bonding was also found to be a versatile process in that it was readily adaptable to the inclusion of titanium interleaves. The ability to utilize local reinforcement had eliminated concern over the inherently low shear transfer and bearing capability of B/Al. On the basis of these encouraging results, work was initiated to develop the production techniques needed to fabricate large, complex boron-aluminum structures. The development of such processing techniques along with the fabrication of complex structures is the subject of Part II of this paper.

Acknowledgement
The research described in this paper was conducted under the direction of the McDonnell Douglas Astronautics Co.-East, utilizing the laboratory facilities of the McDonnell Aircraft Co. The authors thank Mr. F. Pogorzelski of the Advanced Manufacturing Fabrication Facility who supervised specimen fabrication and Mr. M. Russo of the Metallurgical Laboratory for metallography and microprobe analyses.

References
1. Lynch, J. F., Feinstein, L. and Huggins, R. A., "Brazing by the Diffusion — Controlled Formation of a Liquid Intermediate Phase," *Welding Journal,* 38(2), 1959, Research Suppl., pp. 85-s to 89-s.

2. Owczarski, W. A., "Eutectic Brazing of Zircaloy 2 to Type 304 Stainless Steel," *Welding Journal,* 41(2), 1962, Research Suppl., pp. 78-s to 83-s.

3. Mechtly, E. A. "The International System of Units, Physical Constants and Conversion Factors (Revised)" *NASA-SP-7012,* 1969.

4. *Aluminum Vol. I; Properties, Physical Metallurgy and Phase Diagrams,* Van Horn, K. R., ed., American Society of Metals, Metals Park, Ohio, 1967.

5. Darken, L. S. and Gurry, R. W., *Physical Chemistry of Metals,* McGraw-Hill Book Co., New York, 1953.

6. Blucher, D. I., Spencer, W. R., Stuhrke, W. K., "Transmission and Scanning Electron Microscopy of Boron/Aluminum Interfaces," presented to The 17th Refractory Composites Working Group, Williamsburg, Va., June 16-18, 1970.

7. Stuhrke, W. F., "Solid State Compatibility of Boron Aluminum Composite Material" Proc. Symposium of Met. Soc. AIME, May 12-13, 1969, *DMIC Memorandum 243, Metal Matrix Composites,* 1969, p. 43-46.

Soldering and Brazing of Advanced Metal-Matrix Structures

Several processing techniques and filler metals have been developed to realize the low weight, high strength capabilities of boron/aluminum composites

BY A. R. ROBERTSON, M. F. MILLER AND C. R. MAIKISH

ABSTRACT. There is increased awareness of the potential weight savings resulting from the use of advanced composites in aerospace designs. Boron/aluminum, one of these advanced composites, has been shown to be adaptable to a full range of joining techniques. Soldering of boron/aluminum is applicable over a wide range of temperatures, and has no deleterious effect on the boron filaments. Room temperature lap shear strengths for various solders range from 10 ksi (69 MN/m²) to more than 12 ksi (83 MN/m²); certain solders have lap shear strengths as high as 4.5 ksi (31 MN/m²) at 600 F (589 K). Boron/aluminum is brazed using aluminum-silicon alloys, at temperatures up to 1,140 F (880 K), either by dip brazing in a molten bath or fluxless brazing in a vacuum. Even short exposures at these elevated temperatures can cause serious degradation of the boron filaments. In an effort to reduce brazing temperatures, a development program has been conducted using eutectic diffusion brazing. This flux-less process uses an elemental metal interface. Upon heating, a low melting point eutectic alloy forms and diffuses from the interface, leaving a joint with a remelt temperature 200 to 500 F (366 to 533 K) higher than the brazing temperature.

Introduction

There is increased awareness of the potential weight savings resulting from the use of advanced composites in aerospace designs. These materials exhibit high specific stiffness and high strength-to-weight ratios that result in lower weight and size, and greater range and payload for aerospace vehicles. Only one of these advanced composites, boron/aluminum, has been shown to be adaptable to a full range of joining techniques, including welding, brazing, soldering, riveting, and diffusion bonding.

Boron/aluminum is composed of boron fibers interspersed in an aluminum matrix. The fibers are aligned in rows through the thickness. The basic unit of boron/aluminum is the monolayer, which consists of one row of boron fibers in an aluminum matrix. These monolayers (or plies) are stacked, one atop the other, and bonded together to form a continuous material (Fig. 1). All plies may be oriented with the fibers in the same direction (unidirectional) or at various angles to each other (crossplied). Boron/aluminum has the density of aluminum, but strength comparable to steel. The room temperature tensile strength in the longitudinal direction for 50 volume per cent boron in aluminum is 216 ksi (1,490 MN/m²), with a modulus of 31 msi (213 GN/m²). Even at 600 F (589 K), which is considered to be the maximum use temperature, the strength is greater than 150 ksi (1,040 MN/m²) (Ref. 1). The room temperature shear strength, measured by the double-slotted shear test described in Ref. 2, is approximately 23 ksi (159 MN/m²).

Commercial acceptance of boron/aluminum (and other advanced composites) has been hampered by several factors, including the final cost of structural members, the lack of suitable joining techniques, and processing temperature limitations. Application of soldering and brazing to boron/aluminum structures has helped to minimize these factors.

A. R. ROBERTSON and C. R. MAIKISH are Senior Engineers, and M. F. MILLER is Staff Scientist with Convair Aerospace Division of General Dynamics, San Diego, Ca.

Paper was presented at the 2nd International Soldering Conference held in Chicago on April 3, 1973.

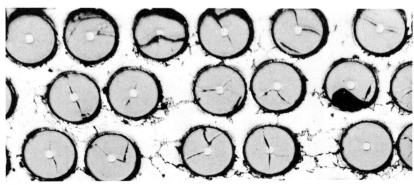

Fig. 1 — Microstructure of Borsic/aluminum braze panel. Borsic fibers interspersed in Al-Si matrix X 100, reduced 40%

Soldering

Soldering is an excellent method for attaching metal-matrix composites to similar or dissimilar metals, such as titanium, since processing temperatures fall below the point where filament degradation occurs. Soldered structures can be used at operating temperatures up to 600 F (589 K) — generally believed to be the upper operating temperature limit for boron/aluminum or Borsic/aluminum structures.

Alloy Selection

Surface preparation of boron/aluminum for soldering has been found to be critical in attaining highly reliable joints. It has been shown that zincate coatings followed by thin electroless nickel plating of about 0.0002 in. (0.05 mm) is effective in promoting soldering ease, alloy flow, and general strength increase (Ref. 2). This result is attributed to higher wettability of the nickel surface, leading to generally better flow conditions. Electroless nickel plating is used in preference to electrolytic processes because it allows the nickel to be plated on exposed boron filaments, which are electrically nonconductive. Depending upon the joint configuration, electroless nickel plated specimens result in joints with 10% to 30% greater strength than those plated with electrolytic nickel systems. The significant difference is attributed to the dewetting action associated with electrolytic nickel systems resulting from any exposed boron filaments.

An electroless nickel plate about 0.0002 in. (0.05 mm) thick on both boron/aluminum and titanium alloys has proven to be a readily solderable combination. Titanium — normally a difficult material to plate — is nickel plated by excluding surface oxide formation with the use of citrate and tartrate coatings soluble in the ammoniacal solution of the plating bath. Standard methods of electroless nickel plating titanium require vacuum heat treatment at 900 F (755 K) for one hour to produce the adherent bond for soldering. Similarly, electroless nickel plated boron/aluminum is baked for one hour at 350 F (450 K).

Solders for both 200 and 600 F (366 and 589 K) applications were selected on the basis of consideration of strength, temperature, and adaptability to use with boron/aluminum composites. Alloys chosen for initial evaluation were 95% cadmium-5% silver; 95% zinc-5% aluminum; 96.5% tin-3.5% silver; and 82.5% cadmium-17.5% zinc. These solders have flow temperatures of 750, 720, 425, and 509 F (672, 656, 492, and 538 K), respectively.

Room temperature and elevated temperature tests were conducted on single overlap shear specimens. The specimens were 0.040 in. (1.02 mm) thick by 0.5 in. (12.7 mm) wide boron/aluminum joined to 0.125 in. (3.2 mm) thick by 0.5 in. (12.7 mm) wide 6061-T6 aluminum with a 0.125 in. (3.2 mm) overlap. Actual joints were made by fluxing both surfaces to be joined with the manufacturers' recommended flux, clamping the overlapping parts in a stainless steel fixture for alignment, and heating with a soft, slightly caburizing oxyacetylene flame. Joint clearance was contact only, and a small piece of solder alloy was preplaced at one end of the joint; upon heating the joints were formed through capillary action.

Table 1 summarizes lap shear test results for three alloys selected for potential applications. Joints made with 82.5% cadmium - 17.5% zinc alloy were found to be extremely brittle and occasionally cracked during cooling, whereas joints made with the other alloys were relatively ductile.

The 95% cadmium-5% silver alloy was chosen for 200 F (366 K) applications. Its shear strength at 200 F of 12.90 ksi (89 MN/m²) satisfies most design requirements for a boron/aluminum structural joint intended for operation at that temperature. Its strength drops off rapidly above 500 F (533 K), making the alloy unsuitable for 600 F (589 K) applications.

The 95% zinc-5% aluminum alloy with a strength of 4.4 ksi (30 MN/m²) at 600 F (589 K) was selected for use at that temperature. Although the alloy is as strong as the cadmium-silver alloy at 200 F (366 K), the latter was selected for the lower temperature application on the basis of its better flow characteristics and the better visual appearance of the finished joint.

Failure modes vary with the test temperature. At low temperatures, the composite fails by interlaminar shear, usually at the outer aluminum foil, exposing the boron filaments. As the useful temperature limit of the solder is reached, the failure surface passes through a mixed mode until the failure is wholly in the solder (Fig. 2). Lap shear joints of boron/aluminum to titanium have the equivalent strength of those between boron/aluminum and boron/aluminum since the failure occurs in the soldered joint or in the composite plies.

Lap shear specimens were also prepared with 0.0002 in. (0.05 mm) of cadmium plated over the electroless nickel. This system was evaluated to determine if it would improve wettability and encourage better flow of

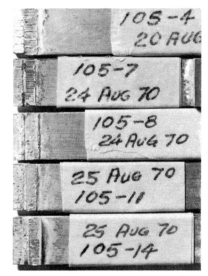

Fig. 2 — Failure surfaces of boron/aluminum with 95% cadmium-5% silver braze alloy

Solder composition	Test temperature		Failure stress		Failure mode [a]
	F	K	ksi	MN/m²	
95% cadmium, 5% silver	70	294	11.68	81	1
	200	366	12.90	89	1
	300	422	10.17	69	1
	400	478	6.79	47	2
	500	533	4.22	29	3
	600	588	0.82	5.6	3
95% zinc, 5% aluminum	70	294	11.60	80	1
	200	366	13.60	94	1
	600	588	4.43	30	2
82.5% cadmium, 17.5% zinc	70	294	10.67	74	1
	200	366	13.31	90	1
	300	422	8.52	59	2

Table 1 — Results of Lap Shear Tests for Soldered Boron/Aluminum Specimens

(a) Failure mode:
   1. Composite interlaminar shear
   2. Combination of 1 and 3
   3. Braze alloy adhesive and cohesive failure.

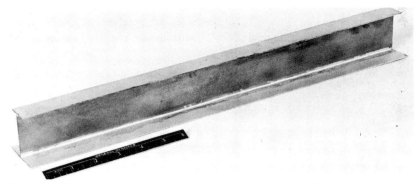

Fig. 3 — Con Braz joined boron/aluminum I section with titanium web

T-TENSION SPECIMEN   T-SHEAR SPECIMEN

Fig. 5 — Loading direction for Con Braz joined Tee sections.

Fig. 4 — Con Braz Tee section fillet, X10, reduced 50%

the solder to yield increased joint strength. The lap shear tests resulted in an average shear strength of 10.3 ksi (71 MN/m²). It was concluded that using cadmium over nickel offered no improvement in brazing ease or joint strength. Further, on the basis of ultrasonic C-scan of the joints, it was found that the system could possibly be detrimental.

Con Braz Joining

One primary application of soldering boron/aluminum is the Con Braz joining process. In this technique, flat sheets and plates of composite (previously consolidated) are cut to size, assembled in a fixture, and soldered into the desired configuration.

Con Braz joining is suitable for making structural sections such as Tee sections, angles, I sections, and hats. Typical I sections can be assembled with unidirectional bor-

on/aluminum caps and a titanium or crossplied boron/aluminum web to give increased shear strength in the web area. In other fabrication processes, this approach is either impossible or requires a complex tape layup procedure. Figure 3 shows a Con Braz joined I beam with boron/aluminum caps and a titanium web.

Con Braz joining results in the formation of a natural radius between the composite and the solder, with complete composite surface wetting. Figure 4 illustrates the natural fillet formed at the joint area. Close examination reveals interdiffusion of the solder and aluminum matrix alloys to approximately the third ply in from the surface.

Before proceeding with development of this process, it was necessary to determine the tension and shear strengths of Tee joints made with the Con Braz process. Tee sec-

tions were assembled from 0.15 in. (3.8 mm) thick diffusion bonded boron/aluminum sheet. The specimens were 1 in. (0.02 m) long with a 2 in. (0.05 m) base and 1.5 in. (0.04 m) leg. Joint surfaces were electroless nickel plated before soldering. The joints were soldered with the 95% cadmium-5% silver alloy and tested at both room temperature and 200 F (366 K). Figure 5 shows the direction of loading. Test results, summarized in Table 2, indicate that the strength of the solder can be fully realized in structures of this kind at both ambient and 200 F (366 K) temperatures. Con Braz joined Tee sections with undirectional boron/aluminum and titanium details have been subjected to thermal cycling between –320 and 200 F (77 and 366 K) with no evidence of cracking or other damage.

If necessary, joint strengths can be increased by using external fillets machined from boron/aluminum or aluminum to increase the joint area. To ensure complete flow of the solder throughout the joint area, solder in the form of 0.005 in. (0.13 mm) thick foil is preplaced between the supplementary fillet and other part details.

To demonstrate the feasibility of using soldering to fabricate long sections suitable for aircraft structure, a combination heating and tooling module was designed and fabricated for Con Braz joining. The module is stationary and the parts to be joined are hand fed through the module by an operator who watches the soldering operation from above. The length of the finished component is limited only by the length of available material and floor space.

The module uses three 1,200 watt T3 quartz radiant heat lamps in three Research Inc. Model 5305A strip heaters to heat the part to the soldering temperature. These units have a 6 in. (0.15 m) long polished aluminum reflector that concentrates the radiant heat over a 1.5 in. (0.04 m) wide by six in. (0.15 m) long target area. The lamp units are watercooled to avoid overheating the reflector and lamp ends. Overheating oxidizes the reflector, thus increasing the emissivity of the reflector surface and reducing the effi-

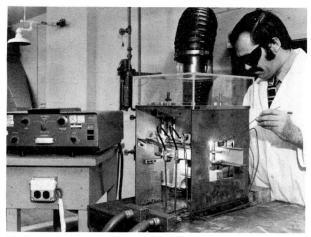

Fig. 6 — Con Braz joining of boron/aluminum I section stiffener

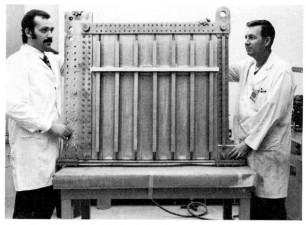

Fig. 7 — Completed boron/aluminum shear beam component showing 15 Con Braz joined I sections

Table 2 — Average Tension and Shear Strengths of Soldered Boron/Aluminum Tee Sections

Test Mode	Test temperature		Failure stress	
	F	K	ksi	MN/m²
Tension	70	294	5.70	39
Shear	70	294	10.60	73
Tension	200	366	5.50	38
Shear	200	366	10.70	75

ciency of the heating unit. Quartz lamp end seal temperatures must be maintained below 600 F (589 K) to ensure a satisfactory lamp life. If the seal temperature exceeds this limit, oxidation of the element at the junction with the quartz envelope is accelerated and the lamp life is considerably reduced.

To try out the module, tooling was initially fabricated to make a 0.5 in. (12.7 mm) thick Tee section, Tooling consisted of two identical stages with spring loaded stainless steel rolls that guided the part through the module and maintained an even pressure on the individual part details to ensure intimate contact at the joint area during soldering. To improve efficiency and reduce stray glare from the lamps, polished aluminum reflectors were added between the lamps forming a chamber with open ends. The top was left partly open to permit visual examination of the joint during soldering.

An extractor system was installed over the unit to remove any fumes generated during the soldering operation. It was constructed from clear acrylic sheet to allow unrestricted visibility of the joint by the operator. Figure 6 shows the Con Braz joining module being used to fabricate an I section stiffener.

Heating and joining tests were conducted using 0.5 in. (12.7 mm) thick 6061 aluminum Tee sections. Static heating tests were conducted to verify the ability of the strip heaters to bring the part to 800 F (700 K), which would satisfy the requirement for the solder alloy(s) to be used. A 0.5 in. (12.7 mm) thick section can be continuously joined at a rate of about 3 in. (0.08 m) per minute. Should production quantities of Con Braz joined sections be required, the maximum speed could be increased by installing a preheating zone to increase the heating rate of the part.

Maximum heating rate and maximum speed could then be attained by installing a closed loop control system, consisting of a radiation pyrometer sighted on the part with a feedback to a temperature controller that transmits, to the power controller, a signal proportional to the temperature deviation from setpoint. The power controller would then regulate input to a drive motor, which would drive the part through the module at a rate sufficient to maintain the area being soldered at the set point temperature.

A 40 by 38 in. (1.02 by 0.96 m) boron/aluminum shear beam component of a Space Shuttle thrust structure shear web beam was designed and fabricated to demonstrate the ability of these boron/aluminum joining methods to withstand both thermal and load cycling (Ref. 1). Twenty-two boron/aluminum sheet metal I section stiffeners were fabricated for this shear beam component. Composite thickness varied from 0.068 to 0.109 in. (0.17 to 0.28 mm) and stiffener length ran from 11 to 40 in. (0.28 to 1.02 m). All stiffeners were joined in the Con Braz joining module, using a 95% cadmium-5% silver alloy. The sections were joined in two passes: the first to make a Tee section, the next to join the second cap to the assembly to form an I section.

The two radiant lamp units at the top of the module were directed at the bottom of the web of the part. The radiant lamp under the base of the section was located close to the part to promote maximum heating. from the bottom. This method reduced the possibility of melting the joint at the top of the part when the second cap was being joined to the Tee to make the I section.

Static heating tests were conducted to determine the power input required to produce an equilibrium temperature of 800 F (700 K) at the joint area. At the optimized heating intensity, overheating of the part and its subsequent detrimental effects did not occur. With a manually fed mode of operation, this rate provides improved control and permits incremental feeding of the part through the unit.

The composite areas adjacent to the joint are protected by brushing on a thin coat of a brazing stopoff agent, which prevented both staining of the surface by excess flux and excessive wetting of the part by unrestricted flow of the solder. The joint area was prefluxed, followed by preplacement of the solder at the joint area. Details were fed through the module and observations made to determine when the alloy melted and flowed through the joint. Any inadequately joined areas were supplemented by hand feeding a prefluxed rod into the required area. The part was fed through the module at a speed consistent with producing a good joint of uniform quality. Ultrasonic C-scan inspection of both joint surfaces indicated excellent joints.

Con Braz joining with the heating/tooling module allowed close temperature control at the joint area. With other brazing techniques this control is difficult when thin gage material is being joined. The equilibrium temperature control system for the Con Braz joining module allows the joint area of the part to be held at the soldering temperature for an indefinite time without risk of overheating. This promotes good flow of the solder and allows careful examination of the joint during soldering.

Two of the stiffeners, with bases 2.25 in. (0.06 m) wider than the others, were soldered using an oxygen-gas torch since the small length of joint involved could not justify making new tooling for the Con Braz module to accommodate these larger details.

Torch soldering is considerably slower than using the Con Braz joining module, since assembling and aligning the part details must be accomplished manually before joining, whereas the spring loaded rolls of the Con Braz joining module align the details automatically. With torch soldering, heating is effected over a much smaller area and controlling the joint temperature is more difficult, requiring considerably more skill and care by the operator than is necessary when using the Con Braz joining module. Also, joining is interrupted for fixture relocation (to provide access to all areas of the joint).

Approximately 80 ft (24.4 m) of joint were made by this process, with

about 6% being rejected due to poor quality materials.

No failures of the radiant quartz lamps in the heating units occured. When using the equilibrium temperature control system and feeding the part manually through the Con Braz joining module, the soldering speed was approximately 1.5 to 2 in. (0.04 to 0.05 m) per minute. Modification of the module to include automatic control and permit full use of the maximum heating capability can increase the feed rate to about two to three feet per minute.

One side of the completed shear beam component is shown in Fig. 7; fifteen of the soldered I beams are evident.

Tube Fittings

OV1 satellites, when launched in the dual configuration by an Atlas launch vehicle, are supported by an aluminum truss structure some 80 in. (2.04 m) high and 30 in. (0.76 m) wide and deep. This support structure was built of boron/aluminum tubes to provide increased design efficiency and a 51% weight reduction (Ref. 3). Boron/aluminum tubing was soldered to aluminum fittings with a 96.5% tin-3.5% silver alloy that had a flow temperature of 425 F (492 K). Tests were run to develop the soldering process and to verify critical tension and compression joints. The structural joints were stronger than adhesively bonded joints and required less overlap. Figure 8 shows the completed truss assembly with the 32 boron/aluminum tubes soldered into the fittings. Three Con Braz joined boron/aluminum channels are also used in the assembly.

Brazing

Boron/aluminum may be brazed with such aluminum-silicon alloys as 713 Al, 718 Al, and 719 Al at brazing temperatures between 1,070 and 1,140 F (850 and 889 K). Short exposure times at temperatures above about 1,025 F (824 K) result in a reaction between the aluminum matrix and the boron filaments with a subsequent loss in filament strength. For that reason, a boron filament coated with 0.0005 in. (0.01 mm) of silicon carbide (trade-named Borsic) is used. The silicon carbide acts as a diffusion barrier between the boron and the aluminum matrix, delaying the onset of any interaction until about 1,100 to 1,125 F (866 to 881 K). This feature allows use of several aluminum-silicon alloys for brazing.

Vacuum Brazing

Two forms of braze alloy Borsic or boron/aluminum monolayer tape are commercially available for vacuum brazing.

In one form, on one side of the

filaments is a surface made up of a 0.002 in. (0.005 mm) 6061 aluminum foil; the other side is a 0.0014 in. (0.04 mm) 6061 aluminum foil and a 0.001 in. (0.03 mm) aluminum-silicon braze foil. This combination avoids placing the brazing filler metal in direct contact with the filaments, thereby reducing the chance of severe filament degradation by the molten braze.

The other commercial tape is a single layer of boron or Borsic filaments sandwiched between a sheet of aluminum-silicon braze foil and plasma-sprayed 6061 aluminum.

Monolayer sheets can be stacked and vacuum brazed in an autoclave or press to produce a flat panel or structural shape. Monolayer Borsic or boron/aluminum tape can also be obtained with a single layer of filaments diffusion bonded between two layers of 6061 or 1100 aluminum. Such sheets can be vacuum brazed by stacking them with alternate layers of an aluminum-silicon braze foil.

Flat panels of the required thickness are made by cleaning and stacking the monolayer Borsic/aluminum tape with or without alternating braze foil, depending upon the type of monolayer used. The assembled stack is placed between two ground flat metal plates, then sealed in a vacuum tight steel envelope. Application of almost any commercial brazing stopoff agent to the platen surfaces before assembly prevents the Borsic/aluminum from bonding to the steel platens during the brazing cycle. The envelope is then evacuated and processed to consolidate the composite panel. Successful parts have been made by either evacuating and sealing the pack before consolidation or by pumping a vacuum throughout the consolidation cycle.

Consolidation is achieved in either an autoclave or a press that heats the part to the brazing temperature [between 1,070 and 1,140 F (850 and 889 K)] while maintaining a bonding pressure between 15 and 200 psi (103 and 1,380 N/m²). Temperature and pressure are interrelated and dependent upon part configuration, tool design, and vacuum envelope design.

Strengths of Borsic/aluminum flat panels containing a nominal 45 volume per cent of filaments have been reported to be between 142 and 187 ksi (978 and 1,290 MN/m²) (Refs. 4 and 5), depending upon the specific consolidation parameters used. These specimens were tested with all filaments running parallel to the direction of the applied test load. The strengths considerably exceed those of the strongest aluminum alloys of the same density as Borsic/aluminum, and compare favorably with steels of about three times that density.

Fig. 8 — Boron/aluminum truss

Autoclave fabrication is more suitable for structural sections because the isostatic gas pressure allows use of simpler tooling to ensure uniform application of the bonding pressure to all part surfaces than would be possible with a single axis loading press.

Pressure application is not as serious a problem in flat panel fabrication. High temperature presses offer the advantage of step pressing techniques that do not limit part length. Step pressing has been used extensively in producing large flat panels. Both open face and matched die tooling concepts can be used, depending upon the configuration of the part. Figure 9 shows open face tooling used for brazing an angle section and matched die tooling used for fabricating a Tee section. Note that in the Tee section a natural space occurs at the center of the cap and leg intersection. To prevent this void, a filler wire of 1100 or 6061 aluminum is added. During consolidation, the aluminum wire consolidates with the aluminum in the tape to form a sound structure. The wire size can be calculated by equating the cross sectional area of the void to that of the wire. The filler wire is first annealed and then formed in a roll die to the shape of the void.

Borsic/aluminum plasma-sprayed braze tape has also been used to fabricate seamless tubes in an autoclave. Bonding pressures used in this process are 1 to 10 ksi (6.9 to 69 MN/m^2) — considerably higher than those used for flat panels or other structural sections.

Figure 10 shows a 4.5 in. (0.11 m) diam brazed Borsic/aluminum tube with a 0.17 in. (4.3 mm) thick wall that was tested in compression at room temperature. The failure load of 456,000 lb (207,000 kg) is equivalent to a stress of 266 ksi (1,830 MN/m^2).

Dip Brazing

Dip brazing has been used to fabricate structural shapes from previously consolidated composite sheet material. Standard aluminum dip brazing techniques are used with aluminum-silicon brazing filler metals.

A typical cycle used to form a Tee section from 0.040 in. (1 mm) thick diffusion bonded sheet material involves preheating the part for 4 min at 1,000 F (810 K), then dip brazing for 45 sec at 1,100 F (866 K) in a salt bath. The structure is quenched in hot water to remove the salt and flux. Time in the braze bath is held to a minimum to avoid solution of the filler metal to a depth sufficient to allow migration of the filaments. One Tee section made using this process failed in crippling at 141% of design ultimate.

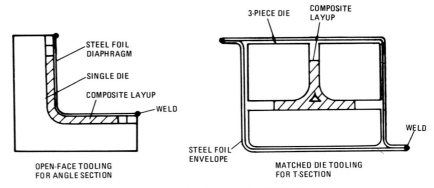

OPEN-FACE TOOLING FOR ANGLE SECTION

MATCHED DIE TOOLING FOR T-SECTION

Fig. 9 — Autoclave tooling for boron/aluminum section fabrication

Table 3 — Results of Crippling Tests on Metal-Matrix Tee Sections.

Material	Ultimate stress	
	ksi	MN/m^2
Borsic/6061 plasma-sprayed material — diffusion bonded	36.2	250
	37.9	261
	44.1	304
	28.3	195
Boron/6061 diffusion bonded material — Con Braz joined	45.0	310
	66.6	459
	54.8	378
Borsic/713 braze/6061 plasma sprayed — vacuum brazed	40.9	282
	34.8	240
	34.1	235

Fig. 10 — Brazed Borsic/aluminum tube, 4.5 in. (0.11 m) diam, after compression testing

Table 3 compares results of this test, along with those for two other 6 in. (0.15 m) long Tee sections dip brazed using different fabrication parameters, with tests of parts fabricated with other processes (Ref. 2). The other processes are (1) layup of boron/6061 aluminum plasma sprayed tape on a tool and diffusion bonding in a high pressure autoclave; and (2) layup of Borsic/aluminum braze plasma sprayed tape on a tool and brazing in a low pressure autoclave.

Dip brazed specimens exhibited a higher crippling strength than all other specimens. The generous fillet radius present in the Con Braz joined section effectively reduces the critical dimension of the individual elements of the section and increases the allowable stress.

Brazed metal-matrix structures have also been used successfully to stiffen aluminum ring structures. In one example, 100 layers of composite tape were dip brazed into a ring structure, then attached to an aluminum structure by low temperature soldering.

Eutectic Diffusion Brazing

To reduce the brazing temperature of boron/aluminum and to avoid filament degradation, considerable attention has been devoted to developing eutectic diffusion brazing. This process is a fluxless technique in which an applied surface coating and the matrix metal diffuse to form a low melting-point eutectic alloy that brazes the joint.

The eutectic phase diffuses away from the bond line into the matrix during prolonged soaking at elevated temperature. This feature raises the remelt temperature to a temperature near the melting point of the base metal, 200 to 500 F (366 to 533 K) higher than the brazing temperature.

Elements potentially suitable for eutectic diffusion brazing aluminum-matrix composites are silver, copper, magnesium, germanium, and zinc; their eutectic temperatures with aluminum are 1,051, 1,018, 820, 795, and 720 F (839, 820, 711, 697, and 655 K), respectively. Silver and copper are unsuitable for 2000 and 7000 series alloys because their

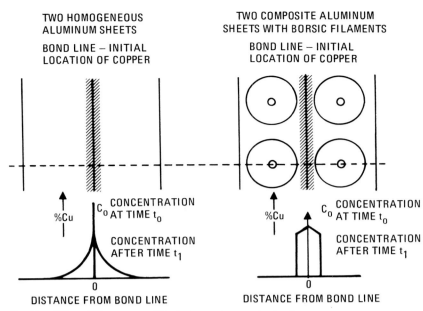

TWO HOMOGENEOUS ALUMINUM SHEETS

BOND LINE – INITIAL LOCATION OF COPPER

TWO COMPOSITE ALUMINUM SHEETS WITH BORSIC FILAMENTS

BOND LINE – INITIAL LOCATION OF COPPER

Fig. 11 — Effect of filaments on diffusion rate of copper in aluminum

brazing temperatures are higher than the incipient melting points of the alloys. Zinc and magnesium oxidize rapidly in air (during layup on the bonding tool), which tends to inhibit bonding.

Eutectic diffusion brazing has been used successfully with non composite materials (e.g., in sealing the caps on nuclear fuel cells), but the presence of filaments in the composites complicates the process. With only the matrix material and no filaments, the coating can diffuse freely down a steep concentration gradient (Fig. 11). In a composite, the filaments impede this diffusion and the homogenization rate of the coating and matrix is drastically reduced.

Mechanical properties of eutectic bonded composite material, particularly when copper is used, are degraded by the inability of the brittle layer to disperse through the matrix. Heat treatment has been used in an attempt to homogenize the structure; but, while successful with 1100 aluminum, it did not prove completely successful with boron/aluminum or Borsic/aluminum. Should continuing investigation lead to resolution of this problem, eutectic diffusion brazing would probably contend technically with other processes for fabricating boron/aluminum composite structures; however, the expense of cleaning and plating the surfaces — especially when vapor deposition is required — places the process at an economic disadvantage compared to braze tape and diffusion bonding techniques.

Discussion

Considerable work has been per-

formed to assess the potential use of soldering and brazing boron/aluminum structures. It has been found that for all types of solder investigated, a protective layer of 0.0002 in. (0.05 mm) of electroless nickel coated on the faying surfaces increases wettability, prevents attack by the flux on the boron fiber/matrix interface, and yields higher allowable joint strengths. For soldered joints seeing service below 200 F (366 K), a 95% cadmium-5% silver alloy, yielding a room temperature lap shear strength of 12 ksi (83 MN/m²) is recommended; for 600 F (589 K) applications, a 95% zinc-5% aluminum alloy, having the same room temperature strength as the cadmium-silver alloy and a strength of 4.5 ksi (31 MN/m²) at 600 F (589 K) is recommended. The cadmium-silver alloy is easier to handle than the zinc-aluminum alloy, and is therefore recommended for applications below 200 F (366 K). Thermal cycling of soldered joints between –320 and 200 F (77 and 366 K) caused no damage to soldered components, indicating that the soldered joint can be used over a wide range of temperatures.

While the zinc-aluminum solder is useful up to 600 F (589 K), there is a vital need for a solder than can be applied in the 950 to 1,000 F (783 to 810 K) temperature range that would have higher properties than the zinc-aluminum system. The shear strength of the composite at 600 F (589 K) is 8 ksi (55 MN/m²); therefore, maximum joint strengths are not achieved with the present system. A secondary benefit of a higher temperature solder would be the possibility of soldering the composite

at the solution treatment temperature for 6061 aluminum: 980 F (799 K). This would mean that a soldered joint could be effectively used in a heat treated structure.

One significant application of boron/aluminum soldering is Con Braz joining of structural elements. The use of Con Braz joining to make Tee sections, I sections, and angle joints has proven to be both economical and efficient. Cross tension tensile strengths of Tee joints made with the cadmium-silver alloy were measured at almost 5.5 ksi (38 MN/m²) at 200 F (366 K). In addition, crippling allowables averaged almost 50% higher for Con Braz Tee sections than similar autoclave bonded sections. An analysis was performed to evaluate the cost of a Con Braz joined Tee section and an autoclave bonded Tee section. The Tees were 12 in. (0.3 m) long and 0.040 in. (1.01 mm) thick. The Con Braz Tee section cost only 44% of the autoclave bonded Tee section. Con Braz parts can be fabricated at less cost, at higher production rates, and will yield stronger structures than similar autoclave bonded parts.

An example of the efficient use of Con Braz joining is the fabrication of a boron/aluminum shear beam. Twenty-two I section stiffeners were made by Con Braz joining, using a specially built module. Approximately 80 lineal feet (24.4 m) of joint were made in this manner, with a rejection rate of only 6% (all attributable to poor quality material).

Brazing boron/aluminum, using aluminum-silicon alloys, is more critical than soldering because of the high temperatures involved. Temperatures typically range from 1,070 to 1,140 F (850 to 889 K). At these temperatures, it is necessary to protect the boron filaments by coating them with a silicon carbide layer that prevents an aluminum/boron filament interaction and subsequent strength degradation. The coating increases the cost of the boron/aluminum by about 20% to 30%, and, therefore, is used only when absolutely necessary.

Dip brazing of boron/aluminum has recently been performed. One dip brazed component was a circumferentially reinforced gimbal housing. A disadvantage of dip brazed components is the increased susceptibility to corrosion that is greater than for non-dip brazed components because of the interaction between the salt and the composite. All salt must be removed from the part immediately after brazing if severe corrosion is to be avoided.

Vacuum brazing of boron/aluminum has been investigated as a potential low cost fabrication method for producing composite structures.

Composite tape material using brazing filler metal matrix can be autoclave bonded at pressures as low as 15 psi (103 N/m²). At these pressures, existing production autoclaves can be used rather than the high cost, high pressure autoclaves or presses currently required for diffusion bonded boron/aluminum.

An alternative low pressure brazing approach is eutectic diffusion brazing in which a plated or vapor deposited alloy forms a bond between aluminun/aluminum interfaces. While this process is promising, strict control is required to minimize fiber degradation and interlaminar embrittlement.

Further development of these low pressure materials is required to increase their strength — which is currently between 20% and 30% below those reported for diffusion bonded boron/aluminum (Refs. 1 and 4) and to decrease the material costs before they can become competitive with current diffusion bonded components. The lower cost capital equipment cannot offset the higher material cost of between $50 and $100 per pound that is due to the requirement for the use of a coated filament. Finally, all major metal-matrix components successfully tested to date have been fabricated using high-pressure diffusion bonded material.

It has been shown that boron/aluminum can be successfully joined by soldering and brazing techniques. Soldering has been shown to not only produce high performance joints in boron/aluminum, but to reduce component fabrication costs. Considerable improvements can be made if a new soldering alloy, compatible with boron/aluminum, can be found for the temperature range of 950 to 1,000 F (783 to 810 K).

References

1. Miller, M. F., Christian, J. L., et al., "Design, Manufacture, Development, Test and Evaluation of Boron/Aluminum Structural Components for Space Shuttle," NAS8-27738, GDCA-DBG73-006, Aug., 1973.

2. Miller, M. F., Schaefer, W. H., Weisinger, M.D., et. al., "Development of Improved Metal Matrix Fabrication Techniques for Aircraft Structures," AFML-TR-71-181, July 1971.

3. Forest, J. D., et. al., "Advanced Composite Applications for Spacecraft and Missiles," AFML-TR71-186, March 1972.

4. Bilow, G. B., "Fabrication Processes for Aluminum Composites, NASA Contract 8-2773, Quarterly Report No. 2, October 1972.

5. Private Communication with J. Dolowy-Webb Associates.

Microstructural Observations of Arc Welded Boron-Aluminum Composites

Investigation indicates it may be possible to join B/Al composites by arc welding

BY J. R. KENNEDY

ABSTRACT. Welding studies were conducted on boron-aluminum (B/Al) composites to observe the effects of gas tungsten-arc welding on the boron reinforcing filaments and aluminum matrix. The objective of this investigation was to determine the basic potential of arc welding B/Al composites for possible structural applications. Microstructural observations after welding revealed matrix fusion without apparent boron filament damage. Analysis of weld metal regions indicated that aluminum filler metal additions intermixed with the matrix and altered its chemical composition. It is concluded that arc welding of B/Al composites may be possible by control of welding energy input.

Introduction

Exposure of boron filaments to molten aluminum raises the question of chemical reactivity and its effect on filament properties. Boron-aluminum interactions are time-temperature dependent and may be sluggish in the solid state, or very rapid in the presence of a superheated liquid aluminum matrix. Thermal treatments

J. R. KENNEDY is with the Research Department, Grumman Aerospace Corporation, Bethpage, New York

such as diffusion welding, casting, and arc welding may induce various interfacial reactions detrimental to filament strength and composite structural efficiency. For example, it has been found that amorphous boron dissolves noticeably in molten aluminum at 1000 C.[1] In other work on cast boron-aluminum composites, several minutes of exposure to molten aluminum at about 740 C caused considerable interaction leading to partial dissolution and edge scalloping of the boron-filaments.[2] It was also determined[2] that exposures up to three minutes at 740 C resulted in minimal observable reaction effects.

In a study of the interaction between boron and aluminum,[3] boron

filaments were placed in molten aluminum at 680 C. After exposure times of 1 min and 15 min, interaction layers were measured at 2-3μm and 5μm, respectively. In a study of B/Al solid state reactions,[4] it was found that a time-temperature dependent incubation period exists, after which a loss of strength occurs. In some cases, tensile strength increased during this period, suggesting some beneficial effects of the interfacial reaction. Studies of B/Al interactions during fusion welding have been generally quite limited. It is reported that gas tungsten-arc, electron beam, and plasma welding usually result in severe weld embrittlement and filament degradation.[5,6] Typical effects on the boron included fil-

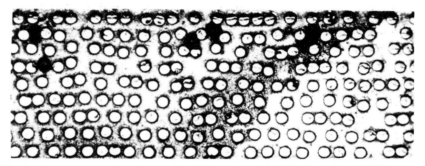

Fig. 1 — Transverse section of arc weld in B/Al, made with 4043 filler wire at 6435 joules. X30, not reduced

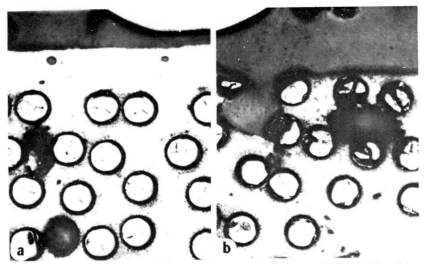

Fig. 2 — Comparison of arc welds made at 6435 joules: (a) 4043 Al filler addition; (b) no filler metal addition, X150, reduced 29%

The most significant qualitative difference between resistance welding and the other fusion processes is the relatively lower thermal energy input of the former. The influence of welding energy input on interfacial reactions has also been demonstrated in welding studies on titanium-tungsten and titanium-graphite composites.[9] As welding energy input was increased, tungsten dissolution became greater and titanium carbide formation around the graphite filaments grew thicker. It was concluded that thermal energy delivered to the composite during welding is a significant factor in controlling the nature of filament-matrix reaction products. High welding heats can increase dissolution between components, producing extensive diffusion zones and larger quantities of additional phases. Quantitative evaluation of these effects and their contributions to composite efficiency will be necessary for practical utilization in future applications.

It is clear that the specific effects of short time, high thermal energy exposure on boron-aluminum composites have not been completely characterized. This program was undertaken

ament cracking, break-up, misorientation, and partial or complete dissolution. In the worst cases, reductions in composite strength at the weld joint have been as high as 90 percent. On the other hand, resistance-spot welding of boron-aluminum, which also requires matrix fusion, has shown more promise in minimizing adverse thermal effects on the boron filaments.[5,8]

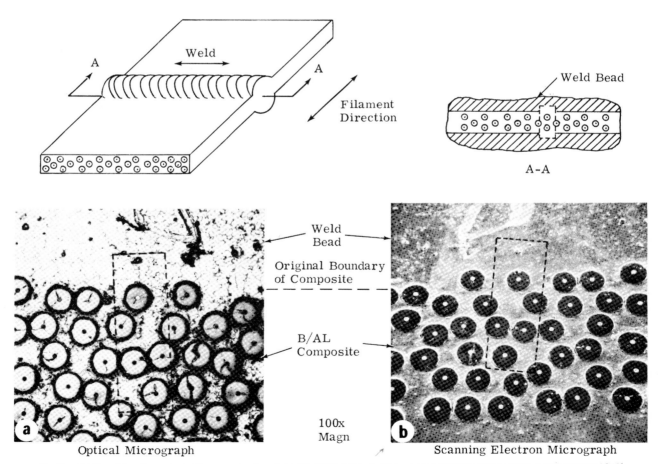

Fig. 3 — B/Al weld region selected for examination: (a) optical micrograph; (b) scanning electron micrograph. X100, reduced 21%

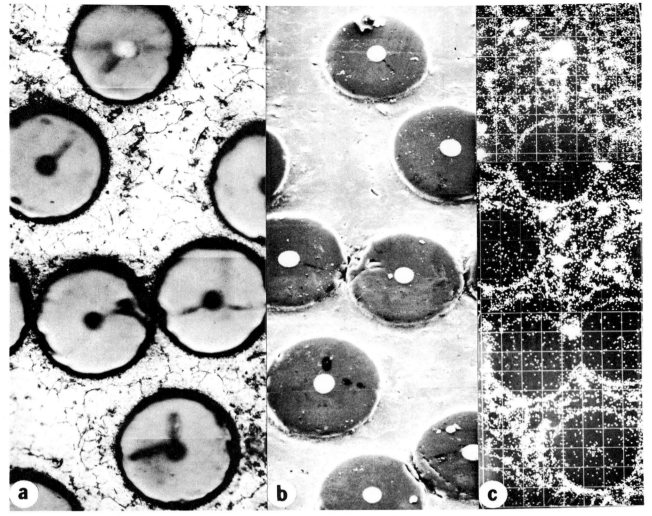

Fig. 4 — Comparison of selected region on arc welded B/Al composite: (a) light micrograph (X500); (b) scanning electron micrograph (X500); (c) silicon distribution electron microprobe (X400). Entire group reduced 35%

to obtain more insight into arc welding of B/Al composites. The present work describes an initial study to observe and to assess qualitatively the condition of boron filaments and aluminum matrix after exposure to gas tungsten-arc welding.

Experimental Procedure

Tests were conducted on 0.025-in. (50%-6 ply) and 0.050-in. (40%-9 ply) thick B/Al composite sheets, made from unidirectionally aligned 0.004-in. diam boron filaments and 6061 aluminum. The composites were fabricated by hot pressing. Arc welding was performed using a standard 300 A ac/dc welding power supply with a manual gas tungsten-arc welding torch. An 0.040 in. diam thoriated tungsten electrode, alternating current, and argon gas shielding were employed during welding. All specimens were bead-on-sheet welds;

when filler metal was added during welding, 1/16-in. diam 4043 aluminum wire was employed.

Immediately prior to welding, the B/Al specimens were degreased by cloth wiping with isopropyl alcohol, followed by a light surface milling in the intended fusion region with a hand draw file. The specimens were clamped in a copper welding fixture with a 1/4-in. space between the hold-down bars. A copper backing bar with a 1/4-in. wide x 1/16-in. deep groove was also used.

Specimens were metallographically prepared by minimizing final polishing time, first with diamond to 1μm, and then with alumina to 0.05μm. Boron filaments were also extracted for examination by leaching the aluminum matrix with dilute hydrochloric acid.

In addition to optical microscope observation of composite microstructures, scanning electron microscopy

(SEM) was performed with a Cambridge Stereo-Scan instrument, and electron microprobe analysis was conducted with a Philips AMR-3 analyzer.

Fig. 5 — B/Al interface in weld region

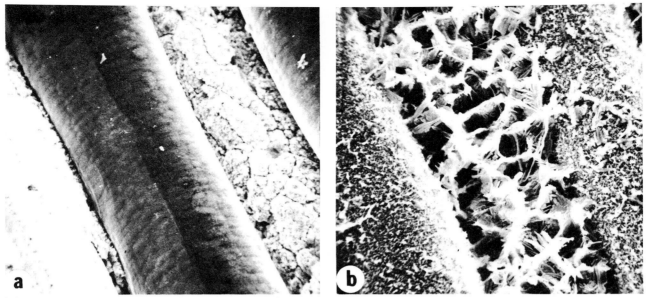

Fig. 6 — Scanning electron micrographs of exposed boron filaments: (a) as received; (b) as welded. Both X400, reduced 5%

Results and Discussion

The effect of increases in welding energy input on 0.050-in. thick B/Al composite was studied. The specimens used for this study were 1/4-in. wide and 2-in. long; they were clamped and welded from one side only under a stationary arc, essentially producing a localized arc spot weld. Under these conditions, the energy input (joules) is the product of arc power and fusion time (A x V x sec).

Welding With and Without Filler Metal

A transverse section of a weld made with 4043 Al filler addition at 6435 joules (26 A, 16.5 V, 15 sec) is shown in Fig. 1. Large pores were formed in the weld metal interior, as contrasted to fine pores in the weld crown. Matrix melting has caused some shifting of the boron filaments in the fusion zone. A comparison of two welds made at 6435 joules with and without filler metal is shown in Fig. 2. These results generally typify the entire group of specimens made in this study that were welded up to energy levels of 18,000 joules. When filler metal was not added, the upper boron filaments were subjected to intense arc heating that caused severe filament fragmentation and dissolution. In addition, the difficulty in maintaining a stable weld puddle increased, usually resulting in the aluminum's being drawn away from the puddle center to its edges, indicative of poor wetting between boron and aluminum.

On the other hand, these problems were largely eliminated when filler metal was present. Apparently, the additional Si-enriched filler metal helps to shield the top layers of boron and promotes metal flow into the matrix, as evidenced by the penetration observed in Fig. 1. Due to the limited scope of this initial work, a full spectrum of fusion effects have not been determined. This includes the effects of higher energy inputs, the welding of thicker sheet, and the effect of longer passes. However, it is known qualitatively that excessive energy inputs will cause significant filament damage and displacement regardless of filler metal additions.

It should be noted that a certain percentage of filament damage, such as radial cracking and contact fragmentation, as shown in Figs. 2 and 4, is considered "normal" and also exists to varying degrees in as-received and welded B/Al composite sheet. This damage is believed to occur during primary fabrication of the composite.

Microstructural Examination

An analysis was made of the microstructure from an arc welded bead-on-sheet specimen of 0.025-in. thick B/Al, shown schematically in Fig. 3. This specimen was welded from both sides (not simultaneously) using 4043 Al filler addition, at an energy level of about 4950 joules/in. (20 A, 16.5 V, 4 ipm). A weld region was randomly selected from this specimen for subsequent examination by various techniques, as shown in Fig. 4. Our interest was to observe particular effects on the boron and the matrix

resulting from this exposure. In Figs. 3 and 4, the SEM region shows the filaments intact within an apparently sound matrix.

The distinct grain boundaries in the matrix are clearly evident in the light microscopy photograph of Fig. 4. The presence of Si in this region is confirmed by the microprobe analysis that shows a definite correlation between the grain boundaries and Si-rich regions. It can also be noted that the Si distribution is relatively uniform through the specimen thickness. The average Si concentration after welding was calculated to be about 4.7%; the Si concentration in unwelded as-received composites was 0.58%. The nominal Si distribution in 4043 Al filler metal is 5% and in 6061 Al about 0.5%.

The matrix appears to have been completely melted and fairly uniformly enriched with filler metal, while the filaments remained apparently undamaged. This microstructure appears to be typical of a Si-rich condition in aluminum welds and castings,[10,11] where relative insolubility of Si in Al causes dispersion of small areas of Al-Si eutectic in the Al matrix and grain boundary enrichment. A Si layer or ring is also seen immediately adjacent to each filament (in some cases, a double Si ring is apparent). It is likely that as freezing progressed in the matrix, a solidification front rich in Si advanced to the filament periphery where the Si then segregated as along a grain boundary.

After being polished and etched, the boron usually stands slightly in relief above the matrix, making sharp planar focusing of both B and Al difficult with the optical microscope; this results in the characteristic shadowing around each filament.

Examination of the B/Al interfacial region at higher SEM magnifications revealed a portion of the interface configuration to be in the form of a small concave fillet between the filament and matrix, as shown in Fig. 5. A somewhat rough or jagged texture was also discernible in the transition zone vicinity of various filaments.

Further SEM observation of B/Al interactions was made on welded specimens in which the matrix was leached to expose boron filaments. Figure 6 is a comparison between as-received and as-welded B/Al composite. In the welded specimen, to which 4043 Al filler was added, the variable mottled surface texture of the boron filaments is evident. The filigreed matrix between the filaments, resulting from a differential etching rate, probably represents a skeletal grain boundary network, enriched in Si as a result of welding.

Subsequent electron microprobe analysis showed these areas to be Si rich relative to the as-received composite. Microprobe scans over the filaments in the welded specimen also showed significant concentrations of Al and Si. The work by Klein[4] on solid state reactions in B/Al composites showed that the interaction

phase has an uneven, acicular appearance with the needlelike protrusions into the matrix. The interaction product in that work has been tentatively identified as AlB_2. The reaction product in the welded condition has not yet been identified, but the presence of a combination of complex intermetallic compounds is not unlikely.

Conclusions

It has been shown that thin-sheet B/Al composites can be subjected to certain arc welding thermal conditions without severely damaging the boron filaments. It is possible to add and to intermix filler metal through the matrix to alter its chemical composition significantly. These results indicate that arc welding of B/Al may be possible by control of welding energy input. Identification of the subsequent fusion reaction products, and the effects of those products on composite mechanical properties, has not yet been determined but is planned in continuing studies. In addition, knowledge of reaction growth rate kinetics and the means to control the reaction products during welding would increase the potential of B/Al for consideration as welded structures.

References

1. Hansen, M., *Constitution of Binary Alloys,* McGraw-Hill Book Co., New York, 1958.

2. Hill, R. et al., "The Development and Properties of Cast Boron/Aluminum Composites," *Advances in Structural Composites,* 12th SAMPE Symposium, Vol. 12, p. AC-18, Western Periodicals Co., N. Hollywood, California, October 1967.

3. Restall, J. et al., *Metals and Materials,* pp. 467-473, November 1970.

4. Klein, M. and Metcalfe, A., *Effect of Interfaces in Metal Matrix Composites on Mechanical Properties,* AFML-TR-71-189, Solar Division of International Harvester Company, San Diego, California, October 1971.

5. Schaefer, W. and Christian, J., *Evaluation of the Structural Behavior of Filament Reinforced Metal-Matrix Composites,* Convair Division of General Dynamics, San Diego, California, AFML-TR-64-36 (Vols. I, II, III), January 1969.

6. *Structural Design Guide for Advanced Composite Applications: Vol. 3 — Manufacturing,* 2nd Ed., pp. 8.4.2-8.4.18, Air Force Materials Laboratory, Wright-Patterson Air Force Base, Ohio, January 1971.

7. Hersh, M., *Welding Journal,* Vol. 47, No. 9, Research Suppl., pp. 404s-409s, 1968.

8. *Development of Improved Metal-Matrix Fabrication Techniques for Aircraft Structure,* Report No. GDC-DBG70-002-1, Convair Division of General Dynamics, San Diego, California, July 1970.

9. Kennedy, J. and Geschwind, G., presented at the 2nd International Conference on Titanium, Boston, Massachusetts, May 1972.

10. Knight, J., *Welding Kaiser Aluminum,* Oakland, California, 1967.

11. Brick, R. and Phillips, A., *Structure and Properties of Alloys,* 2nd Ed., McGraw-Hill Book Co., New York, 1949.

SECTION III
Trimming and Cutting

Cutting Advanced Composite Materials

A 60,000 psi waterjet is an effective tool for cutting certain types of composites. Advantages include no airborne dust and noise levels below 80 dB

ADVANCED COMPOSITE MATERIALS are being considered for use in an increasing variety of applications. The auto industry, for example, is taking a look at the composites as a substitute for metal in its effort to reduce weight for improved fuel economy. And the aerospace industry continues to play a leading role in putting the composites to work as a metal substitute.

A recent announcement from Lockheed Missiles & Space Co., Inc., a subsidiary of Lockheed Aircraft Corp., describes an application on the Navy's Trident I (C4) — the first missile to use advanced graphite/epoxy composite materials in primary structures. Bidirectional woven graphite/epoxy cloth is used for building the missile's equipment section support. It's the first production use anywhere of this particular type of material. Similar applications involving the composites are underway at most of the major aircraft companies.

The advanced composites are complex materials and require some unusual approaches in fabrication, assembly, and machining. The use of a waterjet, *Figure* 1, at pressures up to 60,000 psi (414 MPa) is one of the methods being evaluated as a better means of cutting these materials. This technique was used in a series of laboratory cutting tests performed at Flow Industries, Inc., Kent, WA, and described in SME paper MR77-225 by G. H. Hurlburt, research engineer, and J. B. Cheung, research scientist.

Materials Cut. Materials used in the study included graphite/epoxy, boron/epoxy, Kevlar/epoxy, and fiberglass/epoxy. In addition, the following three hybrids were also used: graphite/epoxy and boron/epoxy, graphite/epoxy and Kevlar/epoxy, and graphite/epoxy and fiberglass/epoxy. The materials in this group are believed to be fairly representative of the fibrous composites and

are among the best now available. Also, other researchers in the field estimated several years ago that over 90% of the high rigidity, high strength composites in use were the epoxy type.

Power for the Waterjet. A Flow Industries 60,000 psi, 1.5-gpm (5.8-L/min) jet cutter unit supplies the high pres-

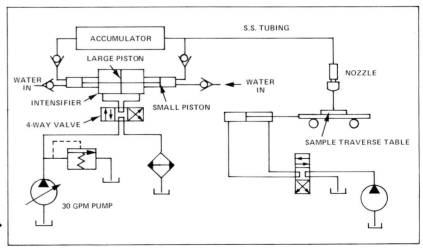

1. CLOSE-UP OF WATERJET cutting sample of advanced composite material. Orifice diameters in test program ranged from 0.006" up to 0.014".

sure water flow to the nozzle. As shown in *Figure* 2, a 30-gpm (113.5-L/min) hydraulic pump delivers hydraulic oil at up to 3000 psi (21 MPa) to an intensifier via a four-way valve. The intensifier is a differential area double-acting piston type in which a large piston is shuttled back and forth by the

2. JET CUTTER system layout. A hydraulic cylinder traverses the sample material under the nozzle during cutting. ▶

ACCUMULATOR
S.S. TUBING
LARGE PISTON
NOZZLE
WATER IN
WATER IN
INTENSIFIER
SMALL PISTON
4-WAY VALVE
SAMPLE TRAVERSE TABLE
30 GPM PUMP

3000 psi oil. Two small pistons attached directly to the large piston each have an area that is 1/20th the area of the large piston, converting the 3000 psi oil to 60,000 psi water.

Compressed water flows out of the high pressure cylinders through a pair of check valves and into a line leading to the nozzle. Here the water is forced through a synthetic sapphire orifice. In these tests, orifice diameters ranging from 0.006″ (0.15 mm) up to 0.014″ (0.36 mm) were used.

A starting set of cutting parameters was chosen from past experience with similar materials. Parameters which were varied during testing included the jet pressure P_o, nozzle orifice diameter d_o, traverse velocity v, and material type and thickness. The nozzle stand-off distance was held constant at about 1/8″ (3 mm). In general, a smaller diameter nozzle produces a finer cut, but a larger diameter is necessary to cut the thicker materials. It has also been found that a higher nozzle pressure and lower traverse velocity produce a better quality of cut.

Following completion of cutting trials on each type of material, the best cut quality was selected on the basis of appearance, jet pressure, orifice diameter, and traverse velocity. In other words, if a group of cuts were generally equal in appearance, the one involving the smallest pressure and orifice diameter, and the highest velocity was selected as the best. The intent was to minimize pumping horsepower while maximizing the cutting rate.

Cutter Performance. All of the materials in the test group were cut successfully, but with varying degrees of effectiveness. For example, materials with the relatively soft Kevlar fibers cut well, leaving a reasonably smooth edge typical of jet cut quality. Those with the hard and strong boron fibers had the roughest cut quality. These fibers tend to break rather than cut, leaving a short stubble of fibers protruding from the cut surface.

Figure 3 shows examples of jet cut slots in 1/8″ thick fiberglass epoxy material. The slots in views A and B were cut with a nozzle pressure of 60,000 psi and nozzle diameter of 0.010″ (0.25 mm), but that in A was cut at 0.10 ips (2.5 mm/sec) and the one in B was cut at 2.0 ips (51.0 mm/sec). The slot in B shows the effect of too high a traverse speed. Some of the water from the jet stagnated in the material and penetrated sideways between the layers of fibers causing some delamination along the length of the slot.

View C shows the reverse (exit) side of the slot in A. Only a very small amount of delamination is evident. View D shows the exit side of a slot cut with a nozzle pressure of 55,000 psi (380 MPa), nozzle diameter of 0.010″,

3. *TYPICAL APPEARANCE of slots cut in 1/8″ thick fiberglass/epoxy.*

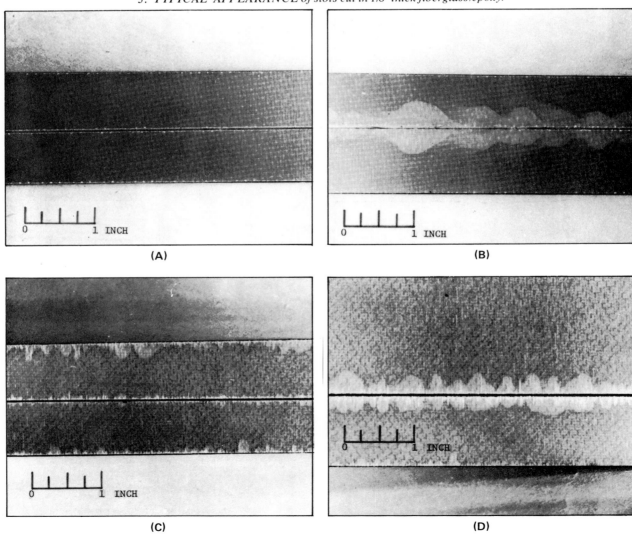

(A)

(B)

(C)

(D)

Material Code	Thickness (in.)	P_o (psi)	d_o (in.)	v (ips)	Remarks
Graphite/Epoxy	0.067	55,000	0.008	1.0	Virtually no separation of the cloth backing.
	0.136	60,000	0.010	0.50	Minor separation of the cloth backing on exit side.
	0.273	60,000	0.014	0.11	Good cut, no separation of the cloth backing.
Boron/Epoxy	0.058	60,000	0.012	2.0	Poor cut, loose fiber left along cut.
	0.136	60,000	0.010	2.0	Poor cut, separation of the surface resin layer, top and bottom.
Kevlar/Epoxy	0.058	55,000	0.006	2.0	Only slight separation of the cloth backing, exit side.
	0.125	55,000	0.010	0.50	Good cut.
Fiberglass/Epoxy	0.139	60,000	0.010	0.10	Small amount of separation of the cloth backing, exit side only.
Graphite/Epoxy and Boron/Epoxy	0.088	60,000	0.012	0.23	Poor cut, loose fibers left along cut.
	0.154	60,000	0.012	2.0	Poor cut, loose fiber and some separation of the cloth backing, bottom side.
	0.321	60,000	0.014	0.15	Some exit side separation of the cloth backing.
Graphite/Epoxy and Kevlar/Epoxy	0.125	60,000	0.010	0.25	Good cut.
	0.250	60,000	0.014	0.08	Very good cut.
Graphite/Epoxy and Fiberglass/Epoxy	0.068	55,000	0.012	0.15	Good cut.
	0.253	60,000	0.012	0.15	Rather severe delamination of bottom fibers of composite.

and traverse velocity of 0.25 ips (6.4 mm/sec). In this case, the higher degree of delamination apparently is due to the lower jet pressure and higher traverse velocity.

A summary of the results of all the cutting tests is presented in the table. In most cases, the materials can be cut at faster feedrates than the figures shown. The rates listed are for the opti-mum, or best, cut.

On materials with a cloth backing, some water from the jet tended to pen-etrate beneath the backing where the jet entered the material. In most cases this was a localized problem that tended to be more cosmetic than real.

The results presented are for cuts in a single direction on the sample. A con-trol cut done on most of the samples at a 90° angle to the others showed no significant difference in cutting perfor-mance. This indicates that orientation of the fibers in the materials does not permit easier cutting in one direction than the other.

The main conclusion drawn from these tests is that the waterjet can cut fibrous advanced composite materials of the epoxy type with no airborne dust and a noise level of less than 80 dB. Materials with the very hard and strong boron fibers were cut completely through at the fairly high feedrate of 2 ips (51.0 mm/sec), but quality of the cut was relatively poor. It would be possi-ble, however, to trim this rough edge by profile routing. The other materials were cut with a generally better edge finish, but at slower feedrates.

The thickness for these types of ma-terials is limited to about 1/4″ (6 mm) for jet cutting. Thicker materials can probably be cut by the jet, but at a slower feedrate.

The action of the jet as a point cutter makes it possible to traverse the nozzle over a stationary sheet of material. Nozzle travel could be placed under numerical control to make repeated cuts along a complicated outline, in-cluding sharp radii. Thus, the jet cutter can be used as a high-production cut-ting tool. ■

WATERJET CUTTING OF ADVANCED COMPOSITE MATERIALS

G. H. Hurlburt
Research Engineer

J. B. Cheung
Research Scientist
Flow Industries, Inc.

ABSTRACT

A waterjet at pressures up to 60,000 psi (414 MPa) can be an effective tool for the cutting of some rather hard materials.[1] Therefore, such a waterjet is being considered in the current search for better means of cutting advanced composite materials.

This paper describes laboratory cutting tests performed on a selection of composite materials. These tests examined the effect of varying the jet pressure, nozzle orifice diameter, material feed rate, material type, and material thickness on cutting effectiveness.

The tests demonstrated that a water jet can cut these materials with practically no airborne dust generated and at the reasonably low noise level of less than 80 db.

INTRODUCTION

Waterjets operating at pressures up to 60,000 psi (414 MPa) are currently in commercial use for the cutting of materials ranging from corrugated paper board to concrete. It follows that this tool now available to industry would be examined for its potential in the cutting of advanced composite materials, some of which present problems in machining due to their strength, and hardness and abrasive qualities of their fibers.

A series of jet cutting tests has been performed at Flow Industries on a collection of composite materials. The purpose of these tests was to determine how well the jet cutter can cut these materials. Because the waterjet was performing a function analogous to that of a bandsaw, the first objective was, of course, to separate the test sample completely at the line of the cut. The second objective was then to provide as smooth a cut (free of surface roughness) as possible with no damage to the adjacent material. The degree of success in reaching these objectives is reported in the following pages

The waterjet cuts by concentrating a very high speed, small diameter stream of water normal to the surface being cut. Speeds of 2600 ft/sec (800 m/sec) and diameters on the order of 0.010 in. (0.25 mm) are typical. The cutting mechanism appears to be a combination of localized erosion and stress failure of material by the jet.

In the tests, a Flow Industries, Inc. intensifier unit was used to deliver the flow to the nozzle. This unit is capable of delivering 1.5 gpm (5.8 li/m) of water at the maximum pressure of 60,000 psi. The nozzle makes use of a sapphire orifice to control wear of the nozzle.

Parameters studied in the tests were nozzle pressure, nozzle orifice diameter material feed rate, and material type and thickness. Orientation of the fibers in the fabric of the composite in relation to the direction of traverse of the jet was considered as well. The test results indicate

---

[1]Flow Industries water jet equipment for the cutting of fiberglass reinforced plastics are now being used in the industry.

that the waterjet can cut composites, but with varying degrees of cut-edge quality depending on material type and thickness.

DESCRIPTION OF MATERIALS

The composite materials used in the tests are listed in Table 1. In the table, the fiber type in each material is given in the beginning of each name with the resin type (epoxy in all cases) following. A convenient abbreviation then follows with the nominal thicknesses of the materials used in testing indicated with an "x". Some of the materials were hybrids of the four basic graphite, boron, Kevlar, and fiberglass epoxy types and are indicated as the last three entries in the table.

Table 1

Material	Material Code	1/16 in.	1/8 in.	1/4 in.
Graphite/Epoxy	GR/EP	x	x	x
Boron/Epoxy	BO/EP	x	x	
Kevlar/Epoxy	KE/EP	x	x	
Fiberglass/Epoxy	FG/EP		x	
Graphite/Epoxy & Boron/Epoxy	GR/EP & BO/EP	x	x	x
Graphite/Epoxy & Kevlar/Epoxy	GR/EP & KE/EP	x		x
Graphite/Epoxy & Fiberglass/Epoxy	GR/EP & FG/EP	x		x

This group of materials is felt to be a fairly representative collection of fibrous composites as the fiber types represented are among the current best available. Also, a few years ago, it was estimated that over 90% of the high rigidity and high strength composites in use were of the epoxy type.[2] This trend seems to have continued to the present time.

TEST APPARATUS

The main feature of the test apparatus is the pumping equipment which supplies the high pressure water flow to the nozzle. It is a Flow Industries, Inc. 60,000 psi (414 MPa), 1.5 gpm (5.8 li/min) Jet Cutter unit. Its operation is described in Fig. 1, where the 30 gpm variable volume, pressure comensated pump delivers its flow of hydraulic oil at up to 3000 psi (21 MPa) to the intensifier via a pilot-operated four-way valve. The intensifier is a differential area double-acting piston type where the large piston is shuttled back and forth by the 3000 psi oil pressure. Attached directly to the large piston are two smaller pistons, each of whose area is 1/20th the area of the large piston. Thus, the 20 to 1 pressure intensification of the 3000 psi oil to the 60,000 psi water is achieved.

---

[2]Fleck, I. N. and Hanby, K. R., "High Performance Fiber Reinforced Polymers," in Composites: State of the Art, The Metallurgical Society of AIME, Detroit, Michigan, 1974, p. 29.

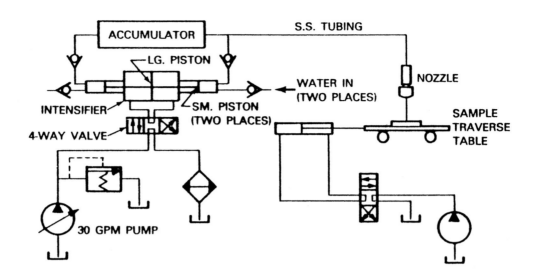

FIGURE 1.   SCHEMATIC DIAGRAM OF JET CUTTER TEST APPARATUS

The compressed water flows out of the high pressure cylinders via a
pair of check valves. The water then enters the stainless steel line.
Included in the water circuit is an accumulator which smoothes out the
pulses produced by the pump. The stainless steel tubing then conveys the
high pressure water to the nozzle where it is forced through the synthetic
sapphire orifice. A 1/4 in. (6.4 mm) inside diameter tube beneath the
nozzle catches the spent water from the nozzle plus all dust from cutting.
Noise level is less than 80 db.

In the diagram of Fig. 1, the sample to be cut is shown resting upon the
traverse table. A fast-acting hydraulic cylinder traverses the sample
under the nozzle as the sample is cut.

A photograph of the equipment is available in Fig. 2, where the traverse
table mechanism is seen on the left and the high pressure pumping unit is
seen on the right. Figure 3 shows the jet cutter nozzle with the water
jet stream penetrating a specimen of composite material.

TEST METHOD

In performing the tests, it was desired to obtain the best, or optimum,
set of cutting parameters for each material and thickness. A starting
set of cutting parameters was chosen from past experience with similar
materials. After the initial trial cut, the fineness of the cut (the
surface roughness of the cut edge) was evaluated qualitatively. A cut
judged to be satisfactory then called for a smaller diameter nozzle
orifice. On succeeding cuts, the nozzle pressure and traverse velocity
were adjusted according to cut quality. (In general, a smaller diameter
nozzle produces a finer cut, but a larger diameter is necessary to cut

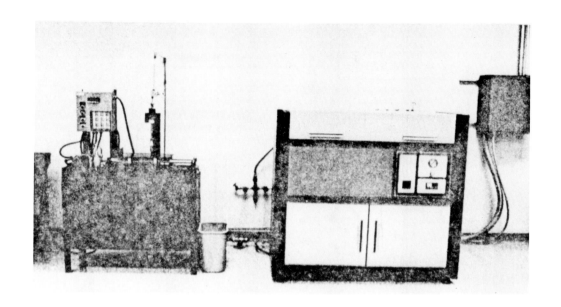

FIGURE 2. JET CUTTER TEST APPARATUS. THE TRAVERSE TABLE UNIT IS ON THE
LEFT, THE HIGH PRESSURE PUMPING UNIT IS ON THE RIGHT.

FIGURE 3. CLOSE-UP OF WATER JET CUTTING SAMPLE MATERIAL.

through the thicker materials. Also, in general, a higher nozzle pressure and a lower traverse velocity produce a better quality of cut.)

The parameters which were varied during the testing were the jet pressure P_0, nozzle orifice diameter d_0, traverse velocity v, and material type and thickness. The nozzle standoff distance was held constant at about 1/8 in. (3 mm). After the series of cutting trials were finished for each material, the best cut quality was subjectively chosen on the basis of the appearance of the cut, the pressure P_0, diameter d_0, and the velocity v, which was used. Stated more plainly, if a group of cuts were subjectively equal in appearance, the one using the smallest P_0 and d_0, and the largest v was chosen as the best. In this way, the pumping horsepower was minimized while the cutting rate was maximized. The above could be said to represent an optimization strategy, albeit of a necessity somewhat unsophisticated.

In actually performing the tests, the proper size nozzle is installed on the traverse table apparatus, the speed of the table set to the desired value, the desired pressure set on the pumping apparatus, the sample put in place and the sample is then ready to cut. The actual speed of the traverse is also measured electronically by the apparatus.

RESULTS

The jet cutter succeeded in cutting all of the materials in the test group. However, some of the materials were cut more effectively than others. For example, the materials with the hardest and strongest fibers (i.e., the boron fiber) left the roughest cut quality. These fibers tended to break under the impact of the jet rather than cut, leaving a short stubble of fibers protruding from the cut surface. However, the softer Kevlar fibers cut well leaving a reasonably smooth cut edge typical of jet cut quality.

Figures 4 through 10 offer examples of the appearance of jet cut slots, both entry and exit side (i.e., the sides where the jet entered and exited, respectively). Figures 4 and 5 show slots on the entry side of 1/8 in. (3 mm) thick fiberglass epoxy material. Both slots were cut with nozzle pressure of 60,000 psi (414 MPa) and nozzle diameter of 0.010 in. (0.25 mm), but the slot in Fig. 4 was cut at 0.1 in./sec (2.5 mm/sec) while the slot in Fig. 5 was cut at 2.0 in./sec (51 mm/sec). The slot in Fig. 5 shows the effect of an improper cut where too high a traverse speed was used. In this case some of the water from the jet stagnated in the material and penetrated sideways between the layers of fibers causing some delamination along the length of the slot. The slot shown in Fig. 4 was cut at the slow rate of 0.1 in./sec but is clean with no delamination at all.

Figures 6 and 7 are the same 1/8 in. thick fiberglass material but the photos were taken from the exit side. The slot in Fig. 6 was cut with the nozzle pressure of 60,000 psi, nozzle diameter of 0.010 in. and traverse velocity of 0.1 in./sec, while that of Fig. 7 was cut with nozzle pressure of 55,000 psi (380 MPa), nozzle diameter of 0.010 in. and traverse velocity of 0.25 in./sec (6.4 mm/sec). A very small amount of delamination is visible along the slot of Fig. 6, but Fig. 7 shows a

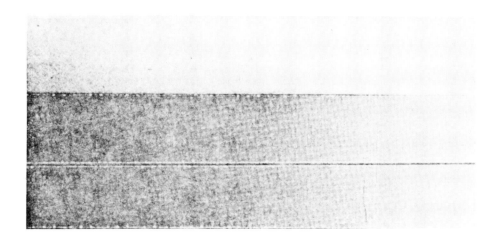

FIGURE 4.   JET CUT SLOT, ENTRY SIDE OF 1/8" THICK FIBERGLASS/EPOXY
            MATERIAL.   CUTTING PARAMETERS WERE P_o = 60,000 psi, d_o =
            0.010 in., V = 0.1 in./sec.   THIS CUT SHOWS THE RESULT OF
            CORRECT JET CUTTING PARAMETERS ON THE APPEARANCE OF THE CUT.

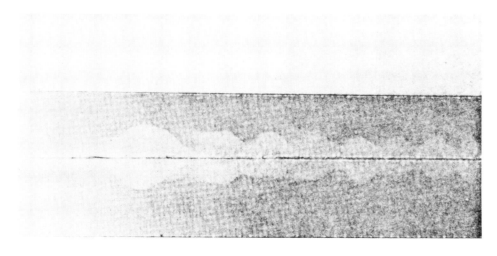

FIGURE 5:   SAME MATERIAL AS FIGURE 4, ENTRY SIDE.   CUTTING PARAMETERS WERE
            P_o = 60,000 psi, d_o = 0.010 in., and V = 2.0 in./sec.   THE
            DELAMINATION OF THE TOP LAYERS, VISIBLE IN THE PHOTO, IS A
            RESULT OF INCORRECT CUTTING PARAMETERS.

FIGURE 6.   SAME MATERIAL AS FIGURES 4 AND 5, EXIT SIDE.   THIS IS A VIEW
OF THE REVERSE SIDE OF THE SLOT SHOWN IN FIGURE 4.

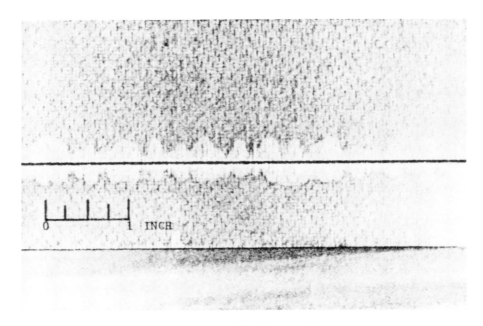

FIGURE 7.   SAME MATERIAL AS FIGURES 4, 5, and 6, EXIT SIDE.   CUTTING
PARAMETERS WERE P_o = 55,000 psi, d_o = 0.010 in., and V =
0.25 in./sec.   THIS PHOTOGRAPH SHOWS THE EFFECT OF USING
INCORRECT CUTTING PARAMETERS ON EXIT SIDE SLOT APPEARANCE.

much higher degree of delamination, apparently caused by the lower jet pressure and the higher traverse velocity.

Figures 8 and 9 show slots cut in a 1/16 in. (1.6 mm) thick hybrid of graphite/epoxy and fiberglass/epoxy, both on the entry side. This material, unlike the fiberglass/epoxy material shown in Figs. 4 and 5, has a fabric backing on both sides. Under some conditions, this fabric backing tended to separate from the composite proper, seen in Fig. 9, under a similar action of the water as described above for Fig. 5. In this case, the separation of the cloth backing on the entry side could be stopped, as seen in Fig. 8, by increasing the nozzle pressure from 55,000 to 60,000 psi, the nozzle orifice diameter from 0.008 to 0.012 in., and while holding the traverse velocity constant at 0.15 in./sec. Not in all cases can the separation of the cloth backing be avoided, as one can see from perusing Table 2. (Table 2 contains only the optimum, or best, cut for each material.)

Lastly, Fig. 10 shows a slot cut in 1/16 in. thick boron/epoxy, entry side. The cutting parameters were nozzle pressure of 60,000 psi, nozzle orifice diameter of 0.010 in., and traverse velocity of 2 in./sec. Seen in the photograph are random, short lines running parallel to the slot. These are small cracks in the rather brittle surface layer of resin due to the "wedging" action of the jet water again as described for Fig. 5 (above). The appearance of this slot was typical for all cutting parameters tried and for entry and exit sides.

Results of the cutting tests are presented in Table 2 where the material code (key in Table 1) is given in the first column with the material thickness, P_o, d_o, and v following. The last column gives a brief description of the cut appearance ("Remarks" column). It should be mentioned that for most of the cases, the materials can be cut at faster feed rates than the figures shown in the table. The figures shown are for the optimum, or best, cut.

The "Remarks" column includes references to a cloth backing which some of the materials had and some of the materials lacked, notably the materials containing any glass fiber. This is apparently due to the fact that the fiberglass is in the form of a woven fabric which creates a finished material with a durable surface. The fibers in some of the other materials were placed in laminae, each of which had fibers oriented in one direction only. These laminae were placed at angles to one another to produce a finished material with good strength properties in all directions. The unidirectional laminae apparently have poor durability qualities at the surface of the finished material, requiring the cloth backing to correct this.

At any rate, some water from the jet did tend to penetrate beneath the cloth backing where the jet entered the material. This phenomenon was very localized in most cases, tending to be more of a cosmetic problem than a damage to the material.

Examination of Table 2 reveals some appearingly anomalous conditions. For example, with the BO/EP materials, the thicker material is shown obtaining the best cut with a smaller diameter nozzle than the thinner piece. This is due to the subjective method of cut evaluation mentioned earlier (a

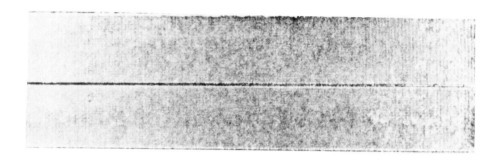

0   1 INCH

FIGURE 8.   SLOT CUT IN 1/16 in. THICK HYBRID OF GRAPHITE/EPOXY AND FIBERGLASS/EPOXY, ENTRY SIDE. CUTTING PARAMETERS WERE P_O = 60,000 psi, d_O = 0.012 in., and V = 0.15 in./sec. ANOTHER EXAMPLE OF THE USE OF CORRECT CUTTING PARAMETERS.

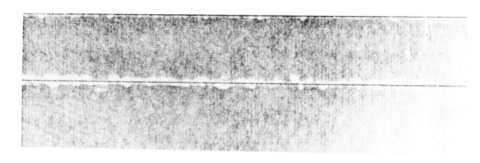

0   1 INCH

FIGURE 9.   SAME MATERIAL AS FIGURE 8, ENTRY SIDE. CUTTING PARAMETERS WERE P_O = 60,000 psi, d_O = 0.008 in., and V = 0.15 in./sec. THE DELAMINATION VISIBLE IS ONLY BETWEEN THE CLOTH BACKING AND THE COMPOSITE PROPER.

244

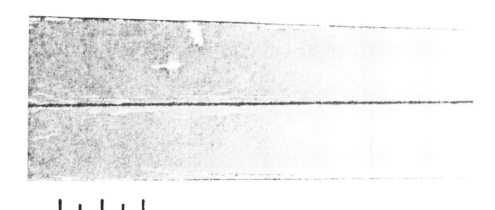

0   1 INCH

FIGURE 10.   SLOT CUT IN 1/16 in. THICK BORON/EPOXY, ENTRY SIDE.   CUTTING
             PARAMETERS WERE P_o = 60,000 psi, d_o = 0.010 in., and V =
             2 in./sec.   THE SHORT, RANDOM LINES RUNNING NEAR TO AND
             PARALLEL TO THE SLOT ARE BREAKS IN THE SURFACE LAYER OF THE
             RESIN DUE TO DISPLACEMENT OF UNDERLYING FIBERS BY THE JET.

Table 2. Results of Jet Cutting Tests

Mat'l. Code	Thickness (in.)	P_o(psi)	d_o(in.)	v(in./sec)	Remarks
GR/EP	.067	55,000	.008	1.0	Virtually no separation of the cloth backing
"	.136	60,000	.010	0.50	Minor separation of the cloth backing on exit side
"	.273	60,000	.014	0.11	Good cut, no separation of the cloth backing
BO/EP	.058	60,000	.012	2.0	Poor cut, loose fiber left along cut
"	.136	60,000	.010	2.0	Poor cut, separation of the surface resin layer, top & bottom
KE/EP	.058	55,000	.006	2.0	Only slight separation of the cloth backing, exit side
"	.125	55,000	.010	0.50	Good cut
FG/EP	.139	60,000	.010	0.10	Small amount of separation of the cloth backing, exit side only
GR/EP & BO/EP	.088	60,000	.012	0.23	Poor cut, loose fibers left along cut
"	.154	60,000	.012	2.0	Poor cut, loose fiber and some separation of the cloth backing, bottom side
"	.321	60,000	.014	0.15	Some exit side separation of the cloth backing
GR/EP & KE/EP	.125	60,000	.010	0.25	Good cut
"	.250	60,000	.014	0.08	Very good cut
GR/EP & FG/EP	.068	55,000	.012	0.15	Good cut
"	.253	60,000	.012	0.15	Rather severe delamination of bottom fibers of composite

GR/EP : Graphite/Epoxy

BO/EP : Boron/Epoxy

KE/EP : Kevlar/Epoxy

FG/EP : Fiberglass/Epoxy

simple visual inspection). Thus, with the thicker material, the 0.012 in. diameter nozzle did not appear to improve the cut quality, but with the thinner material it did. Also, the rather high traverse velocity of 2 in./sec and the resulting poor cut quality would seem to suggest the efficiency of using a slower speed. However, slower traverse speeds did not improve the cut quality with this material.

The results shown in Table 2 are for cuts in a single direction on the sample. A control cut done on most of the samples at a 90° angle to the others showed no significant difference in cutting performance. This was accepted as reasonable assurance that the orientation of the fibers in the materials did not permit easier cutting in one direction than the other.

OBSERVATIONS AND CONCLUSIONS

The main conclusion arising from the cutting tests was that the waterjet can cut fibrous advanced composite materials of the epoxy type with no airborne dust and a noize level of less than 80 db. The materials with the very hard and strong fibers (boron) were cut completely through at the fairly high feed rate of 2 in./sec, but the quality of the cut was relatively poor. It would be possible to trim this rough edge by profile routing, however. The other materials were cut with a better edge finish, in general, but at generally slower feed rates.

It would appear that material thickness for these types of materials would be limited to about 1/4 in. (6 mm) for jet cutting. Thicker materials might be successfully cut by the jet but at a slow feed rate.

It is possible to provide apparatus such that the jet cutter nozzle can be traversed over a stationary sheet of material. The action of the jet as a point cutter makes this practical. This situation readily lends itself to numerical control of the nozzle (or by simply following a template) to make repeated cuts along a complicated outline, including sharp radii. Thus, the jet cutter can readily serve as a high-production cutting tool.

ACKNOWLEDGEMENTS

This study was partially supported through a subcontract from Grumman Aerospace Corporation under sponsorship of Manufacturing Technology Division, Air Force Materials Laboratory, Wright-Patterson Air Force Base, Ohio.

The authors would like to express their thanks to Mr. Mark Marvin who did all the test cutting in this study.

APPENDIX: CUTTING OF ADVANCED COMPOSITE PREPREG MATERIALS

At the time of this writing, jet cutting tests were being performed at Flow Industries to determine the ability of the water jet to cut prepreg materials. Preliminary results on single layers of graphite/epoxy and Kevlar/epoxy prepreg materials indicate that they are very successfully cut at very high feed rates (see the table below). The fiberglass/epoxy prepreg materials were cut well but the cut edge showed that the waterjet affected the material adjacent to the cut for a distance of about 1/32 in. (0.8 mm). The effect was a lightened color of the material, presumably due to wetting of the resin. This effect was absent in the graphite and Kevlar/epoxy materials.

The boron/epoxy prepreg materials in single thickness did not cut well. The water jet tended to dislodge the fibers from their paper backing causing a generally disrupted appearance of the material along the cut.

Preliminary Results of Jet Cutting Tests
Single Plies Prepreg Materials

Mat'l. Code*	P_o (Kpsi)	d_o (in.)	v (in./sec)	Remarks
GR/EP	50	.003	50	Very clean cut
BO/EP	60	.014	10	Poor cut
KE/EP	60	.003	50	Very clean cut
FG/EP	60	.006	50	Some apparent wetting along cut

*Key in Table 1, in text.

Water cuts composite aircraft parts

At Sikorsky, using a water jet to trim molded and cured parts
made of advanced composites has increased productivity 80%

ROUTING, the traditional method of trimming aluminum-sheetmetal parts in the aircraft industry, doesn't work too well on advanced composites, which have been replacing aluminum for various components. Bandsawing also leaves much to be desired.

These methods are noisy, create critical dust problems, and are labor-intensive—after trimming, the cut edge must be deburred in a secondary manual operation. Aramid-reinforced epoxy, which is rather widely used at Sikorsky, is especially difficult to rout or bandsaw. The Kevlar aramid fibers simply do not shear or fracture cleanly, and frayed edges almost always result.

Thus, in 1979, when it was obvious that the company's use of advanced composites for military and commercial helicopters would continue to expand, a water-jet cutting system was purchased from Flow Systems (Kent, Wash.). Water-jet cutting, also called hydrodynamic machining, harnesses the power of water pressurized to as much as 60,000 psi and focuses it into a controllable needle-like jet capable of cutting today's most widely used organic-matrix advanced composites: glass-, aramid-, graphite-, or hybrid-reinforced epoxy. And there's no reason to suspect that it wouldn't work equally well on the emerging polyimide-matrix materials.

The initial system cost about $35,000 and included a single water-jet table model, a 60-hp compressor, a pressure intensifier, and various accessories. This price was just about one-third of what the inevitable expansion of the dust-collection system would cost. Since the initial purchase, a portable, hand-held model, three fixed C-frame stations (Fig 1), another compressor (50 hp), and two more intensifiers have been added. The

By Mel M. Schwartz, manager, Manufacturing Technology, Sikorsky Aircraft, division of United Technologies Corp, Stratford, Conn

system is shown schematically in Fig 2.

Although several aircraft companies are using the water jet to cut the periphery of flat, advanced-composite plies in their prepreg form, Sikorsky applies the process solely to trimming and producing cutouts in molded and cured parts. Prepreg plies are cut by steel-rule dies, reciprocating knives, or hand scissors.

Most parts are nested in a trim fixture and trimmed at a C-frame station, as shown in Fig 1. The worker holding the fixtured part butts the fixture against the guide-post nozzle of the water-jet assembly. Large fixtured parts are often trimmed with the hand-held water jet (Fig 3), and some unfixtured parts are trimmed with the table model. In such a case, the worker butts the workpiece against a special guide ring, which may be attached to the table or to the guide-post nozzle of the water-jet assembly, as shown in Fig 4 and 5.

At the time that water-jet cutting was

1. Worker trims part nested in fixture at one of three C-frame water-jet cutting stations at Sikorsky plant. Most composite parts that can be hand held are trimmed this way

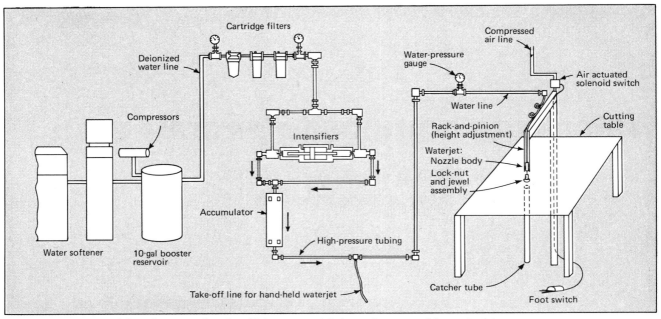

2. Inlet water enters at left and emerges in a fine, high-pressure jet stream from any of five water jets

introduced at the company, more than 400 trim fixtures were being used for routing and bandsawing composite parts. To accommodate the standard guide-post nozzle of the water jet, trim fixtures require a 0.156-in. setback. Instead of modifying this multitude of fixtures, at considerable expense, the company developed special guide-post nozzles with zero setback (Fig 6A). Trim fixtures for new parts entering production were, of course, designed with the 0.156-in. setback (Fig 6B).

During the past few years, virtually all composite parts made at Sikorsky have been trimmed by water jet. Although laminate thicknesses up to ¼ in. or more have been cut elsewhere, most of our composite parts are only 0.025-0.030 in. thick. Fiber orientation in the plies may be all-longitudinal (0°), all-transverse (90°), or a combination of these with or without +45°, −45° intermediate plies. Fiber orientation has little or no effect on the quality of the cut edge. Increasing water-jet pressure or jet diameter or slowing the cutting pass permits thicker sections to be cut and generally enhances the quality of the cut edge.

In almost every case, there has been no evidence of delamination and/or fraying and, thus, no need for secondary finishing. Only on rare occasions have even the aramid-reinforced parts required subsequent edge refinement. As a result, compared with the former routing and bandsawing methods, productivity has improved 80%.

There are several other benefits besides improved edge quality. Because of a phenomenon that has been termed

3. Large parts are set on work table and trimmed with portable unit

"jet entrainment," the cutting operation is dust-free. As the high-velocity water mixes with the air, a vacuum is created, drawing the air downward.

Also, since the cutting stream does not impose any stress on the trim fixture, essentially any material that can be formed to nest the part can be used for these tools. Obviously, the lowest-cost materials get the nod. Because lightweight fixtures are desirable, ordinary glass-reinforced epoxy (fiberglass) is the most common material, although low-cost aluminum sheetmetal is also used.

Because of the high pressure, the water of the jet stream is hot, though not hot enough to degrade properties of the cured material: there's no heat-affected

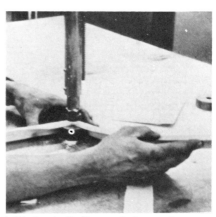

4. Worker butts flange of close-out member against guide to trim other flange

250

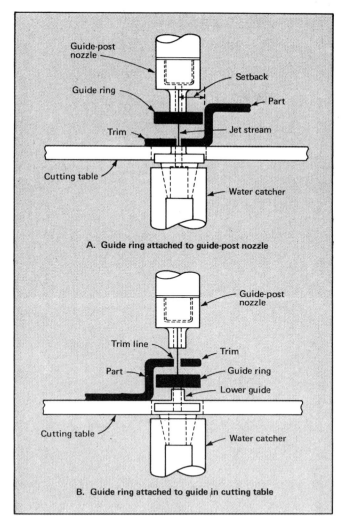

5. Two guide-ring arrangements aid in trimming unfixtured parts

Guide-post nozzle
Guide ring
Setback
Part
Trim
Jet stream
Cutting table
Water catcher

A. Guide ring attached to guide-post nozzle

Guide-post nozzle
Trim line
Trim
Part
Guide ring
Lower guide
Cutting table
Water catcher

B. Guide ring attached to guide in cutting table

Guide-post nozzle modified to zero setback
Trim tool
Jet stream
Part
Water catcher

A. Guide-post nozzle modified for routing fixtures

Guide-post nozzle
Trim tool
0.156-in. setback
Jet stream
Part
Water catcher

B. Guide-post nozzle for new tools

6. Cutting nozzle back (A) precluded modifying existing tools

zone, for example. Nor does the momentary exposure of water to the edge being cut degrade properties: strength retention has been substantiated by tensile and short-beam-shear tests.

Besides trimming external part peripheries, the water jet can also produce internal cutouts. Both internal and external cuts can be produced with corner radii as tight as 1/16-1/8 in. Also, no starting hole has to be drilled for internal cuts. Such cuts, however, must be initiated away from the trim line because starting them at the trim line can cause delamination. This limitation also precludes use of the water jet as an alternative to drilling small holes.

Generally, the water jet cuts best at a standoff distance of 1/8-3/16 in. from the workpiece. This is advantageous from the standpoint of safety in that it essentially eliminates the danger of placing a finger in the jet stream.

Unlike with routing and bandsawing, there are, of course, no cutters to buy, periodically clean, and replace when

worn. The high-pressure water does, however, take its toll on the synthetic sapphire "jewel" in the water-jet nozzle assembly, despite the exceptional wear resistance of this material.

The orifice, through which the high-pressure water passes, is quite small and will erode. Erosion stems not solely from the high pressure of the water but also, and possibly mainly, from the mineral content of the water and the accumulation of minute dirt particles. "Fan out" of the water-jet stream indicates a worn or dirty orifice.

Because of the high mineral content of inlet water at Sikorsky, the company had to add water-deionization as well as filtration equipment in the hydraulic system (Fig 2). Still, jewel life is on the order of 200-400 hr. The jewels, however, cost only $7 or so and can be replaced in only 2-3 minutes.

Orifice size and pressure depend on the number of cutting stations operating. The maximum orifice size that can be used if one station is run at 55,000 psi

and only one of the compressors and a single pressure intensifier are used is 0.012 in. With two stations operating at this pressure and only one compressor and intensifier used, maximum orifice size is 0.008 in.

To operate all five units—the three C-frames, table model, and portable unit—at 55,000 psi with 0.008-in. orifices, both compressors and the three intensifiers must be used. To prolong orifice life, however, as well as the life of the high-pressure seals in the intensifiers, the system is normally operated in the range of 35,000-50,000 psi.

Although water-jet cutting is currently a manual operation, Sikorsky is looking into automating it. Within the next few years, any or all of the C-frame stations and table model could be replaced or supplemented by robots, which would eliminate the need for trim fixtures, as well as reduce labor costs. At the very least, the automated equipment would have stand-alone controls and might be tied to DNC. ■

SECTION IV
Painting

CORROSION PROTECTION METHODS FOR GRAPHITE
FIBER REINFORCED ALUMINUM ALLOYS

J. H. Payer
Corrosion and Electrochemical Technology Section
BATTELLE
Columbus Laboratories
Columbus, Ohio 43201

———

P. G. Sullivan
NETCO
Long Beach, California 90802

Abstract

Corrosion protection methods for graphite/aluminum composites
were studied for application in hostile environments where
protection of bulk aluminum alloys is required. Organic
coatings, metallic coatings, and metal claddings on graphite/
aluminum composites were exposed to marine atmosphere, salt
spray, and alternate immersion in seawater. Several systems
studied were readily applied and provided corrosion protection for Gr/Al composite.

1. INTRODUCTION

Since potential uses of metal-matrix composites expose materials to a broad range of corrosive environments, the importance of characterizing corrosion behavior of graphite fiber-reinforced aluminum alloys (Gr/Al) was recognized early in the development of this advanced material system. In addition, the feasibility of applying corrosion protection methods commonly applied to bulk aluminum alloys in hostile environments to Gr/Al composites is under investigation.

In this paper, emphasis is placed on methods for additional corrosion protection for Gr/Al composites. Other work under way is focused on optimizing the inherent corrosion resistance of Gr/Al composite. Organic coatings, metallic coatings, and metal claddings were applied to Gr/Al composite and exposed to (1) marine atmosphere at Battelle's Daytona Beach Test Facility, (2) salt spray cabinet, and (3) alternate immersion in seawater.

2. EXPERIMENTAL PROCEDURE

2.1 GR/AL COMPOSITE

Continuous length, eight-strand, Thornel-50 graphite fiber-reinforced aluminum alloy composite wire (~0.060-in. diam) obtained from Fiber Materials, Inc., and from Aerospace Corporation was consolidated into approximately 6-inch by 6-inch panels using commercially available state-of-the-art hot-pressing technology. The precursor wire matrix alloys utilized consisted of 201 and 202 aluminum containing 25 to 30 vol % Thornel-50 graphite fibers. The wire bundles were consolidated by hot pressing using face and interior foils of 1100 aluminum alloy. Representative photomicrographs of these materials are shown in Figure 1. Throughout this paper, the first number in each composite identification refers to the aluminum alloy in the wire configuration and the second number to the aluminum alloy used for the interlayer foils.

The nominal composition of the 201 and 202 alloys is shown below:

201 Alloy
Cu – 4.7%   Ag – 0.6%
Ti – 0.3%   Si – 0.1%
Mn – 0.3%   Fe – 0.15%

202 Alloy
Same as above, with the addition of ~0.4% Cr.

All composites evaluated were characterized by radiography and metallographic examination prior to exposure.

2.2 COATING AND CLAD SYSTEMS

The coating and clad systems investigated are listed in Table 1.

Specimens of 201-1100 Gr/Al composite for coating were cut to approximately 1/4-inch wide by 3-inch long by 0.1-inch thick.

TABLE 1.   CORROSION PROTECTION SYSTEMS INVESTIGATED

COATING SYSTEMS ON Gr/Al COMPOSITES

METALLIC

Electroplated Nickel
Chemically Vapor Deposited Nickel — nominal 1 mil
Chemically Vapor Deposited Nickel — nominal 3 mil
Chemically Vapor Deposited Chromium Carbide
Physically Vapor Deposited Nickel + Chromium
Physically Vapor Deposited Titanium

ORGANIC

Polyurethane — nominal 5 mil
Chlorinated Rubber — nominal 5 mil
Epoxy — nominal 5 mil

CLAD SYSTEMS ON Gr/Al COMPOSITES

Aluminum Clad
Titanium Clad
Nickel Clad

Metallic coatings were applied by electroplating, chemical vapor deposition, and physical vapor deposition. Electroplating of nickel was preceded by a zinc displacement layer and copper flash layer. For purposes of coating, the Gr/Al composite was handled as bulk aluminum. In all cases, this represents the first attempt at coating Gr/Al composite by these techniques, and particularly for vapor deposited coatings better quality can be anticipated as experience is gained.

The organic coatings selected all have exhibited good service in marine environments. All were applied with primer and procedures recommended by the manufacturer. Flow coating and air cure were used for organic coatings.

Clads of aluminum, titanium, or nickel were applied during initial panel consolidation from wire. For titanium and nickel clad, this again represented the first attempt to fabricate the comibinations and some defects were observed during characterization.

2.3 CORROSION EXPOSURE

Coated specimens were exposed under conditions described below. Visual inspections were made throughout the exposure, followed by visual, optical metallographic, and scanning electron microscopic examination after exposure.

2.3.1 Marine Atmospheric Exposure

Specimens were exposed at the atmospheric corrosion site of Battelle's Daytona Beach, Florida, Marine Laboratory. Racks containing the specimens were mounted at 45 degrees facing the ocean in accordance with standard ASTM procedures. A set of racks at installation are shown in Figure 2. Black silicone sealant was applied to secure specimens in the racks and to protect exposed ends.

2.3.2 Alternate Immersion Exposure

Specimens were exposed in the laboratory to alternate immersion in seawater. A cycle consisted of a 10-minute immersion period followed by a 50-minute drying period. Seawater was replenished each week.

2.3.3 Salt Spray Cabinet

Salt spray tests were run in accordance with ASTM Standard B117-73. Test conditions were relative humidity 95-98%, temperature 95 F, and volume of condensed spray 1-2 ml/hr per 80 cm^2 of collecting surface. As in the above exposures, visual inspections were made throughout the test and detailed analysis after exposure.

3. RESULTS

While this paper is concerned primarily with additional corrosion protection methods for Gr/Al composite, a summary of the current status of corrosion characterization studies is included for reference. In the past year, significant improvement in the inherent corrosion resistance of Gr/Al composite has been observed. Increased corrosion resistance is attributed to improved quality of precursor wire and advancement in hot pressing consolidation technology. Current programs are focused on further improvement of corrosion resistance through matrix and process optimization.

The corrosion of Gr/Al composite, as bulk aluminum alloys, is strongly structure dependent, and therefore subject to control by alloy selection and consolidation practice. It is of interest to note that the graphite fiber-aluminum interface is not subject to accelerated attack as might be expected. An illustration of this is presented in Figure 3 which shows a schematic diagram and photomicrograph of a cross section through a partially drilled hole in Gr/Al composite. The photomicrograph taken after 5 weeks' exposure to alternate immersion in seawater shows that no preferential corrosion has occurred along graphite fibers where they intersect the surface. Similar behavior was observed along edges with exposed fibers. This is not to say that the galvanic action between graphite and aluminum does not enhance corrosion somewhat, but it does indicate that the interface reactivity is low.

In development of Gr/Al composite, as a role of macro- and microstructural features is more fully understood, composites can be produced with corrosion resistance approaching that of bulk aluminum alloys. For hostile environments which readily corrode aluminum alloys, further corrosion protection is required. For comparison with coated specimens, photomacrographs of an uncoated Gr/Al composite sample and a 3-inch by 3-inch Gr/Al panel with edges coated

are shown in Figure 4. The specimens were exposed for 14 weeks and 1 year, respectively. Surface corrosion was observed with no significant penetration into the composite.

An evaluation of some candidate corrosion protection systems is presented below.

3.1 METALLIC COATING

Metallic coatings for Gr/Al composite were chosen for their demonstrated corrosion resistance. Three application procedures were used: electroplating, chemical vapor deposition, and physical vapor deposition. In all cases, coating parameters chosen represent a first attempt with no prior experience in treating Gr/Al composite. Characterization of coated specimens prior to exposure indicated good quality for electroplated nickel and numerous flaws in physically vapor deposited and chemically vapor deposited coatings. Thus, the capability of applying electroplated nickel coatings to Gr/Al composite was demonstrated, as was the need for some further experimentation to improve CVD and PVD coating parameters. The probability of identifying suitable parameters is high.

Specimens of each coating system were exposed for 14 weeks to marine atmosphere, 4 weeks to alternate immersion in seawater, and 4 weeks to salt spray cabinet.

Electroplated nickel provided excellent corrosion resistance in all three exposures. While exposure of the overall coating group was terminated in marine exposure at the end of 14 weeks, an electroplated nickel specimen was left on test for 8-1/2 months. A photomacrograph of the specimen is shown in Figure 4. Except for a slight surface stain, no evidence of corrosion was observed. Flaws in the nickel plate coating on a few specimens resulted in isolated corrosion sites toward the end of the exposure periods. Electroplated nickel is seen as a presently available coating system which can be applied to complex shapes by state-of-the-art procedures. Gr/Al can be treated as bulk aluminum with this system.

Results for CVD coatings of nickel and chromium carbide are similar. Corrosion initiated at flaws present prior to exposure. These flaws were pores which penetrated through the coating. As exposure continued severe corrosion led to loss of coating. All three exposures resulted in the same morphology of corrosion. Adhesion of CVD coatings to Gr/Al substrate was good, and adjustment of coating conditions can eliminate these flaws.

Nickel and chromium coating and titanium coating were applied by PVD. In all three exposures corrosion initiated at flaws in the coatings and resulted in loss of coating as exposure continued. As with CVD coatings, adhesion to Gr/Al substrate was good and adjustment of coating condition can eliminate flaws. Some experimental work is necessary to identify coating conditions for application of high quality CVD and PVD coatings; however, Gr/Al composite can be treated as bulk aluminum for both processes.

3.2 ORGANIC COATINGS

Three organic coatings were selected for evaluation on the basis of their documented, good service performance in seawater. Each coating cures in air and is, therefore, suitable for maintenance coating in the field. The coatings were polyurethane, chlorinated rubber, and epoxy. The former two coatings require a primer, while the latter is self-priming. All coatings were applied to a nominal thickness of 5 mils.

Organic coatings provided excellent corrosion protection for all three exposures. An epoxy-coated specimen was left on test in marine atmosphere for 1 year and showed no signs of deterioration or corrosion. A photomacrograph of the specimen is shown in Figure 4. No corrosion was observed after 4 weeks' exposure to alternate immersion and to salt spray cabinet.

No difference between Gr/Al composite and aluminum alloys was observed with respect to organic coatings. The composite can be treated as bulk aluminum with equivalent corrosion protection anticipated. Several well-established coating systems are now available for protection in hostile environments; no difficulties are foreseen in their application to Gr/Al composites.

3.3 METAL CLADDING

Metal cladding is a method to protect large surface areas of Gr/Al composite samples. The feasibility of cladding with aluminum alloys, titanium, and nickel was demonstrated. It is convenient to apply cladding during the hot pressing consolidation of panels.

Aluminum facesheets are routinely used in the fabrication of Gr/Al shapes and bond well to the composite substrate. Bonding was not found to be influenced by composition of aluminum alloy cladding, hence facesheet composition can be selected to provide maximum protection. Some aluminum alloys which have been applied to Gr/Al composite are Al-1100, Al-2024, Al-3003, Al-5056, and Al-6061. With edge protection these clad composites exhibit identical corrosion behavior as their aluminum alloy counterpart.

Titanium has excellent corrosion resistance in marine environments and can be clad to Gr/Al composite. Limited experience with titanium cladding indicates that some improvement is required to increase the bond between clad and substrate. Exposure of Ti-clad material resulted in delamination along the interface, initiating at exposed edges. Edge protection would alleviate this behavior.

As with titanium, limited experience has been gained with nickel cladding of Gr/Al composites. Here again some improvement in the clad/substrate bond is needed. Exposure of Ni-clad material to alternate immersion in seawater resulted in some delamination at exposed edges. Edge protection would, also, alleviate this behavior.

4. CONCLUSIONS

The necessity for and selection of corrosion protection for Gr/Al composite are determined by specific application. For mildly corrosive conditions, Gr/Al composite can be used as fabricated and requires no additional corrosion protection. As with aluminum alloys, additional corrosion protection is required for more hostile environments. Several methods were used successfully to protect Gr/Al composite from corrosion. These methods have been applied to aluminum alloys and the technology is directly transferable to Gr/Al composite. The feasibility of other methods was demonstrated and some areas for further development identified.

Specific conclusions of the study are
(1) Gr/Al composite can be used in many applications with no additional corrosion protection.
(2) For organic coatings, Gr/Al can be processed as aluminum alloys with equivalent protection provided.
(3) Electroplate nickel coating is readily applied and provides excellent corrosion protection.
(4) Application of metallic coatings by physical or chemical vapor deposition is feasible.
(5) Aluminum alloy clad can be

applied with any composition desired that is available in sheet form. (6) Titanium and nickel cladding is feasible.

The results reported here are part of an ongoing study of the corrosion behavior of Gr/Al composite. In addition to the study of corrosion protection methods, other aspects of the program are focused on characterizing the corrosion behavior and improving the inherent corrosion resistance of Gr/Al composite.

ACKNOWLEDGMENTS

This work was performed as a part of the Launch Vehicle Materials Program sponsored by the Naval Sea Systems Command and managed technically by Battelle's Columbus Laboratories. Marlin A. Kinna is NAVSEA Technical Monitor and Jack M. Snyder is the Battelle Launch Vehicle Materials Program Manager. The LVMT program is directed toward development and application of new materials to meet the needs of future, advanced submarine launched ballistic missile systems.

BIOGRAPHIES

Joe H. Payer is a research metallurgist at Battelle's Columbus Laboratories in Columbus, Ohio. He holds a B.S. degree in Metallurgical Engineering and a PhD from The Ohio State University. From 1971 to 1974, he was a research engineer in the Coated Products Division of Inland Steel Company's Research Laboratory in East Chicago, Indiana. He directs research in the areas of corrosion and electrochemistry including studies of mechanisms of corrosion processes, corrosion prevention, materials selection and evaluation, and materials development.

Pamela Sullivan is a Project Engineer at NETCO responsible for all aspects of materials testing including mechanical, thermal, chemical, and corrosion. She is currently Principal Investigator on a program for NASA on The Elevated Temperature Properties of Boron/Aluminum. She graduated with a B.S.M.E. from California State University -- Long Beach, in 1971 and expects to complete a masters degree in 1977.

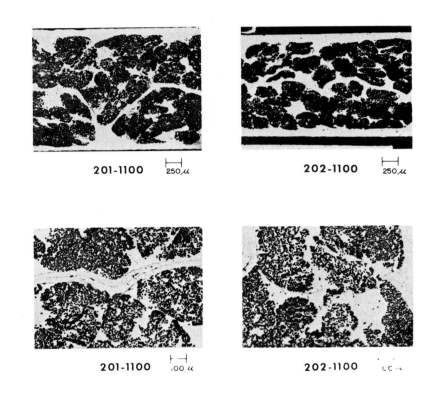

201-1100 ⊢⊣ 250μ   202-1100 ⊢⊣ 250μ

201-1100 ⊢⊣ .00 μ   202-1100 ⊢⊣ cc ≺

FIGURE 1.   PHOTOMICROGRAPHS OF GR/AL COMPOSITES 201-1100 AND 202-1100

(The 201 refers to the aluminum alloy in the wire bundle;
the 1100 to the aluminum in the interlayer foil.)

FIGURE 2.   SPECIMENS IN RACKS AT BATTELLE'S ATMOSPHERIC
EXPOSURE SITE IN DAYTONA BEACH, FLORIDA

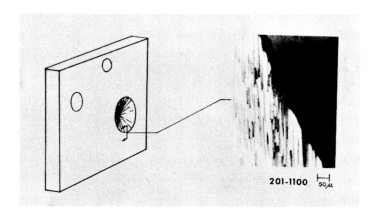

201-1100   50μ

FIGURE 3.   PHOTOMICROGRAPH AND SCHEMATIC DIAGRAM OF
CROSS SECTION THROUGH A PARTIALLY DRILLED
HOLE IN GR/AL COMPOSITE AFTER 5 WEEKS'
ALTERNATE IMMERSION IN SEAWATER.   NO
CORROSION OBSERVED ALONG GRAPHITE-ALUMINUM
INTERFACES.

DAYTONA BEACH MARINE ATMOSPHERIC EXPOSURE

14 Weeks	1 Year	8½ Months	1 Year

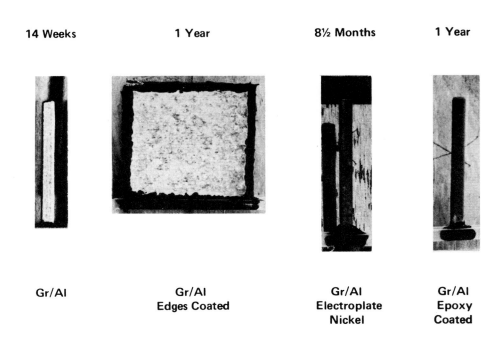

Gr/Al	Gr/Al Edges Coated	Gr/Al Electroplate Nickel	Gr/Al Epoxy Coated

FIGURE 4.   MONTAGE OF GR/AL COMPOSITE SAMPLES AFTER MARINE
ATMOSPHERIC EXPOSURE: UNCOATED, EDGE COATED,
ELECTROPLATE NICKEL COATED, EPOXY COATED

EXPERIMENTAL EVALUATION

OF

LIGHTNING PROTECTIVE COATINGS FOR BORON/EPOXY COMPOSITES

by

E. H. Schulte and D. W. Clifford
McDonnell Aircraft Company
St. Louis, Missouri

Abstract

A number of candidate lightning protective coatings for composite structural materials have been evaluated in the McDonnell Aircraft Company (MCAIR) Lightning Simulation Laboratory. Emphasis has been placed on coating parameter optimization and elucidation of damage mechanisms. Coated and uncoated aluminum foils, plasma sprayed aluminum, aluminum meshes, and bronze meshes, on dielectric substrates were tested by exposure to simulated high current lightning strikes.

INTRODUCTION

Lightweight composite materials offer many advantages to the aircraft designer, and their use is expected to increase dramatically in the future as increased confidence in their reliability is established. One problem associated with composite materials is their vulnerability to lightning strike damage. Early reports[1,2,3] indicated that catastrophic damage to composites might be expected from a strike; and since most all-weather aircraft are struck by lightning several times during their lifetime, it might be assumed that all composite surfaces on such aircraft would have to be protected against the lightning threat.

However, a closer examination of the lightning/aircraft interaction reveals that although a particular aircraft may receive several lightning strikes, the extremities of the aircraft receive most of the strikes, and some regions of the aircraft are very unlikely ever to experience a strike at all. FAA Advisory Circular No. 20-53[4] defines three regions of strike probability in terms of Strike Zones: Zone 1--Direct Strike, Zone 2--Swept-Stroke, and Zone 3--Low Probability of Any Strike. In current aircraft designs almost all composite structures are located in either Zones 2 or 3. Therefore, before assuming the weight penalty and cost of protecting the composite areas, an analysis of the localized strike probability should be conducted. High-voltage attach-point tests are used to determine all Zone 1 strike points (Figure 1). Zone 2 regions then are inferred from the primary attach points. Any composite structures in Zone 1 (strike points) and Zone 2 (swept-stroke regions) may indeed require protection, especially if they are essential to safe flight of the aircraft. Using the criteria established by the FAA, Zone 3 composites need not be protected, and non-"safety of flight" composites in Zones 1 and 2 should be analyzed from a cost-effectiveness standpoint to determine the desirability of protecting them.

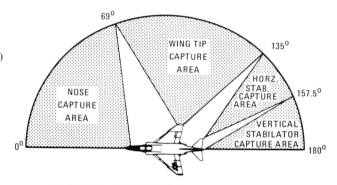

FIGURE 1 LIGHTNING CAPTURE AREAS

Further, differentiation must be made between composites which are purely structural in nature and those which are also used for other purposes, such as wing skin panels used for fuel containment ("wet-wing"). The difference arises because some structural damage can be allowed if the safety of the aircraft is not endangered and because repair techniques have been developed[5] for damaged composite structures. However, the same damage over a wet wing raises questions concerning fuel ignition and adequacy of fire suppression systems. Consequently, special problems such as those on the wet-wing must be assessed separately. The scope of this paper is restricted to a discussion of test results on coatings for composites used in purely structural applications, and studies reported address the problem of protecting composite structural materials from high peak current catastrophic damage.

LIGHTNING THREAT

The nature of the lightning strike in Zone 2 (swept-stroke regions) is not well defined by Military Specifications such as MIL-B-5087B[6] which defines a "high-level" high current strike only. Other specifications[7,8] include a continuing low-level current component which is fundamental to the swept-stroke phenomenon. Table I[9] summarizes the characteristics of lightning strikes in terms of average and maximum values.

TABLE I CHARACTERISTICS OF NATURAL LIGHTNING DISCHARGES[9]

Characteristics	Probable Average	Observed Maximum
Current (Amperes)	30,000	300,000
Rate of Rise (Amperes/Microsecond)	20,000	200,000
Voltage (Megavolts)	100	500
Charge Transfer (Coulombs)	100	1000

DAMAGE MECHANISMS

Lightning damage to resin matrix composites results from electrical heating effects as currents are conducted through the boron/tungsten or graphite fibers. The high peak current occurs in just a few microseconds and does most of its catastrophic damage by explosive heating effects (Figure 2). These fast rising current "spikes" can result in thermal-shock-induced cracking and vaporization of boron fibers (Figure 3) at lower current levels with no external evidence of damage.[10] The low-level continuing currents produce localized burn-through of boron/epoxy laminate panels (Figure 4) but less of an explosive effect.

An extensive study was inaugurated by MCAIR to determine the effects of electrical currents in boron/epoxy materials and to identify the lightning induced damage thresholds and mechanisms. It has been determined that by applying thin conductive coatings to composite materials, the type of explosive damage shown in Figure 2 can be avoided. However, the following observations from the study concerning boron/epoxy composites are believed important to the coating development program:

(1) Very significant structural damage can be sustained with no visible evidence of degradation. Test samples degraded as much as 30% of their original strength after simulated lightning strikes showed no visible discoloration or delamination. The data shown in Figure 5 show the induced current per filament to reduction of tensile strength in boron/epoxy strips 0.5 inch wide.

(2) The extent of hidden damage can be determined by microscopic examination of small section cuts or core samples as in Figure 3.

(3) The level of current passing through the material can be approximately determined by the same microscopic examination.

1 INCH

FIGURE 2 DAMAGE TO UNPROTECTED BORON/EPOXY PANEL FROM HIGH CURRENT UNIPOLAR SIMULATED LIGHTNING STRIKE

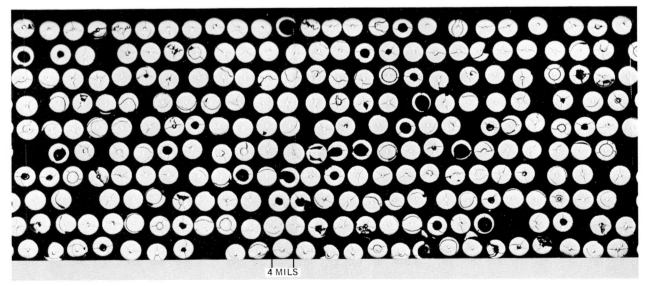

4 MILS

FIGURE 3 BORON-EPOXY TEST STRIP CROSS SECTION SHOWING SEVERE DAMAGE RESULTING FROM SIMULATED LIGHTNING CURRENT

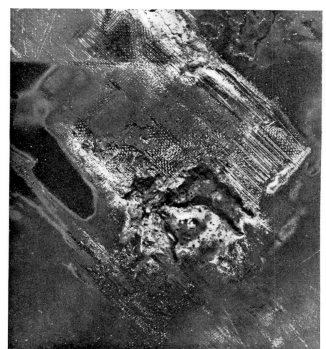

FIGURE 4 DAMAGE TO BORON/EPOXY PANEL FROM COMBINED, HIGH COULOMB HIGH CURRENT, SIMULATED LIGHTNING STRIKE

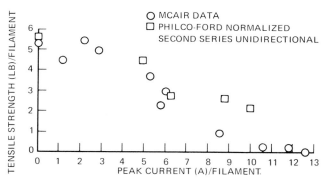

FIGURE 5 DEGRADATION OF BORON-EPOXY CAUSED BY HIGH PEAK CURRENTS

EXPERIMENTAL FACILITIES

The MCAIR Lightning Simulation Laboratory[11] can accurately simulate high-energy electrical discharges representative of actual lightning strikes. The Laboratory's Multi-Component Lightning Simulator (MCLS), which produces a unipolar high peak current pulse closely approximating the MIL-B-5087B pulse has been used for composite coatings development testing.

To test the coated panels a return electrode was clamped to one end of the panel and the high energy discharge was applied at the desired point on the panel by positioning the 1/2-inch copper probe a fraction of an inch above the panel surface. A typical high current pulse oscillogram is shown in Figure 6 (pulse width is measured, in accordance with the requirements of Mil Spec MIL-B-5087B, at one-half the peak amplitude).

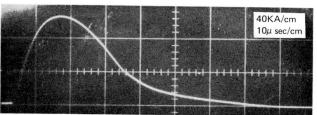

FIGURE 6 HIGH CURRENT WAVEFORM PRODUCED BY MULTI-COMPONENT LIGHTNING SIMULATOR (MCLS)

COATING EVALUATIONS

Several candidate protective coatings for composite materials have been suggested, but Air Force sponsored studies[12] have served to screen out many of them. MCAIR is now concentrating upon optimizing the more promising systems.

The MCAIR testing philosophy has been to evaluate the protective coatings on a dielectric substrate. If the coating is optimized on a partially conductive substrate (such as a composite panel), the coating may be undamaged, yet a portion of the energy may have been shared with the substrate. By optimizing the coating parameters (thickness, mesh size, conductive fillings, etc.) on a dielectric, all of the deposited energy must be absorbed by the coating itself, and there is no doubt that the coating can handle all of the deposited energy. In that case, it is less likely to share the strike current with a conductive substrate when applied to it. The coatings being evaluated at MCAIR are aluminum foils, wire fabrics (aluminum and bronze meshes), plasma-sprayed aluminum, and conductive epoxies.

Tables II and III list the coating systems which have been evaluated and their approximate weight per hundred square feet. The coatings were applied to 9 by 12-inch panels for testing. Panels were of fiberglass about 3/32-inch thick, of plexiglass 1/8-inch thick, and of 10-ply laminated boron/epoxy.

TABLE II CHARACTERISTICS OF CONTINUOUS COATINGS

Material	Thickness (inches)	Weight (lb/100 ft^2)
Foils: 1100 Al	0.001	1.4
1100 Al	0.002	2.8
1100 Al	0.004	5.65
1100 Al	0.006	8.5
1100 Al	0.008	11.3
Plasma-Sprayed Al:	0.006	7.6 (Est.)

TABLE III CHARACTERISTICS OF WIRE CLOTHS

Material	Wire Diameter (inches)	Mesh	Weight (lb/100 ft^2)
5056 Al	0.010	40 x 40	9.4
5056 Al	0.009	50 x 50	9.7
1100 Al	0.004	120 x 120	4.47
5056 Al	0.0021	200 x 200	2.11
Commercial Bronze	0.0114	40 x 40	42.0
Phosphor Bronze	0.0071	60 x 60	23.1
Phosphor Bronze	0.0045	100 x 100	16.2
Phosphor Bronze	0.0030	140 x 140	9.3
Phosphor Bronze	0.0021	200 x 200	7.0

TEST RESULTS AND DISCUSSION

Aluminum Foils

Table IV summarizes the available test results for aluminum foils. Typical test panels are shown in Figures 7 through 10. The criteria for evaluation were based on the area of the substrate which was exposed by vaporization or "blow-off" of the coating.

Some general observations from the tests are:

(1) An 8 mil (.008 inch) aluminum coating will provide complete protection for high peak current strikes and may be used as a baseline system for comparison utilizing the MIL-B-5087B criteria.

(2) The application of a sprayed layer of polyurethane paint over the protective coating significantly affects the performance of the coating, and the effect on each coating scheme must be considered individually. For example, on the aluminum foils two competing mechanisms seem to be working. First the energy consumed in vaporizing the paint layer reduces the amount of energy available to vaporize the protective coating, and in that respect, the effect of the paint layer is favorable. However, the presence of the dielectric paint surface also acts to constrain the arc attachment to a smaller area, thereby increasing the chances of the substrate being exposed. Without the paint it has been observed that the arc current tends to fan out, possibly following the expansion of the ionized plasma sheath.

A preliminary evaluation of aluminum foils on boron/epoxy panels appears to show the same results as on the fiberglass panels (Table IV). A final evaluation of the effect of the paint layer will not be possible until more extensive comparative tests are conducted on composite panels.

TABLE IV TEST RESULTS ON ALUMINUM FOILS

Foil Thickness (mils)	Panel Material	Bonding Agent	Painted (X)	Peak Current (KA)	Pulse Width (μsec)	Damaged Area (inches²)	Notes
2	Fiberglass	FM400		197	18.0	1.18	
2				101.5	25.9	.523	
2			X	193	18.9	.664	
2			X	100	26.3	.196	
4				204	19.2	.207	
4				104.5	26.2	.012	
4				100.8	26.0	.441	Oscillatory Current
4			X	197	19.1	.196	
4			X	106	25.9	.050	
1		FM400	X	93	27.2	.784	
4		Tape		196	16.8	1.00	
4		Tape		104	25.7	.393	
8		FM400		101	25.8	(No Damage)	
8				204	21.2	(No Damage)	
8			X	198	21.2	(No Damage)	
8			X	104	27.6	(No Damage)	
6				101	26.9	(No Damage)	
6				208	22.1	.196	
6				208	20.3	(No Damage)	Negative Probe
6			X	177	18.5	.024	
6			X	206	21.4	.098	
6	Fiberglass		X	105	25.8	.012	
2	Boron/Epoxy			195	21.5	1.76	6" Wide Panel
2			X	191	20.3	.865	
6			X	195	21	.238	
6	Boron/Epoxy	FM400		207	22	.508	6" Wide Panel

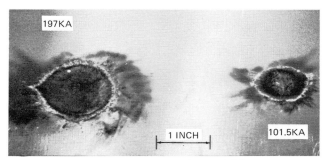

FIGURE 7 2 MIL ALUMINUM FOIL ON FIBERGLASS (UNPAINTED)

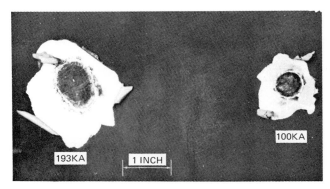

FIGURE 8 2 MIL ALUMINUM FOIL ON FIBERGLASS (PAINTED)

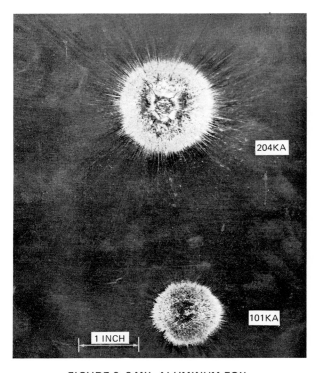

FIGURE 9 8 MIL ALUMINUM FOIL ON FIBERGLASS (UNPAINTED)

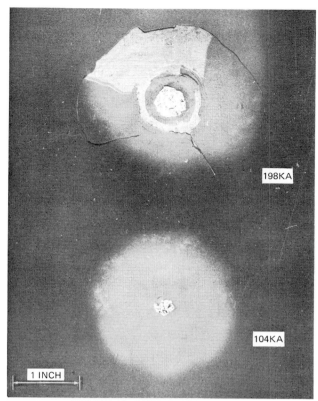

FIGURE 10 8 MIL ALUMINUM FOIL
ON FIBERGLASS (PAINTED)

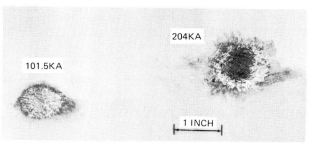

FIGURE 11 6 MIL PLASMA-SPRAYED ALUMINUM
ON FIBERGLASS (UNPAINTED)

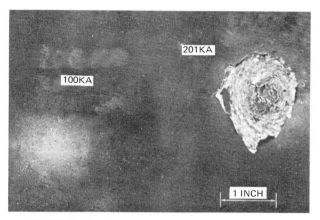

FIGURE 12 6 MIL PLASMA-SPRAYED ALUMINUM
ON FIBERGLASS (PAINTED)

PLASMA-SPRAYED ALUMINUM

Two plasma-sprayed aluminum panels were tested but did not appear to offer any advantages over the aluminum foils and testing was discontinued. An equal weight of plasma-sprayed aluminum appeared to offer approximately the same protection as aluminum foil but costs considerably more. Further, it is difficult to control the thickness of the plasma-sprayed coatings. Test results are shown in Table V, and test panels are shown in Figures 11 and 12.

TABLE V TEST RESULTS ON PLASMA SPRAYED ALUMINUM

Coating Thickness (mils)	Panel Material	Bonding Agent	Painted (X)	Peak Current (KA)	Pulse Width (μsec)	Damaged Area (inches²)	
6	Fiberglass	None		204	18.7	.441	7″ Wide Panel
				101.5	26.8	(No Damage)	
			X	201	19.2	1.76	
6	Fiberglass	None	X	100	26.2	(No Damage)	7″ Wide Panel

WIRE FABRICS

Table VI summarizes the test results on wire cloths. The fabrics used in this study were first bonded to fiberglass panels using FM-400*. The coating was bonded under pressure and heat. After applying high current strikes to a few of these panels, it

*Bloomingdale Dept., American Cyanamid Co.,
Havre-de-Grace, Md.

was obvious that an accurate visual determination of the damage would not be possible, for the bonding techniques had resulted in the wire mesh being completely surrounded by epoxy, and therefore, generally invisible to the naked eye or with a low-power microscope. As seen in Figures 13 and 14, the test panels show carbonized areas which probably approximate the damaged regions. However, the actual amount of wire material removed by the stroke cannot be determined visually, even with a high-power microscope.

It was found that phosphor-bronze fabrics could be examined radiographically. Figures 15 and 16 show typical x-rays of such 60-mesh phosphor-bronze fabrics. These x-rays exhibit an unusual-appearing damage pattern resembling a clover-leaf with the leaves connected but with the wires perpendicular to the connecting wires gone. Surprisingly, the center of the strike area appears undamaged. This damage pattern has been found to be typical for encapsulated metal meshes. Aluminum meshes bonded to fiberglass could not be photographed radiographically. The technique which was adopted for examination of aluminum mesh consisted of applying the fabrics to 1/8-inch transparent plexiglass panels by heat-pressing the material into the surface of the plexiglass, thus maintaining the panel transparency. After the panels were exposed to high current strikes, they were backlighted and examined with a high-power microscope. During a cursory visual examination of the first such

TABLE VI TEST RESULTS ON WIRE CLOTHS

Material: Wire Mesh	Panel Material	Bonding Agent	Painted (X)	Peak Current (KA)	Pulse Width (μsec)	Damaged Area (inches²)	Notes
Bronze 200	Fiberglass	FM400		158	19.1	> 4.*	
200			X	147	20.7	> 5.*	
140				173	16.6	2.26	
140				94	26.5	.787	
140			X	175	19.6	3.53	
140			X	95	27.0	1.25	
140			X	93.4	26.8	1.02	
100				160	16	.503	
100				190	18	1.13	
60				136	18.6	**	
60				195	19	5.03	
100				188	21	1.33	
60	Fiberglass	FM400		200	20.5	.695	
Aluminum 200	Fiberglass	FM400	X	187	16.5	+	Paint Burned for 3 sec
120		Tape		197	21.9	1.20	
120		Tape		107	25.0	.444	
120		FM400		98.4	25.8	+	
120				203	21.4	+	
120			X	197	20.7	+	
120	Fiberglass	FM400	X	102	26.1	+	
200	Plexiglass	None		93.5	24.0	2.55	
200	Plexiglass	None				>10.	90 Coul in .286 sec
200	Fiberglass	FM400				+	100 Coul in .312 sec
120	Plexiglass	None		210	19	++	Oscillatory
40				190	19	(No Damage)	
40						1.09	70 Coul in .215 sec
40				21.6	19	++	Oscillatory
40				216	17	++	Oscillatory
40				204	19	.82	Oscillatory
200			X	178	21	3.14+	
40			X	195		1.07+	
50			X	200	20	.75++	
50	Plexiglass	None	X	193		.75	

* - Damage extended through edges of panel.
** - Only a few slightly damaged wires.
+ - Damage could not be determined because aluminum mesh could not be radio-graphically examined.
++ - High-speed-photo sequence tests. No measurements made of damage.
+ - Panel cracked.
++ - Hole in mesh and in plexiglass substrate.

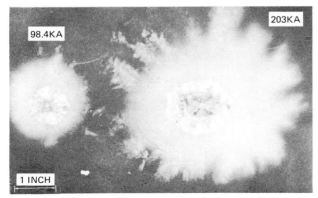

FIGURE 13 120 MESH ALUMINUM WIRE CLOTH ON FIBERGLASS (UNPAINTED)

FIGURE 14 120 MESH ALUMINUM WIRE CLOTH ON FIBERGLASS (PAINTED)

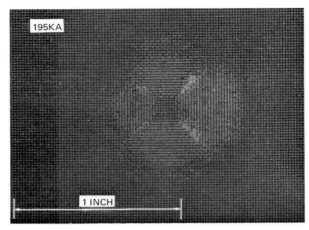

FIGURE 15 RADIOGRAPH OF 60 MESH PHOSPHOR BRONZE WIRE CLOTH ON FIBERGLASS (UNPAINTED)

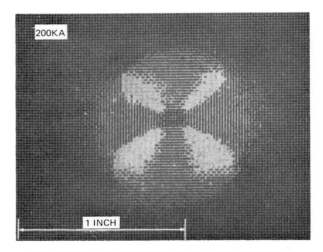

FIGURE 16 RADIOGRAPH OF 60 MESH PHOSPHOR BRONZE WIRE CLOTH ON FIBERGLASS (PAINTED)

panel (Figure 17) after exposure to a 93.5 Kiloamp strike, it appeared that the wire mesh was still intact and essentially undamaged. A more careful examination revealed that an area about 2 inches in diameter had been damaged, displaying the characteristic clover-leaf damage pattern. Though the wire had vaporized, the grid pattern in the plexiglass remained, giving the initial visual impression that the mesh was still intact.

To help understand the clover-leaf pattern, a series of ultra-high-speed photographs was taken with the objective of defining the time history of the damage phenomena. Figure 18 illustrates a sequence taken at the rate of 200,000 frames/second (interframe interval of 5 μsec) of a strike on 40-mesh aluminum fabric on plexiglass. The photographs were taken through the panel, which was backlighted by the arc so that the screen was silhouetted against the arc. In Frames 3 and 4 the arc plasma seems to propagate along the perpendicular directions of the weave. Because the mesh failed to vaporize due to the heavier gauge (10 mils) of the aluminum used in this test, no pattern appeared.

A similar sequence taken with a finer fabric (4 mil, 120 mesh) also failed to yield the desired results. In this test, however, the metal did vaporize, causing the characteristic pattern. During the test what appeared to be an opaque plasma obscured the vaporization sequence. Further effort is required to obtain photographs through the plasma curtain and to extend the viewing time to define more completely the damage time history.

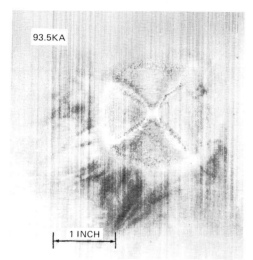

93.5KA

1 INCH

FIGURE 17 200 MESH ALUMINUM WIRE CLOTH ON PLEXIGLASS (UNPAINTED)

One aluminum mesh panel was tested to ascertain whether the near total encapsulation of the mesh in the bonding agent might be responsible for the clover-leaf pattern. In this instance the 120-mesh aluminum was bonded to the fiberglass panel using double-backed pressure sensitive tape and was tested with high current pulses. The results are shown in Figure 19, in which the clover-leaf pattern in missing.

Some general observations of the protective meshes on fiberglass panels are:

(1) If the outer perimeter of the damage is considered as the damage boundary, then the aluminum meshes do not offer any better protection than a continuous aluminum foil having the same weight. Though it is realized that the outer perimeter may not be the best damage criterion, it will be used until actual damage to boron/epoxy can be determined.

(2) On a weight-per-protection basis, the bronze meshes are the least attractive of the various coating schemes mentioned. This fact was recognized early in the testing program, but testing continued because many different mesh sizes were readily available and the damage analysis technique (x-ray) was more suitable to the bronzes. It is believed that a correlation between the aluminum and bronze mesh damage can be found, simplifying damage analysis of future testing of aluminum-mesh-coated boron/epoxy panels.

(3) In general, the meshes suffered greater damage when painted.

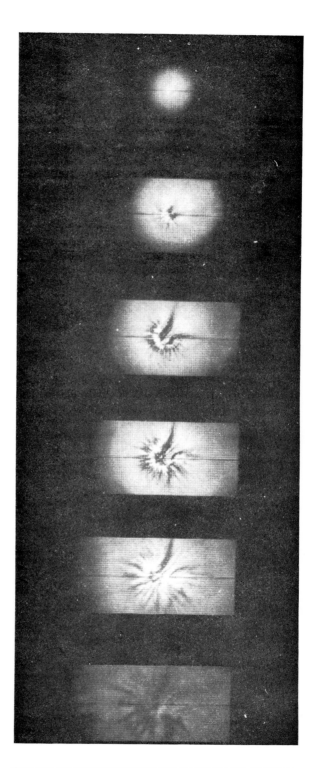

FIGURE 18 TIME HISTORY OF DAMAGE TO 40-MESH ALUMINUM WIRE CLOTH BONDED TO PLEXIGLASS (200 KA PEAK) (5 μSEC INTERVAL BETWEEN PHOTOS)

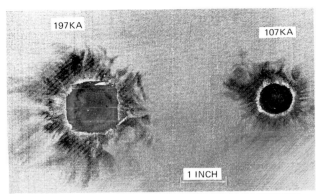

FIGURE 19 120 MESH ALUMINUM WIRE CLOTH ON
FIBERGLASS (UNPAINTED) (BONDED WITH DOUBLE
SIDED PRESSURE-SENSITIVE TAPE)

FIGURE 20 HIGH COULOMB STRIKE (90 COULOMB)
TO 200 MESH ALUMINUM WIRE CLOTH
ON PLEXIGLASS (UNPAINTED)

(4) Although some of the finer meshes on fiberglass and plexiglass were severely damaged, it is not possible at this time to determine whether they would protect a composite panel such as boron/epoxy against a high current strike.

(5) Even though a fine mesh screen may possibly protect a composite panel from a fast high current peak, a moderate, continuing current (high-coulomb transfer) can severely damage the wire fabric (Figure 20) to the extent that the current may choose a path into the composite material rather than continue in the surface mesh.

(6) It is too early to attempt to explain the clover-leaf damage pattern because experiments and analyses in this area are continuing.

CONCLUSION

The work reported here is not complete and should be considered as a progress report. Effort is currently being conducted to determine the effects of the coating systems on boron/epoxy panels and to determine if it is feasible to allow a small amount of current to enter the substrate. The results of the later tests will be reported in the future.

ACKNOWLEDGEMENT

The authors wish to acknowledge the efforts of Mr. William Butters for his efforts in obtaining the high-speed photographs and to thank the MCAIR Materials and Processes Department personnel for their assistance in providing necessary support and technical assistance for the program.

REFERENCES

1. H. S. Schwartz and C. R. Austin, "Effects of Lightning Strikes on Fiber-Reinforced Structural Plastic Composites," Tech. Report AFML-TR-68-196, September 1968.

2. L. G. Kelly and H. S. Schwartz, "Investigation of Lightning Strike Damage to Epoxy Laminates Reinforced with Boron and High Modulus Graphite Fibers," Lightning and Static Electricity Conference, 3-5 December 1968, in Tech. Report AFAL-TR-68-290, Part II, p 485.

3. J. D. Robb, "Mechanisms of Lightning Damage to Composite Materials," AFAL-TR-68-290, Part II, December 1970, p 520.

4. FAA Advisory Circular No. 20-53, "Protection of Aircraft Fuel Systems Against Lightning," October, 1967.

5. G. Lubin, "Effect of Lightning Strikes on Boron-Epoxy Single Skin and Honeycomb Sandwich Panels," Grumman Aerospace Corp. Report No. ADR-02-06-70.3, December 1970.

6. MIL-B-5087B, "Bonding, Electrical, and Lightning Protection, For Aerospace Systems," 6 February 1968.

7. MIL-A-9094D, "Arrester, Lightning, General Specification For," September 1958.

8. MIL-F-38363A (USAF), "Fuel System, Aircraft, Design, Performance, Installation, Testing, and Data Requirements, September 1966.

9. Air Force Systems Command Design Handbook DH1-4, DN 7A1 First Edition, 10 January 1969.

10. A. P. Penton, J. L. Perry and K. Lloyd, "The Effects of High Intensity Electrical Currents on Advanced Composite Materials," Philco Ford Corp. Report U-4866, September 1970.

11. D. W. Clifford and E. H. Schulte, "A Lightning Simulation Facility for Aerospace Testing," Proceedings of 10th National Conference on Environmental Effects on Aircraft and Propulsion Systems, 4 May 1971.

12. J. Quinlivan, C. Kuo, R. Brick, "Coatings for Lightning Protection of Structural Reinforced Plastics," March 1971, AFML-TR-70-303, Part I.

Today's conductive plastics combine shielding plus strength

Novel fillers and conductive combos deliver EMI shielding plus improved properties

By Robert H. Wehrenberg II, Senior Editor

Electromagnetic or radio-frequency interference is all around us. We can't escape it, indoors or out, and neither can senstitive electronic equipment.

Computers, business machines, communications equipment, and even onboard microprocessors that control today's auto engines can all malfunction because of electromagnetic, radio-frequency interference, or EMI/RFI. Many of these devices are housed in plastics, which normally provide no shielding from electromagnetic pollution.

Fortunately, conductive fillers can be added to these plastics to transform them from mere electrical insulators to functional conductors, creating new materials that can be used to shield electronics from electromagnetic pollution. The electrically conductive additives allow plastics housings to reflect and absorb electromagnetic energy, thus combining the advantages of plastics with the shielding ability of metals.

Current research in conductive-plastics technology is aimed at determining the right type and amount of conductive filler required to shield against EMI/RFI without sacrificing part performance — particularly mechanical properties. Perhaps more important is the need to provide shielding at a cost comparable to or less than conventional methods, which include conductive coatings, metallization, and plating.

Conductive plastics is still a new technology. One look at the number of systems now under development for EMI shielding confirms it. While research organizations, material suppliers, molders, and even OEM manufacturers still continue their efforts to develop better conductive plastics, some systems have already received the "go ahead" for production.

Conductive composites

Reinforced conductive plastics offer a number of advantages over coated alternatives. Built-in shielding, for instance, cuts component cost by eliminating secondary operations involved with coatings, and at the same time produces an EMI shield that's less prone to damage. Best of all, the fabrication of complex conductive parts lends itself readily to a variety of plastics processes including injection molding, compression molding of BMC/SMC, structural foam molding, and thermoforming of sheet.

Conductive composites are finding their niche in business equipment. Here, compression-molded SMC (sheet molding compound) parts provide EMI shielding while maintaining much-needed mechanical properties. Conductive SMC is now appearing in a number of such applications.

For example, the base housing of a Xerox computer printer incorporates both carbon-fiber mat and conductive carbon pigments in a 22% glass-fiber-reinforced SMC. Molded by Zehrco Plastics of Ashtabula, Ohio, the conductive base delivers a resistivity of 1 ohm-cm. Compared to nonconductive SMC, the company reports both higher tensile and flexural properties for the flame-retardant part. At the SPI 1982 Reinforced Plastics/Composites Institute, the part won first prize in the business-equipment category.

Conductive pigment alone couldn't provide satisfactory shielding, says Jim McDarment, vice president and technical director at Zehrco. Loadings of 5 to 7% or greater would be needed to approach an effective shielding conductivity, but unfortunately, such levels of conductive filler would have an extremely thixotropic effect on the compound. Adding the fibrous form of carbon results in better circuit continuity so that the conductive path does not solely rely on the point-to-point contact of a conductive pigment.

To date, the company has molded about 400 of the base units for Xerox, and expects to be producing conductive bases for new models. According to McDarment, the company is also exploring the possibility of molding a prototype, conductive SMC hood for Chevrolet's Camaro.

Five fillers perform

Soon expected to be in full production is another conductive SMC part, the middle support base for the new Xerox intelligent

Reprinted with permission from Materials Engineering, Vol. 95, March 1982, 37-43, © 1982 Penton/IPC

Injection molded of polycarbonate, a prototype cover for a computer disc drive achieves conductivity and built-in shielding by incorporating Transmet aluminum flakes.

terminal. Molded by Premix of North Kingsville, Ohio, the 4-lb part (1.8 Kg) provides molded-in EMI shielding by incorporating five separate conductive additives. "No one conductive ingredient added to a conventional business-equipment SMC is going to provide the necessary shielding characteristics," says Frank Bradish, vice president and technical director. Both carbon black and carbon-fiber mat are part of the conductive system, while "other ingredients synergistically combine to provide very subtle shielding characteristics as well as complement other critical properties," says Bradish.

Why polyester SMC? Primarily because the materials can be tailored to fit shielding characteristics of each specific application. "There's no sense in paying for more shielding than necessary," says Bradish. In addition, SMCs can easily be made to meet other property requirements, such as flammability ratings, shrinkage, dimensional control, and cosmetic appearance.

Although only about 60 of the support bases have been produced so far, the unit will soon be a full-production item — 8 to 10,000 pieces per year. Another similar application is already in the works, says Bradish. This one is a 20-lb part (9.1 Kg) for Data Products Inc. Initial production of 500 high-speed computer printer bases is underway. Ultimate production is slated for 30,000 parts per year.

Make way for mat

Conductive nonwoven fiber mats work well in SMC. Mats made of carbon fibers, aluminized glass fibers, or blends of conductive fibers with nonconductive glass fibers can be used. By putting conductive fibers in mat form on the surface, lower amounts of conductive additives can be used since the fibers don't have to be uniformly distributed throughout the material to achieve shielding.

One advantage of using the mat form is that the shielding levels can be easily adjusted by varying the mat weight. And compared to conductive fibers dispersed throughout an SMC, mat forms of conductive fibers require significantly lower levels.

Although research to date at International Paper Co. has emphasized carbon-fiber mats, says Jim Quick, the company is also doing work with aluminized glass-fiber mats. Both types are available in developmental quantities. Successful applications have also included hybrid mats. Work is now in progress, says Quick, to demonstrate the capabilities of conductive mat in other molding processes.

While most applications for conductive mat are in business equipment, at least one automotive application exists. It's a limited production hood scoop for the 1982 Mustang. Here, a carbon-fiber mat imparts EMI shielding properties to an SMC hood scoop. Without provision for shielding, electromagnetic radiation from the engine, distributor and spark plugs could reach the radio antenna and affect the reception. Quick points out that demanding business-machine applications require higher levels of shielding. While a business machine may require 70 gm/m² of carbon fiber to be effective, the auto scoop succeeds with about 3 gm/m² or only minimal shielding. Some shielding applications will adopt the conductive-filler mats, but certain applications will still use conductive paints, plating, and even small metal enclosures around individual components, Quick comments.

Metallized-glass-fiber mat from Lundy Technical Center provides built-in EMI shielding in Diacon System desktop computer for the medical industry. The 3-piece cabinet is compression molded of polyester using one layer of 1-oz conductive mat and multiple layers of 2-oz glass mat.

Glass mat costs less

Conductive mats made of aluminized glass fibers are being considered for use in SMC, primarily because of lower cost. "In some cases, cost savings over carbon are greater than 50%," says Dennis Young of Lundy Technical Center. "Our Rohmoglas material can provide adequate shielding for approximately 30 cents per square foot."

Conductive glass mat is being used commercially, too. Made for Diacon Systems Inc., a desktop computer cabinet for the medical industry incorporates one layer of 1-oz metallized mat to provide higher shielding than a conductive coating, according to Young. Another conductive cabinet also uses metallized glass. The

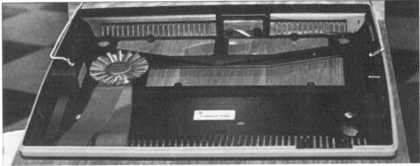

Carbon-fiber mat and carbon pigment provide EMI shielding for this 8.2-lb base for a Xerox computer printer. The 22% glass-fiber-reinforced SMC is compression molded by Zehrco Plastics and delivers 54, 40, and 51 dB attenuation at 3, 100, and 1000 MHz, respectively.

EMI shielding for onboard microprocessors in automobiles can be provided by conductive plastics. Prototype injection-molded polycarbonate housing from Wilson-Fiberfil is reinforced with aluminized glass fibers.

While SMCs are capturing most of the shielding applications that currently exist for conductive plastics, development is continuing in bulk molding compounds (BMCs). So far only Premix Inc. offers conductive BMC for shielding. Typical shielding performance is 20 to 30 dB, and the better the shielding, the higher the price. According to the company, compounds can be formulated to meet specific shielding needs.

American Cyanamid and Fiberite are both continuing research in BMCs. A variety of fillers have been tried, but in BMCs, where strength is not really a consideration, flake reinforcements may be the best bet. Adding pure-aluminum flake results in more conductivity than aluminized glass fibers on a pound-for-pound basis, says MBAssociates' Don Davenport. While aluminum-coated glass can produce some strengthening (as in SMC), it's not really necessary in BMC. Here pure aluminum ribbon or flake has an advantage in conductivity.

New fillers hold the key
New types of conductive fillers may hold the key to better performing conductive plastics. New conductive fillers are now being evaluated in injection-molding compounds to provide consistent shielding at lower loadings, maintain properties in the finished part, and at the same time be economical.

Just introduced by Bekaert Steel Wire Corp., Pittsburgh, are stainless steel fibers, drawn from 316 L stainless rod, that promise EMI shielding performance at lower loadings than other conductive fillers. According to Pierre Veys, product manager, loadings of 4 to 6% (1% by volume) are sufficient to impart the necessary conductivity for shielding without adversely affecting mechanical properties. The company claims that the low

new commercial application is in production at Roller Reinforced Plastics.

While the best material for EMI shielding is the mat form with less than 1-ohm resistance across a 12-in. span (0.3 m), Lundy also supplies aluminized glass in other forms, including roving, milled and chopped fibers, as well as woven fabrics.

Aluminized glass fibers are economical enough to be used as chopped fibers in SMC, says Don Davenport of MBAssociates. Because they are more conductive than carbon fiber, concentrations of 5 to 7% of Metafil G fibers can provide shielding at lower cost than carbon fiber. The mat form of aluminized glass is not really necessary, says Davenport.

Woven fabrics conduct, too
Woven into fabrics, aluminum-coated glass fibers have been used to shield aircraft from EMI. E-glass fibers, chemically bonded with a coating of pure aluminum by MBAssociates, are woven into fabric by Hexcel, Corp., Dublin, California. For EMI shielding in aircraft, the conductive fabric (called Thorstrand) is used in wing-to-body fairings, wing coverings, wing-tip coverings and other areas where both EMI shielding and static bleed are necessary to protect sensitive instrumentation. The woven conductor usually forms the outside layer in a typical aircraft composite structure, which is laid up, then vacuum bagged.

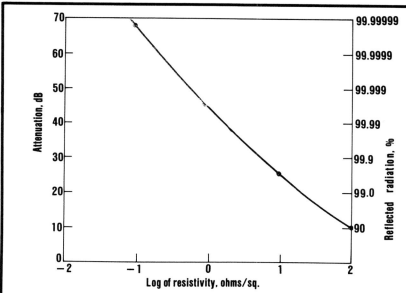

Higher surface resistivity produces better EMI shielding. A conductive plastics system rated at 40 dB would reflect 99.99% of incident electromagnetic radiation. Conductive coatings have higher shielding values, says Wilson-Fiberfil's Steve Gerteisen, because they have lower surface resistivity (higher conductivity). For example, conductive coatings range from 10^{-2} to 10 ohm/sq, while conductive plastics have resistivities that range from 10 to 10^6 ohm/sq. In conductive plastics, however, surface-resistivity values can be erroneously low when a resin-rich surface layer exists.

Zinc arc spray can provide a plastic part with 70 dB shielding, but other conductive coatings produce less. In contrast, shielding values for today's conductive, injection-moldable plastics, says Gerteisen, fall in the 20 to 40 dB range. Those in the 30 to 40 dB category could perform effectively in 95% of today's shielding applications.

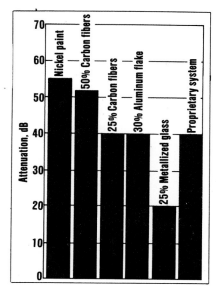

Shielding effectiveness of a new, proprietary, conductive plastic is compared with other conductive systems at 1000 MHz. Developed by Wilson-Fiberfil, the injection-moldable polycarbonate contains metal fibers.

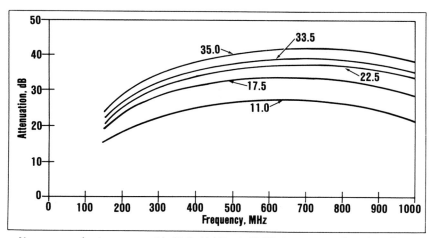

Nonwoven conductive mats provide EMI shielding in SMC. Attenuation values depend on the carbon-fiber content, which is indicated on the graph in gm/m^2. The conductive mats, made by International Paper Co., are 50/50 hybrids of carbon fibers and glass fibers.

concentration has no effect on molded-in color, surface texture, or chemical additives.

Both continuous strands and chopped fibers are available in diameters down to 2 microns. The high-aspect-ratio fibers re-flect energy, says Veys, but unlike many other conductive fillers, also absorb significant electromagnetic radiation by acting as dipole antennas. The high-conductivity fibers are being evaluated by a number of companies, and to date, one new propri-etary, injection-molding compound is based in part on this conductive filler. Mat forms of the fibers, now used in filtration, may soon be available.

Electroplated carbon fibers are yet another alternative to standard conductive fillers. Basically, a metallic coating, such as silver, nickel, copper, or brass, is electroplated onto graphite fibers. With an aspect ratio of 50 to 1, nickel-coated fibers yield electrical properties equal to nickel, but have much lower specific gravity. Silver-plated fibers produce similar results.

In research studies recently conducted at Fiberite, Steve Kidd reports nickel-coated fibers with conductivities 100 times greater than graphite fibers. Although the plated-carbon fibers are currently proprietary, they are expected to become commercially available in six to nine months.

Rapidly solidified metals in flake form are produced by Transmet Inc., which has just gone into full commercial production of aluminum flake. The rapid-solidification process quenches the molten metal so rapidly that it changes the surface structure, leaving passive flakes that do not oxidize as a standard rolled sheet does. The company has introduced a surface-treated series that improves adhesion between the flakes and plastic matrices.

In addition to the standard, uncoated

Comparison of injection-moldable polycarbonates for EMI shielding

Property	Unfilled	Proprietary system[a]	25% Carbon fiber	25% Metallized glass	30% Al flake
Tensile str., psi	9,000	8,000	20,000	12,000	6,000
Elong., %	6—8	7	1.1	2.0	2.9
Izod impact, ft-lb./in.	2.4	1.7	2.0	1.6	2.0
Flex. str., psi	13,000	13,500	28,000	21,000	11,000
Flex. mod., 10^5 psi	3.3	3.3	20.0	9.8	6.4
DTUL[b], °F	270	288	295	290	285
Specific gravity	1.20	1.27	1.31	1.40	1.44
Attenuation at 1.0 GHz, dB	0	40	30	20	30

a. Conductive additives include stainless steel fibers. b. At 264 psi. Data, courtesy of Wilson-Fiberfil International.

BMC and SMC for EMI shielding[a]

Property	BMC	SMC
Tensile str., psi	4,500	7,000
Flexural str., psi	12,000	14,000
Flex. mod., 10^6 psi	1.4	1.5
Notched Izod, ft-lb/in.	5.0	8.0
Comp. str., psi	18,000	20,000
HDT @ 264 psi, °F	400+	395
Specific gravity	1.80	1.75
Shrinkage, mils/in.	1.0	1.0
UL 94 flammability[b]	Pass	Pass
Surf. res., ohm/sq	1–2K	1K
Shielding perf.[c], dB	20–30	30–40

a. Data shown for Premi-Glas compounds from Premix, Inc. b. At 0.100 in. thick. c. Insertion loss attenuation at 1 Hz to 1 GHz (plane wave).

flake, nine coated flakes are now commercial, all of which are treated with silanes, wetting agents, or titanates to provide better properties in specific polymer systems. According to Bob Simon, vice president of the Columbus company, a new flake material, based on a change in the alloy itself, will be released within a few months.

Injection moldables: on the way?
Injection-moldable plastics for EMI shielding are available. There are high-priced carbon-fiber-reinforced versions or lower-priced plastics that use a less expensive conductive filler. Unfortunately, the high loadings (typically 25 to 40%, depending on type of filler) with the less costly types of conductive additives such as aluminized glass, carbon powders, and metallic pigments often result in significant drops in important mechanical properties. So, development in injection-moldable conductive composites is aimed at keeping cost down and performance up.

Where both high strength and shielding are necessary, thermoplastics reinforced with 30 to 50% carbon fibers by weight are commercially available from a number of materials suppliers, such as Fiberite, LNP Corp., and Wilson-Fiberfil. A wide variety of plastic matrices is available, including nylon, polycarbonate, polypropylene, PBT (polybutylene terephthalate) ABS, and most recently, PEEK (polyether-etherketone), according to Wilson-Fiberfil's Steve Gerteisen. But for many applications these materials are uneconomical.

High-performance molding compounds from Wilson-Fiberfil are carbon-fiber-reinforced plastics, both pitch and PAN types. The company also offers compounds that incorporate aluminum flake and metallized glass fiber, which comes strictly in a long-fiber version.

"Our best material for EMI shielding is a 30% carbon-fiber-reinforced polycarbonate," says LNP Corporation's Dick Burns. According to the company, customers are buying carbon-fiber compounds for shielding components that are relatively small in size. LNP is compounding several prototype compounds for evaluation as well as testing new conductive fillers.

"What we need," says Burns, "is a filler system that will provide proper impact resistance as well as shielding capabilities, and still be reasonably priced." Burns sees conductive nylon and polycarbonate as the two prime materials, because, when reinforced, they provide the best impact strength. Presently, injection-moldable plastics are still below an acceptable minimum impact resistance for housings. The company is continuing research in this area and is now evaluating new conductive additives. Presently under evaluation are

steel fibers, which when used as continuous rovings, should require lower loadings than chopped steel fibers.

Fiberite sells a 40% graphite-fiber material that has been used in EMI applications. Attenuation is about 35 dB or about 10 to 20 dB less than a plated thermoplastic. "We're working toward an injection-moldable thermoplastic that will provide shielding in the 55 to 60 dB range," says Steve Kidd. "A 40% graphite-fiber plastic is around 40 dB. Today, graphite fibers seem to be best."

Fiberite conductive materials are being used primarily in computer parts. Most activity is in conductive polypropylene and nylon 6 materials, according to Hugh Miller. A new conductive plastic, designated as a high-electrical-conductivity material, is a special carbon-filled material having a volume resistivity of 0.2 ohm-cm. "Carbon is involved," says Miller, "but it's put through a special process."

Another division of the company has done research with conductive phenolics. A two-stage phenolic compound, says Larry Rupprecht, has been granted an experimental number, but is not yet commercial. The company plans to develop a variety of compounds with various strengths and attenuation characteristics. The experimental phenolic, which can be easily compression or injection molded, is currently undergoing molding trials. Re-

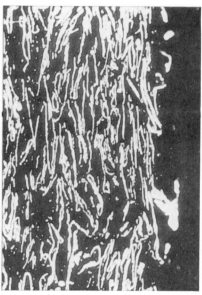

Cross-section of an injection-molded part shows typical distribution and orientation of aluminum flakes. Photo, courtesy Transmet Corp.

Aluminum-coated glass fibers from MBAssociates are woven and resin-impregnated by Hexcel Corp. the resulting conductive fabric is used in EMI-shielding applications for aircraft.

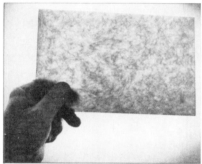

New conductive fillers such as stainless steel fibers form a conductive network within a clear plastic. According to Bekaert Steel Wire Corp., the fibers provide shielding at much lower loadings than other conductive additives.

search is continuing in both epoxy and BMC.

Injection-moldable systems based on new types of conductive fillers may be the economical answer to stronger conductive composites. For example, a new proprietary polycarbonate (PC), developed by Wilson-Fiberfil, uses stainless-steel fibers as part of the conductive network. The compound achieves 40 dB of attenuation at 1000 MHz or 99.99% signal loss. Compared to a 25% carbon-fiber reinforced PC, shielding is about 10 dB better.

The biggest advantage of the material is that the conductive system does not significantly modify the properties of the base resin as do other conductive agents such as carbon blacks, aluminum flakes, and metallized glass fibers. Both impact strength and ductility are maintained in the conductive PC, which is available in developmental quantities.

Both aluminized glass fibers and aluminum flake can easily provide conductivity for effective shielding. Keeping impact strength and other critical properties at required levels is another matter, however. One approach to maintaining properties and conductivity is to use surface modifiers to enhance adhesion between the polymer matrix and the conductive filler. But so far, results from material compounders show only moderate gains in impact resistance.

Longer conductive fibers help increase strength and improve conductivity. With both carbon and aluminized glass fibers, the longer the fiber in the molded part, the better the conductivity. Wilson-Fiberfil, for instance, has reported a 10 dB increase in attenuation with the use of long aluminized-glass fibers.

Complementary filler combos

Already proven successful in SMC applications, the use of synergistic combinations of conductive agents is yet another way to improve injection-molding compounds. At Wilson-Fiberfil, Gerteisen

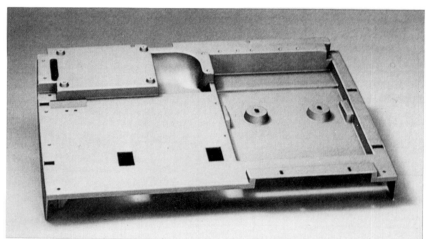

Molded of Premi-Glas conductive SMC, the middle support base for the Xerox 515 Small Business Computer is compression molded in matched-metal dies. The conductive part not only shields the "intelligent terminal" from EMI, but also provides excellent static discharge.

sees the possibility of cost savings by using two different types of fillers, for instance, aluminum flakes and carbon black or a combination of metallized glass fibers with carbon fibers. At Fiberite, combinations of carbon black, nickel powders, and carbon fibers are being investigated. Other suppliers are exploring the same route.

"For a pigment or filler to be suitable in a molding compound," says McDarment of Zehrco Plastics, "each particle has to be discretely wetted by the resin. You're wetting out a conductive material with a non-conductor. It's almost self-defeating — the better the wetout, the better the molding, but conductivity is worse. What you have is a compromise, and in SMC, the fibrous form of carbon makes up for the conductive pigment."

First of a whole family of conductive plastics to be commercialized by Diamond Shamrock is a PVC compound. Based on aluminum flakes, the conductive compound provides from 35 to 50 dB of shielding, depending on the design of the injection-molded part. The UL 94 V-0 material has a volume resistivity of 0.04 ohm-cm. Another new compound made conductive with aluminum flakes is an ABS/polycarbonate alloy from Mobay Chemical.

Developed overseas is a conductive polypropylene, which is being tested in video-camera housings as well as by auto companies to shield their microprocessors, reports Transmet's Bob Simon. Meanwhile, Japanese companies are planning to introduce conductive versions of ABS in about six months. Bayer AG in Germany has developed conductive versions of nylon 6/6 and ABS/polycarbonate.

Why not structural foam?

Research continues to develop conductive structural foams for business machine applications. Wilson-Fiberfil, for instance, has been exploring the use of structural foam to make conductive cases for personal computers. According to Don Davenport, MBAssociates' aluminized glass fibers have been compounded by Fiberfil and molded into some "very interesting" parts. While no one is in commercial production at this time, everybody is evaluating them, he says.

"Shielding is not much different in structural foam than it is with solid material," says Davenport. At first, foam was thought to be easier to shield, but actually, 25 to 30% conductive filler (by weight) is still needed for shielding.

Aluminum flake doesn't seem to work very well in structural foam. Bubbles in structural foam usually nucleate on one or two percent of milled glass or talc, Bob Simon of Transmet explains. But in the case of a flake-modified material, bubbles form on the flakes and blow the flakes apart. The end result is no conductive path for shielding.

Results of earlier work on structural foam — sponsored by the Army Material and Mechanics Research Center — show that carbon/graphite fibers can provide 40 dB and 20 dB of shielding at 1000 and 10 MHz, respectively. Feasibility was demonstrated, but shielding effectiveness was not high enough for military applications.

According to one material supplier, if carbon/graphite fiber prices had declined as far as predicted a few years ago, carbon-fiber-reinforced foam would be a cost-effective shielding material today.

Seeking conductive sheet

Conductive polypropylene sheet for thermoforming has been researched in the past, but now a German company is compounding aluminum flake into ABS sheet. Since the conductive sheet will be a little rough, the company is planning on laminating and coextruding it with nonfilled ABS to produce a multilayer product. One possibility is a three-layer sandwich, with the center layer conductive. Resulting thermoformed parts should then have a nice class-A surface and built-in conductivity.

While conductive RIM (reaction-injection-molded) parts have been talked about for shielding applications, there's essentially no development in progress. One of the reasons may be the fact that similar to injection molding, fiber lengths breakdown during processing. "In the work we've done, ⅛-in. fibers (3.2 mm) may enter the mixing chambers, but what actually comes out are very short fibers," says Davenport. While a few fibers may be ⅛-in. long (3.2 mm), most are probably down below 50 mils (1.3 mm). Therefore, it takes a high concentration, probably up to 40% to achieve shielding. ∎

SECTION V
Drilling

NEW DEVELOPMENTS IN DRILLING AND CONTOURING COMPOSITES CONTAINING KEVLAR® ARAMID FIBER

by
Prof. Dr.-Ing. Dr. h.c. W. König, FHG, Aachen
P. Grass, FHG, Aachen
A. Heintze, Du Pont Co., Geneva
F. Okcu, MBB-UT, Hamburg
Cl. Schmitz-Justen, FHG, Aachen

Subjects of this paper are research and development activities on drilling, countersinking, and contouring of aramid fiber reinforced laminates. It covers results of experimental investigations as well as the technological principles realized in the respective tools. In this regard material properties essentially related to machinability are also being illustrated.

Machining quality and tool life attainable by use of optimized tools and the required preconditions are presented, and cutting data recommendations are given for these tools. A comparative process study covering routing and water jet cutting eventually points out the respective specific and advantageous ranges of application.

INTRODUCTION

Emerging from the aerospace and automotive industry, lightweight design and engineering has gained an increasing importance in the recent past. The prime objective of these efforts is to improve the ratio of performance and weight. An essential contribution has been made by the use of fiber reinforced plastics due to their characteristic properties such as high specific tensile strength and modulus. While these materials were mainly used in fairing parts at the beginning of their application, the introduction of carbon and aramid fibers allowed the design of primary structures as they are common on recently developed aircraft. The function of these parts consequently requires higher quality in the mechanical machining of such materials.

BASIC ASPECTS OF MACHINING FIBER REINFORCED PLASTICS

The machining of fiber reinforced plastics implies requirements and conditions essentially different from those of metal cutting:

— FRPs are machinable only in a limited range of temperature. Even though the fibers will withstand higher temperatures, the curing temperature should not be exceeded to avoid material disintegration.

— The low thermal conductivity favors heat build-up in the cutting zone during machining operations, since there is only little dissipation into the material. The greatest part of the heat has to be carried away through the tool.

— The difference in coefficients of thermal expansion between matrix (highly positive) and fiber (slightly negative for carbon and aramid) is favorable for the formation of residual stress and makes it difficult to attain high dimensional accuracy. Drilled holes, for example, later often show a smaller diameter than the drill used.

— Reinforced fibers and fillings may cause a highly abrasive tool wear which restricts the selection of utilizable cutting material.

— The change of physical properties by absorption of fluids is only insufficiently known, this has to be considered when deciding to use a coolant.

The mechanical machinability of FRPs is essentially determined by the individual characteristics of the respective fiber type.

Figure 1 gives a general account of some physical properties for glass, carbon, and aramid fibers and shows the typical failure mode of a filament. Different values of the properties covered are given by some authors, yet the table is representative for commercially available fibers. The brittle fracture of the inorganic glass and carbon fiber is clearly identified as compared to the ductile failure of the organic aramid fiber. The axial splitting of the fiber is typical for aramid, caused by a weak molecular bond transverse to the fiber axis which allows relative motion of molecule chains in sliding planes. This also accounts for the relatively low loading capacity in compression of the aramid fiber, where buckling of the fibrils and consequently a contraction of the fiber occurs. In contrast, carbon fibers which are also anisotropic show a brittle fracture as they consist of axially oriented, intertwisted graphite layer tapes, which lack a comparable sliding mechanism.

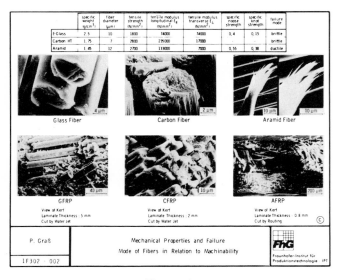

	specific weight (g/cm³)	fiber diameter (μm)	tensile strength (N/mm²)	tensile modulus longitudinal $E_\parallel$ (N/mm²)	tensile modulus transverse $E_\perp$ (N/mm²)	specific noose strength	specific knot strength	failure mode
E-Glass	2.5	10	1800	74000	74000	0,4	0,15	brittle
Carbon HT	1.75	7	2800	235000	17000			brittle
Aramid	1.45	12	2700	133000	7000	0,55	0,38	ductile

Glass Fiber   Carbon Fiber   Aramid Fiber

GFRP   CFRP   AFRP

View of Kerf   Laminate Thickness : 5 mm   Cut by Water Jet   View of Kerf   Laminate Thickness : 2 mm   Cut by Water Jet   View of Kerf   Laminate Thickness : 0.8 mm   Cut by Routing

P. Graß	Mechanical Properties and Failure Mode of Fibers in Relation to Machinability	FhG Fraunhofer-Institut für Produktionstechnologie · IPT
1 F 302 · 002		

Figure 1

The features of the individual filament are immediately reflected by the problems in machining reinforced material as the photographs in Figure 1 show.

In this connection it also becomes evident that the special tool requirements in machining of materials containing aramid fibers such as KEVLAR® 49 are different from those concerning glass or carbon fibers.

Due to its relatively low compressive strength this fiber has a tendency to recede within the matrix in the machining process instead of being sheared off. In this regard the influence of the laminate structure, the orientation of fibers and fiber layers, the volumetric fiber percentage and the filament processing is of great significance for the attainable machining quality and tool life.

Basically, laminates made out of angular layers are easier to machine than unidirectional ones, and plain weaves easier than satin. Even within the individual weaves differences are found: dense fabrics generally give better qualities than loose ones.

Additionally the resin type does affect the machinability of laminates. Generally, epoxy is favorable compared to phenolic resin of which the use has grown in the recent past. Furthermore, it has been noticed that higher curing temperatures seem to be advantageous in terms of machinability. This, however, has not yet been investigated in detail and systematically.

Special problems arise when machining of hybrid composites is required that consist of glass or carbon in addition to aramid fibers. The specific properties of the individual fibers have to be taken into account to obtain an acceptable tool life, especially the highly abrasive effect of glass and carbon fibers. Most important, however, in terms of machining quality on hybrids is the top layer material. At these layers, quality problems are faced in terms of delamination and burrs. Further difficulties are caused by the melting of resin, burn patterns, and — particularly with plastics reinforced with KEVLAR® 49 aramid fiber (AFRP) — the fuzzing of the cut surface.

From the previously described material properties of aramid fibers, certain requirements for the cutting tool geometry can be derived. To attain high machining quality and especially to avoid fuzzing, the cutting process has to proceed in such a way that the fibers are being preloaded by tensile stress and then cut in a shearing action. For a rotating tool this means that fibers have to be pulled from the outside diameter towards the center. A keen cutting edge and a comparatively high cutting speed are necessary to put this motion into effect and to avoid the receding of the fiber into the matrix. These aspects as well as the highly abrasive effect especially of hybrid laminates have to be considered in the selection of cutting material.

Even on unhybridized AFRP, high-speed steel tools are unsatisfactory in terms of tool life due to excessive wear. Better results are given by the application of coated HSS tools. Yet these physical vapor deposition coatings — TiN is mostly used — result in a rounded shape of the cutting edge. Nevertheless it is very important to say that present results concerning coated high-speed steel tools must not be generalized. This special coating technique has only recently been developed and efforts are made to further improve layer composition and processing.

As for tool life, carbide tools clearly are superior, especially if carbide grades of fine grain size and thus high cutting edge stability are used. Tool cost is considerably higher though, and more expensive equipment is needed such as diamond wheels for regrinding.

Polycrystalline diamond tools (PCD) which are extensively used for machining of glass and carbon laminates due to their extremely high wear resistance are used for AFRP in special cases only, as the manufacturing of the required cutting edge geometry is very difficult or even technically impossible. Furthermore, tool cost is about ten times that of a carbide tool.

Several research and development programs were carried out by the Fraunhofer-Institut für Produktionstechnologie (IPT), Aachen, West Germany, to study the problems presented above and to find feasible solutions. The activities were jointly sponsored by Du Pont de Nemours International SA, Geneva, Switzerland, and MBB – Unternehmensbereich Transport – und Verkehrsflugzeuge, Hamburg, FRG. The objective of the projects was a reproduceable improvement of machining quality on aramid and hybrid laminates for drilling, countersinking and routing operations. Furthermore, tool life was to be increased by application of appropriate cutting material and cutting data.

RESULTS IN DRILLING

Drilling and countersinking tests were performed on the same material. Chosen was a flat 17-layer A/C hybrid laminate with epoxy matrix. The sample thickness was 3.2 mm. Both top layers were KEVLAR®, style 120. At the tool exit side a TEDLAR® protective foil was used additionally. The sample plates were drilled without backing; neither a drill press nor any coolants were applied.

The machining tests were based on the following quality criteria emerging from production requirements:
— grade of delamination at tool entry and exit,
— fuzzing at the outside diameter,
— burrs,
— roughness of the drilled hole wall or countersunk cone surface.

The investigation was based on a compilation of commercially available tools. These were partly in production use and partly in a prototype stage.

Beyond the cutting material, tools can be distinguished by their geometrical characteristics which also lay down the principle of function. Figure 2 shows the drilling quality for the respective tool designs as well as the recommended cutting data.

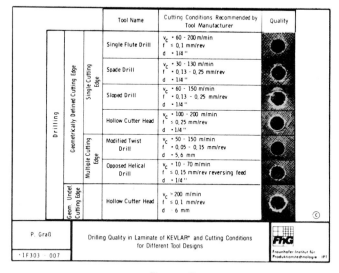

Figure 2

A very significant criterion is whether the cutting edge is geometrically defined or undefined. The former tools have one or more cutting edges to perform the cutting process. For each of those the angles are exactly fixed. Among the single-edged tools, the design can be differentiated by shape and position of the edge and by the way of chip removal. The multi-edged drilling tools are mainly represented by twist drills and moreover by straight flute designs. Serrated cutting edges are found on drills as well as on countersinks.

When using abrasive tools with geometrically undefined cutting edges, a great number of separate edges is constantly in contact with the workpiece. The angles, however, follow a statistical distribution for the individual edges.

Within the machining program all commercially available tools were tested under the cutting conditions recommended by the respective manufacturers. A comparative analysis of the machining results showed substantial quality differences. It turned out that part of the tools did not fulfill the mentioned geometrical requirements and thus are not suitable for machining the given composite reinforced with KEVLAR.® Abrasive tools with geometrically undefined cutting edges in particular gave a bad performance due to the mainly negative rake angle of the individual cutting particles and the very small chip space between the grains. Intensive heat and smoke build-up was noticed when these tools were used. The drilled holes and countersinkings showed clear traces of overheating. After one machining operation, the tools were so heavily loaded with fiber particles and molten resin that extensive cleaning was necessary before any further use.

The cutting mechanism specifically necessary with AFRP as presented above requires a radially C-shaped cutting edge for drills and countersinks, with its tip at the outermost periphery which forms a positive radial rake angle (Figure 3). These tips also have the first axial cutting contact with the workpiece.

A centering point is necessary to obtain good drilling quality in a manual operation. If a drill press is used that provides a firm guidance, this is not necessary. Additionally it should be considered that the major part of the heat generated in the machining process has to be carried away through the tool, as the aramid fiber shows an extremely low thermal conductivity. This, as well as the changing percentage of abrasive carbon and glass fibers, requires a stable cutting edge. In any case, however, the nominal axial rake angle should be neutral at least, or better yet a positive angle to generate the least cutting heat possible. Good chip removal additionally is ensured, and clogging of the tool avoided. The positive rake angle also forms the keen cutting edge that is necessary to achieve a smooth cut without crushing or tearing out of individual fibers.

The twist drill and single flute drill design gave the best results of all tools tested and thus were selected for further optimization. By systematical variation of the rake angle, the relief angle and the cutting edge profile, the best machining quality obtainable was determined.

Figure 3 shows a drill prototype made by Klenk, Balzheim, West Germany, and developed by MBB, and the increase in quality at the tool entry compared to a conventional tool.

For the nominal rake or helix angle, a range of 29° to 35° turned out favorable; and for the relief angle, a range of 12° to 20°. Additionally the point clearance was reduced to avoid pushing away top layers at the tool exit.

The single flute drill was included in the tool life investigations without optimizing modifications. The drilling quality attainable by this tool is below that of the twist drill. The advantage is, however, that it can be reused much easier. It is only the plane flank that has to be reground, whereas regrinding of the twist drill requires a profiled diamond grinding wheel.

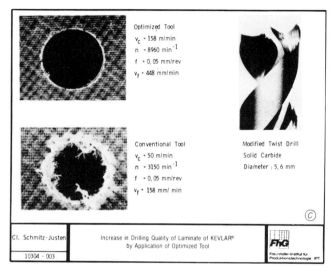

Figure 3

Both tools were made out of solid cemented carbide. This material is characterized by a high cutting edge strength and wear resistance up to a moderate cutting temperature. It consists of 92% WC, 6 to 8% Co and 0 to 2% TiC + TaC.

Meanwhile a tool similar to the single flute drill is available with a PCD bit, as the relatively simple edge geometry can also be manufactured in this material. A considerable extension of tool life is to be expected from this modification.

In addition to tool geometry and cutting material, the tool life and machining quality are substantially influenced by the cutting conditions. As far as drilling is concerned, these are cutting speed v_c and feed per revolution f. High drilling and countersinking qualities, however, will only be obtained within a limited range of these parameters. The upper limit of cutting speed is given by the risk of thermal damage primarily in the workpiece material. The lower limit is defined by a quality decrease based mainly on the receding of the individual fibers in front of the penetrating cutting edge. This is also favored by a low feed rate. Below a certain feed per revolution, the individual filaments are not cut any more and heat-intensive crushing occurs. Additionally the top layer at the tool exit is not cut completely – especially at a low cutting speed – so that caps are formed which cling strongly to the workpiece. With a rising feed the process forces also increase, as Figure 4 shows. Excessive forces will cause delamination at the tool entry and exit. The quality limit acceptable is specific and different with each material. Generally, however, it can be assumed at a feed rate of 0.1 to 0.15 mm per revolution and per cutting edge. For a double-edged tool, as depicted in Figure 4, this gives 0.2 to 0.3 mm/revolution as the limit. The major part of the tests was carried out using a controlled feed unit to ensure reproduceable conditions.

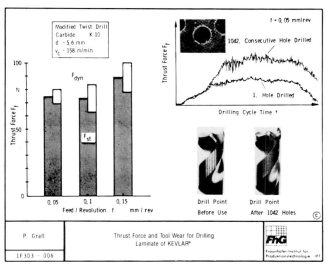

Figure 4

The right side of the figure shows a comparative diagram of the thrust force for the first and the 1042nd consecutive hole drilled, and the corresponding states of the cutting edge showing increasing wear. There is a distinct rise in the static and especially in the dynamic component of the thrust force which results in a slight quality decrease.

Tool life tests were performed using the Klenk modified twist drill originally developed by MBB. The cutting speed varied from 50 m/min to 150 m/min, the feed per revolution from 0.05 mm to 0.15 mm. As the end of tool life was not reached, the tests were terminated since the planned increase in tool life beyond the previously normal 100 drilled holes was exceeded by far. After drilling 1270 holes (a drilling length of over 4000 mm) at 160 m/min cutting speed

($n = 8960$ min^{-1}) and 0.05 mm feed per revolution, the tool showed a flank wear of only 0.15 mm.

The single flute drill (Starlite, Rosemont, Pa.) produced over 700 holes under optimized cutting conditions (2240 mm drilling length) until a slight rounding of the point was noticed. An essential prerequisite for this extended tool life, however, is a rigid and vibration-free clamping of the workpiece.

Furthermore, it emerged from the investigations that only a controlled feed will give a continuously high machining quality over a large amount of workpieces and ensure the increase in tool life of the individual tools.

RESULTS IN COUNTERSINKING

The process mechanisms previously described for drilling operations are basically the same for countersinking. Yet an important additional machining quality criterion is the even and flawless cone surface.

To meet these requirements a sickle-shaped cutting edge is used here too, which pulls the individual fibers towards the center and thus avoids fuzzing especially at the outside perimeter. Various single- and multi-edged tools commercially available were included in preliminary tests. Serrated tools proved to be unsuited due to the large negative rake angle of the serrations. Tools with geometrically undefined cutting edges turned out to be equally useless as the heat generated causes a melting of the resin and thus a clogging of the tool. The technological requirements were widely fulfilled by a double-edged

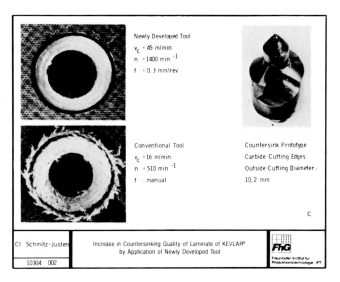

Figure 5

tool, as made for IPT by Gühring, Albstadt, West Germany (Figure 5).

To allow higher cutting speed the axial rake angle should be positive, the favorable range covers 6° to 15°. Carbide cutting bits of the same grade as for drilling tools were brazed onto a steel tool base. A pilot is used for a better concentric guiding of the countersink.

Due to the manufacturing process of the tool bits, the required C-shape is realized only in a limited range of the diameter. This means that qualitatively flawless countersinkings can only be produced close to the nominal tool diameter as Figure 6 shows by the successive stages of the countersinking operation.

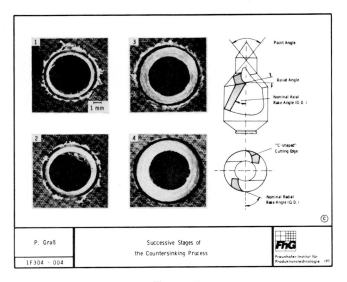

Figure 6

It is obvious that fuzzing at the outermost perimeter is only avoided for a positive radial rake angle. This restriction requires the application of a micro-stop in manual operations. Latest tool developments partly avoid this effect by a deeper offset of the cutting edge curve.

To obtain a smooth cone surface the feed has to be set relatively high. At f = 0.3 mm per revolution, more than 1500 countersinkings were made without a loss of quality whereas conventional tools achieved hardly more than 80 countersinkings per grind. The cutting speed was 45 m/min (n = 1400 min^{-1}). This is an increase compared to conventional speed of more than 400% at a similar quality.

RESULTS IN ROUTING

Routing tests were mainly carried out on an aramid/glass hybrid laminate with an epoxy matrix. Material thickness was 3 mm, both top layers were KEVLAR®, style 281. At the bottom side the material was backed by foil of TEDLAR® PVF film.

Preliminary tests showed that conventional tool geometries, as used in metal cutting as well as routers for GFRP or CFRP, do not give satisfactory results in terms of cut surface quality on aramid laminates.

Figure 7 shows different router types. Due to the specific properties of the KEVLAR fiber as previously explained, tools with unidirectionally twisted flutes (top left) produce fuzzing on that top layer of the material which is not backed by adjacent layers. This is caused by the unidirectional force parallel to the rotational axis of the tool.

A straight flute design (top right) at first gives high machining quality, yet after a very short cutting length intensive clogging of the tool is noticed resulting in a decrease of quality. From this it can be assumed that the compressed material causes a deflecting force upon the top layer fibers which leads to fuzzing.

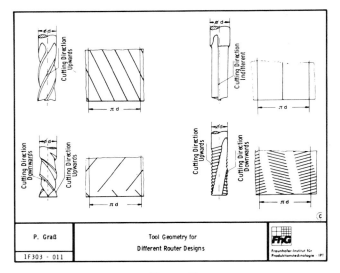

Figure 7

On the other hand, opposed helical tools either produce an alternating force across the entire cut surface (bottom right) or such forces that point towards the material from both top layers (bottom left). Both of these last types give a flawless machining quality even for considerably longer cutting lengths than the other tool designs.

The developed view on the right side shows the alternating orientation of the cutting edges which draw the fibers up and down and thus ensure a backed cut of the top layer.

As for the other tool design (bottom left) the top layer fibers on both sides are permanently being drawn towards the center of the workpiece which results in lower dynamic process forces. The application of this tool, however, is limited to mechanized or guided operations on thick laminates or sandwich structures, as the narrow working zone requires a very delicate positioning of the tool against the workpiece. Due to this restriction, the further study and tests focused upon opposed helical routers as shown on the bottom right side of the figure.

In machining tests the following phenomena were noticed which result in a quality decrease of the cut workpiece:

— clogging
— deposition of resolidified molten resin
— curling of fibers
— rounding of the cutting edge
— flank wear.

Clogging, resin deposition, and curling of fibers are no criteria which define the ultimate end of tool life, as it can be reused after more or less extensive cleaning.

From the routing tests it became obvious that the cut surface on the "upmilling side" generally shows a better quality independent of the way of workpiece clamping. This is likely to be caused by the increas-

ing flute congestion over one revolution of the tool which obstructs the cut on the "downmilling side".

Figure 8 shows the machining quality attained in tool life tests as well as the corresponding state of the tools after a certain distance of cutting. The tools were not cleaned in the course of the test. As for the commercial tool, intensive clogging and deposition of resolidified resin and burnt fibers in the flute were noticed which caused a fracture of the tool after a cutting length of only 4000 mm.

On the other hand the prototype router made by Dixi, Le Locle, Switzerland, attained a cutting length of more than 70 m until a slight quality decrease was perceived. A more favorable cutting edge geometry may account for this result and furthermore a flute surface which is significantly smoother and thus reduces friction as the cut material is carried away through the flute.

Figure 9 gives a general account of this tool design and specifies the geometry data for different tool samples. The prototype is marked as tool "D".

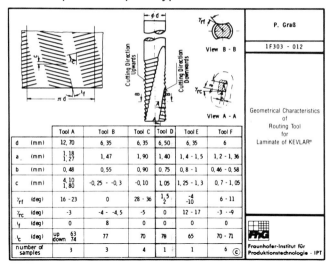

	Tool A	Tool B	Tool C	Tool D	Tool E	Tool F	P. Graß
							1F303 - 012
d (mm)	12,70	6,35	6,35	6,50	6,35	6	
a (mm)	1,18 / 1,27	1,47	1,90	1,40	1,4 - 1,5	1,2 - 1,36	
b (mm)	0,48	0,55	0,90	0,75	0,8 - 1	0,46 - 0,58	Geometrical Characteristics of Routing Tool for Laminate of KEVLAR®
c (mm)	4,10 / 1,80	-0,25 - -0,3	-0,10	1,05	1,25 - 1,3	0,7 - 1,05	
γ_{rf} (deg)	16 - 23	0	28 - 36	1,5 / 2	-4 / -10	6 - 11	
γ_{rc} (deg)	-3	-4 - -4,5	-5	0	12 - 17	-3 - -9	
l_f (deg)	0	8	0	0	0	0	
l_c (deg)	up 63 / down 74	77	70	78	65	70 - 71	Fraunhofer-Institut für Produktionstechnologie · IPT
number of samples	3	3	4	1	1	6	

Figure 9

Yet it happened that a second tool of the same type did not reproduce the former results in terms of quality and cutting length. As the dimensions of all tools included in the testing program were measured, it turned out that each tool of the same type showed a considerable deviation against the specified values. As, however, a slight variation of the tool geometry may already cause a major decrease in machining quality, the insufficient reproduceability in tool manufacturing represents a serious problem for their use in large-scale production.

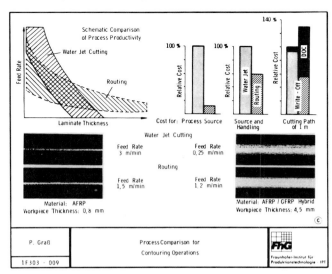

P. Graß	Process Comparison for Contouring Operations	FhG
1F303 - 009		Fraunhofer-Institut für Produktionstechnologie · IPT

Figure 8

In this regard it should be emphasized that a diamond wheel is necessary for grinding or regrinding carbide tools. This is the only way to obtain the required flawless and reproduceable shape of the cutting edge. In any case the grinding direction has to point towards the cutting edge to avoid chipping and thus a loss of keenness.

The workpiece was continuously clamped on one side in all tool life tests. For this setup a periodical short-time clogging of the tool and a corresponding local decrease of quality was noticed during the cutting operation. This process, however, is highly influenced by the way of clamping. Figure 10 shows the feed force F_v, the deflection force $\bar{F}$ and the axial force F_z as a function of four different clamping setups.

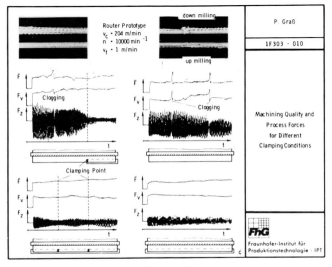

Figure 10

The force F_z, alternating normal to the workpiece plane, excites oscillations of the workpiece and especially of the scrap strip as it is cut off. For one-sided clamping (top right side) the scrap strip performs free bending oscillations. Repeated clogging of the tool occurs, accompanied by significant, short-time peaks of the feed force and the deflecting force as well as a decrease in quality (right photograph). As the tool ejects the congested material immediately after, the cutting process then continues unobstructed. The top left diagram of Figure 10 shows that one of the clogging cycles could be avoided by partial backing and clamping of the scrap strip. There is an essential decrease of the dynamic force F_z to ca., 20% of the initial value, and a corresponding improvement in quality (left photograph). A backing of the free workpiece side across the entire

length results in a continuously steady routing process without any significant clogging. If the scrap strip is clamped additionally, only a slight further decrease of the dynamic force F_z is noticed at the clamping points (bottom left side).

Especially for the machining of thin laminates, which usually requires high rotational speed, these relations between clamping and process performance may gain a major importance so that a rigid and appropriate setup is necessary.

COMPARISON OF CONTOURING PROCESSES

In addition to routing, the high pressure water jet technology is firmly established in contouring operations on FRPs. Comparative tests employing both processes were carried out on identical laminates to differentiate the respective specific and advantageous fields of application. Laser beam cutting was not included in this survey due to its nature as a thermal cutting process.

An essential advantage of water jet cutting is the absorption and removal of dust by the fluid, an important aspect in terms of health safety and equipment lifetime. Moreover, the narrow kerf allows cutting of minimal radii and thus contributes to scrap minimization. Due to the symmetrical jet structure, both cut surfaces are equivalent in quality, whereas for routing – especially on AFRP – a distinct quality difference between up and down milling is noticed. While routing gives a cut rectangular to the material surface, water jet cutting generally shows a sloped curve due to the specific erosion mechanism. This effect, however, can be minimized by the application of a newly developed nozzle geometry. The necessity to catch the water jet as it leaves the workpiece requires free access to both sides of the part. The synchronized motion of a catching device makes it difficult to cut intricate three-dimensional shapes.

Figure 11 gives a comparison of water jet cutting and routing. As for thin laminates a higher cutting speed is attained by water jet which, however, rapidly falls with increasing material thickness. The ultimate thickness to be cut by water jet amounts to 15 mm for GFRP, 8 mm for CFRP and 6 mm for AFRP, whereas routing covers a much wider dimensional range. The machining quality obtained by both processes is exemplified for two different laminate thicknesses.

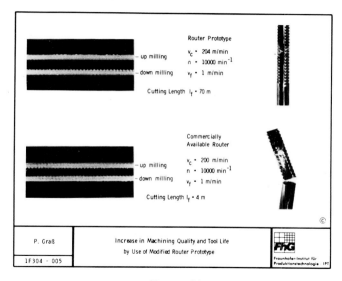

Figure 11

While the pressure unit as a process source is a much higher investment than a high speed spindle, this relation is shifted already for a complete operational system comprising a workpiece handling device and control. The direct operational cost per meter cut — excluding write-off for depreciation —· as shown here for the cutting of an AFRP laminate at equal cutting speed is considerably lower for the water jet, primarily due to the nonexistence of tool cost. The relation, however, is strongly affected by the respective cutting speed attainable. From the mentioned individual process characteristics, criteria for the selection of an appropriate contouring operation can be developed:

- Thin laminates are favorably cut by water jet, parts of higher thickness by routing.

- If the laminate structure is symmetrical and the same quality is required on the top and the bottom side of the cut, routing is superior as well as it is for sandwich material.

- Workpieces which are thermally delicate and mechanically unstable are usually better cut by water jet.

- Long, continuous cuts as well as cutting of narrow radii and angles is facilitated by water jet cutting.

The selection criteria as presented here will be helpful in a decision which, however, still has to consider the individual quality and productivity requirements of the specific part to be machined.

SUMMARY

To attain high quality in the mechanical machining of fiber reinforced material it is an essential prerequisite to consider the characteristic properties of the individual fiber types. With regard to aramid laminates the technological requirements concerning tool geometry and cutting material were laid down for drilling, countersinking and routing.

A sickle-shaped cutting edge is necessary with drills and countersinks. Such tools were developed within the program, and optimized cutting data were pointed out. Consequently the commercial introduction of these tools was another project activity.

Improved tool geometries were presented and discussed for routing. Cemented carbide turned out to be the cutting material best suited for all three individual techniques. In a comparative study of routing and water jet cutting, assistance is given in the selection of the appropriate contouring process.

INNOVATIVE MANUFACTURING FOR AUTOMATED

DRILLING OPERATIONS

Carl Micillo and John Huber

Grumman Aerospace Corporation

ABSTRACT

The major cost drivers in airframe fabrication are identified. In
the assembly area, drilling for various fastening systems is described
both from economic and quality aspects. The Five-Axis Automated Assembly
Fixture can automatically locate the substructure of a part using a digital
scanning process and form a precise map of the location of the parts. The
digitized information is used to drill and coutersink through skins and sub-
structures without costly templates. The evolution of the system is followed
from the development stage into production, and economic analyses and projec-
tions for aircraft structures at various learning curves and production rates
are given, including the aluminum A-6E wing and A-10 horizontal stabilizer,
and advanced composite B-1 horizontal stabilizer.

INTRODUCTION

In the brief history of aviation, a continuous and rapid evolution of
new materials and design concepts has occured. In the quest to fly faster
and further with ever-increasing mission requirements, stronger, lighter
and more corrosion-resistant materials have been developed,

ranging from wood and simple fabrics to aluminum, titanium and advanced composites. Technological advances in airframe structures, however, have increased production costs which must be offset by innovative manufacturing approaches. Some of the major cost drivers identified in the acquisition cost of today's aircraft are:

- Cost of new raw materials
- Design complexity
- High part count
- High fastener-to-weight ratio
- Non-optimum utilization of equipment and facilities

It is interesting to note that assembly and subassembly labor account for as much as 50% of the total manufacturing cost of current airframes. Production labor is almost directly proportional to the number of detail parts, holes for joining, and fasteners. Since a fighter may have 250,000 to 400,000 holes and a bomber or transport 1,000,000 to 2,000,000 holes, drilling has become a major production cost factor.

DISCUSSION

Design/Material Impact on Process Selection

Automated drilling and riveting can drastically reduce subassembly costs. This approach should be considered whenever design criteria, weight considerations or raw material costs warrant its use. For example, an efficient design concept for a small, high-performance wing structure involves use of a skin-stringer structure that must be

subassembled to produce a skin assembly before it can be installed on the
wing box substructure (Fig. 1). A five-axis, numerically controlled
(N/C) drilling/riveting machine (Fig. 2) is being used to produce titanium wing structures for the F-14A air-superiority fighter. This machine
is capable of oversqueezing a rivet or slug at a predetermined stress
level to produce a specified amount of fastener-to-hole interference or
installing precision attachment hardware in high interference ranges in
titanium structures. This approach is also being used to produce large
aluminum wing structures for transport and commercial airliners
(McDonnell Douglas DC-10, Lockheed 1011 and Boeing 727, 737 and 747) for
which the most cost-effective configurations are skins stiffened with
riveted stringers with access to both sides of the structures.

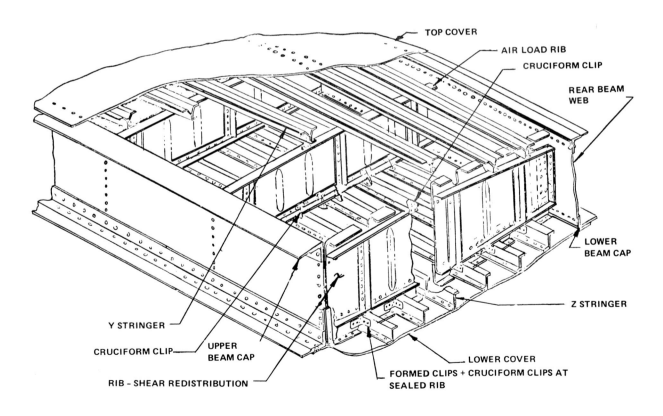

Fig. 1 Multi-Rib Wing Box Configuration

Fig. 2 Five-Axis, Numerically Controlled Drilling/Riveting Machine

For other types of wing structures, such as multi-spar wing box configurations (Fig. 3), however, automated systems are either unavailable or simply have not yet been developed. In this case, attaching hardware must be located and installed on assembly fixtures. As a result, preparation of the proper drilling operation can be the most costly aspect of the total mechanical fastening operation (Fig. 4). Positioning of stringers, beams and ribs beneath a skin can be off by as much as 6.4 mm (0.250 in.). Normally, pilot drilling, back drilling, countersinking and templates are required for such "blind" drilling operations.

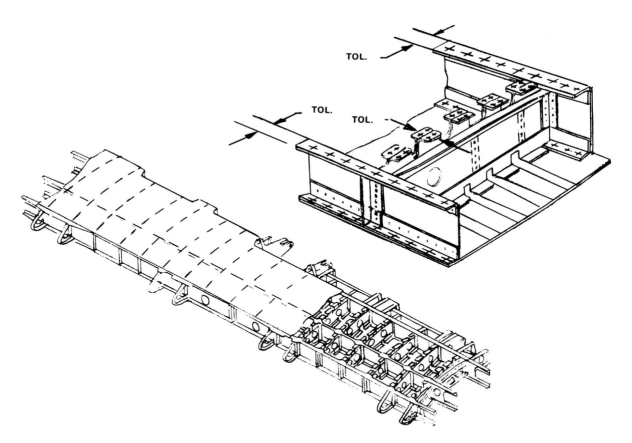

Fig. 3 Multi-Spar Wing Box Configuration

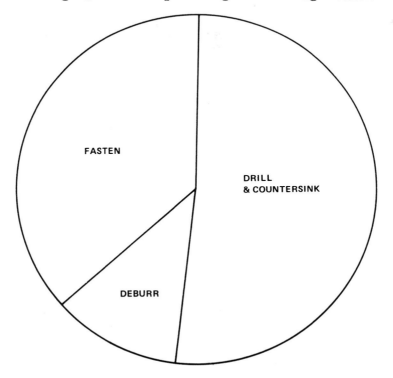

Fig. 4 Breakdown of Mechanical Fastening Operation

As a case in point, let us examine a typical manual fastening
operation on an aluminum fighter wing structure for which an average
manufacturing tolerance of 0.8 mm (0.032 in.) is required. The procedure
would normally involve placement of the substructure without its covers
(Fig. 5) in an assembly fixture, followed by manual positioning of a
series of templates on the substructure (Fig. 6) and then laboriously
moving the templates by hand to obtain the proper edge distance (as
measured from the hole to the edge of the individual spar, rib or
stringer). These time-consuming operations, for which all misalignments
and tolerance build-ups must be accounted for and corrected, are un-
avoidable unless we are fortunate enough to be working with oversized
details (at a significant weight penalty) to get the proper edge

Fig. 5 Aircraft Wing Substructure in Assembly Fixture

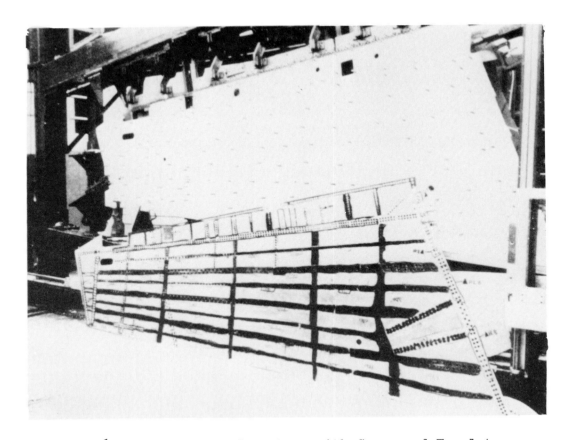

Fig. 6   Aircraft Wing Structure with Cover and Templates

distances. After all dimensional corrections have been indicated, holes

can be drilled through the wing cover and substructure. Since these are

manual operations, we are all too painfully aware of the potential for

improperly positioned holes.

Selectivity of Automation in Aerospace Manufacturing

When the purchase and deployment of automated equipment are being

considered, the nature of the aerospace manufacturing community must be

examined. We are essentially batch-type producers, often manufacturing

various details in relatively small quantities. As a result, the ad-

vantages of full automation, such as that found in the automotive industry,

are not applicable to the aerospace industry. If design and procurement criteria warrant large capital investment and "upfront" money, then full automation is the answer.

In all other cases, the obvious challenge is a system providing production cost savings and reliable hole quality with nominal capital expenditures. The system we have developed, the Five-Axis Automated Assembly Fixture (Fig. 7), is capable of automatically locating and correcting for variations in the positions of the substructure components by means of a digital scanning process, and then automatically positioning and operating a portable drilling device on an assembly fixture. The unique and most attractive feature of this system is the capability to create a precise map of the locations of the various sub-

Fig. 7 A-6 Right and Left Wing Outer Panel Drilling Fixtures

structure components without the need for costly templates. The digitized information obtained in the scanning process is computerized and subsequently used by the Five-Axis Assembly Fixture to automatically drill and countersink through the cover and substructure in one operation -- rapidly, accurately and economically. This system is being used in production on Grumman A-6E wings and has been evaluated for use with the Fairchild A-10 stabilizers.

Five-Axis Automated Assembly Fixture

Modular Features

Since the Five-Axis Automated Assembly Fixture is modular in nature, it can be readily adapted to many other applications. The system consists of a horizontal fixture that holds the wing structure, a vertical gantry that holds the drill head which houses either the scanning camera or drill unit, a central controller, a remote interface box, and an operator control pendant (Table I). The system has five axes of motion (Fig. 8). Four axes of motion are operated electrically by drive motors. These are:

- X-Axis - horizontal movement (left to right) of the gantry across the wing structure [(8.8 m/min) (350 in./min)]

- Y-Axis - vertical movement (up and down) of the scanning camera/ drill head unit on the gantry [(3.0 m/min) (120 in./min)]

- α-Axis - rotation around the X-axis that allows the drill spindle or camera lens to pitch up or down at various angles [(1.4 rad/min) (78 deg/min)]

Table I Automated Assembly Drilling Fixture - Modular Components

ITEM	MODULE	KEY COMPONENTS	FUNCTION
A	CENTRAL CONTROLLER	• COMPUTER • DISC • REAL-TIME CLOCK • POWER SUPPLIES • INTERFACE ELECTRONICS • LOGIC RACKS AND CABINET • TELETYPE	CONTROLS THE MOTION AND OPERATION OF THE SCANNING CAMERA AND DRILL MOTOR; CAN ALSO BE USED FOR EDITING PARTS PROGRAMMER'S SOURCE FILES
B	DRILL HEAD ASSEMBLY	• MECHANICAL ASSEMBLIES INCLUDING ALPHA, BETA AND Y-DRIVE MOTORS, ENCODERS, Z-AXIS PNEUMATIC CYLINDERS, POTENTIOMETER OR HYDROCHECK • FLUIDIC CONTROLS • POSITION DISPLAYS	THIS UNIT, WHICH MOUNTS ON THE GANTRY, HOLDS THE DRILL UNITS OR THE SCANNING CAMERA, SETS THE ANGLE TO BE NORMAL TO THE WORK SURFACE AND MOVES THE DRILL OR SCAN CAMERA INTO THE WORK SURFACE
C	GANTRY	• MECHANICAL ASSEMBLY INCLUDING X-DRIVE MOTOR AND ENCODER, VERTICAL POWER TRACK • DRIVE MOTOR TRANSLATOR CARDS • CABLES, CONNECTOR AND CABLE DISCONNECT PLATE • POWER SUPPLIES AND POWER SWITCHING CIRCUITS	HOLDS THE DRILL HEAD AND MOVES THE HEAD IN THE X-AXIS; IS MOUNTED TO THE FIXTURE AND CAN BE MOVED FROM FIXTURE TO FIXTURE
D	REMOTE INTERFACE BOX	• INTERFACE ELECTRONICS • POWER SUPPLIES • CABINET	THIS UNIT, WHICH RESIDES AT THE FIXTURE IT IS CONTROLLING, RECEIVES DATA AND COMMANDS FROM THE COMPUTER, PROCESSES THESE AND SENDS AND RECEIVES SIGNALS TO THE DRILL HEAD, GANTRY AND OPERATOR'S CONTROL PENDANT
E	OPERATOR'S CONTROL PENDANT	• CONTROL PENDANT WITH SWITCHES, LIGHTS, MESSAGE DISPLAY, THUMBWHEELS	THIS UNIT, WHICH HANGS FROM THE GANTRY, PROVIDES COMMUNICATIONS BETWEEN OPERATOR AND CENTRAL OR REMOTE CONTROLLER. MACHINE CAN BE CONTROLLED MANUALLY WITHOUT THE COMPUTER, USING THE CONTROL PENDANT
F	SCANNING SYSTEMS	• SCAN CAMERA • SCAN CONTROLLER • CAMERA ROTATE MECHANISM	ALLOWS SCANNING OF THE SUBSTRUCTURE AND DIGITIZING OF THE VIDEO DATA. THE CAMERA IS MOUNTED INSIDE THE DRILL HEAD CYLINDER IN PLACE OF THE DRILL UNIT. THE SCAN CONTROLLER IS MOUNTED IN THE REMOTE INTERFACE CONTROL BOX
G	DRILLING UNIT	• DRILL UNIT	DRILLING, COUNTERSINKING, COUNTERBORING
H	FIXTURE ITEMS	• HORIZONTAL POWER TRACK • CABLES • POWER SUPPLIES FOR X, Y, ALPHA AND BETA MOTORS	SET OF CABLES TO ROUTE SIGNALS AND POWER TO THE GANTRY FROM THE REMOTE CONTROL UNIT. DC POWER FOR STEPPER MOTORS

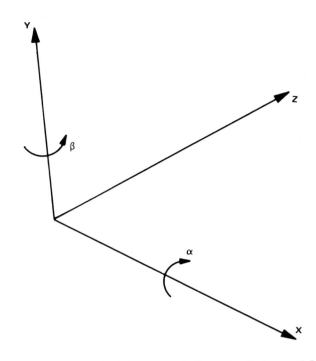

Fig. 8 Five Axes of Motion of Automated Assembly Fixture

- <u>β-Axis</u>-rotation around the Y-axis that allows the drill spindle
  or camera lens to yaw left or right at various angles.

The fifth axis of motion (Z axis) is operated by a pneumatic cylinder
for drill feed [(2.1 m/min) (84 in./min)] and camera focus control.

Parts Programming

The sequence of events for parts programming is shown in Fig. 9.
Parts programming is initiated after finalization of the clamping-and-
drilling sequence for the assembly to be drilled. Using marked-up
drawings, each hole is digitized on the Orthomat drawing machine in the
actual sequence to be drilled using a zero reference point (X=0 and Y=0).
The X and Y coordinates are then put on magnetic tape. This information
is fed into a FORTRAN program to convert it from the vehicle coordinate
system to the assembly fixture system and to compute surface normals and

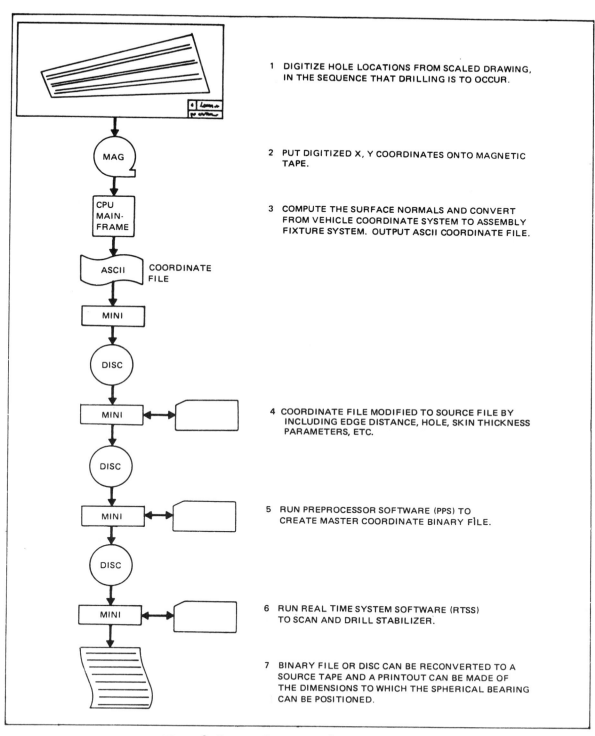

1 DIGITIZE HOLE LOCATIONS FROM SCALED DRAWING, IN THE SEQUENCE THAT DRILLING IS TO OCCUR.

2 PUT DIGITIZED X, Y COORDINATES ONTO MAGNETIC TAPE.

3 COMPUTE THE SURFACE NORMALS AND CONVERT FROM VEHICLE COORDINATE SYSTEM TO ASSEMBLY FIXTURE SYSTEM. OUTPUT ASCII COORDINATE FILE.

4 COORDINATE FILE MODIFIED TO SOURCE FILE BY INCLUDING EDGE DISTANCE, HOLE, SKIN THICKNESS PARAMETERS, ETC.

5 RUN PREPROCESSOR SOFTWARE (PPS) TO CREATE MASTER COORDINATE BINARY FILE.

6 RUN REAL TIME SYSTEM SOFTWARE (RTSS) TO SCAN AND DRILL STABILIZER.

7 BINARY FILE OR DISC CAN BE RECONVERTED TO A SOURCE TAPE AND A PRINTOUT CAN BE MADE OF THE DIMENSIONS TO WHICH THE SPHERICAL BEARING CAN BE POSITIONED.

Fig. 9 Parts Programming

Z coordinates This file is then edited to include hole treatment, structure angle, scan points, edge distance, skin thickness and control moves. The edited file becomes the source file for the part. The substructure is scanned and marked using the Rockwell drill unit fitted with a lead pencil. All the holes are checked in this manner for proper location and number. The "debugging" process of correcting hole location, tool path and scan points is accomplished in several scan-and-mark cycles. Once the source file is corrected, subsequent structures are scanned and drilled without a marking cycle. When the program has been established, a change can easily be made in path, sequence or function to optimize the scan/drill operation.

Hole Position Calculation

Through the control pendant, the operator can direct the computer to issue electrical camera-positioning signals for the drive motors in the four axes to start the scanning operation. The digital scanning camera (Fig. 10), which is housed in the drill head assembly, locates the edge of the structural members. The computer then calculates the offset distance for hole placement from the individual electrical pulses generated through detection of the reflected light by a linear array of photodiodes as a function of light intensity and exposure. Since the scanning camera is positioned so that it is centered at the edge of the substructure, half of the photodiodes respond to the reflected light. The operational range of the camera is 256 pulses (plus 6.5 mm (0.256 in.)) because there are 512 photodiodes in the array. If the scanning camera is more than 6.5 mm (0.256 in.) from the center while seeking the edge of a substructure component, the computer run-time software drives

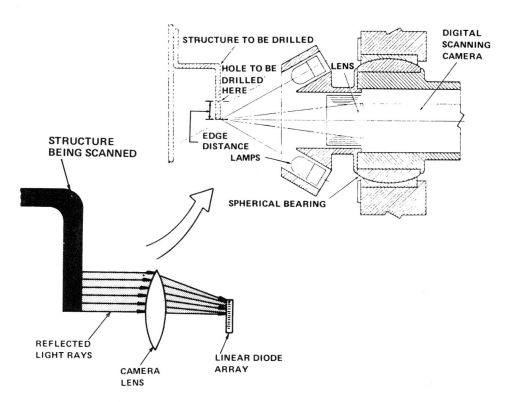

Fig. 10 Operation of Scanning Camera

the camera to its computed midpoint so that it can rescan the hole after
taking into account any overshoot or undershoot. If the camera finds the
edge of the substructure component within its operational range of
6.5 ±1.3 mm (0.256 ±0.050 in.), it is not moved. Hole position is
calculated by the computer based on the camera reading once the reading
is within 1.3 mm (0.050 in.).

Scanning

The scanning cycle is the first operation in the sequence, since
holes must be scanned before being drilled. In the case of the wing
structure assembly (described previously for manual operation), the
substructure without the cover is placed in an assembly fixture, the
scanning camera is placed in the drill head assembly, and all five axes
are at the zero position. Through the control pendant, the operator
activates the program for the scanning cycle, which turns on the camera

lights and sends back signals to the computer to identify the actual position of the substructure. The computer corrects the nominal hole position to the desired position as a function of the actual substructure position and sends this information to the disk file. The scanning program, based on a file of master coordinates taken from engineering drawings, generates a modified set of hole coordinates to drill holes for temporary fasteners to hold the cover to the substructure and then another set of hole coordinates to drill and countersink all other holes.

Drilling

Upon completion of the scanning program, all axes are returned to the zero position, a drill head replaces the camera in the drill head assembly, and the wing cover is temporarily clamped to the substructure. Drilling and installation of temporary Cleco fasteners precedes the drilling of finished holes in accordance with the modified program. Normality of the drill head to the wing cover surface is maintained by programming. A residual force of 91 kg (200 lb) is maintained at the nosepiece to clamp the cover to the substructure. The minimum-to-maximum diameter differences are within 0.05 mm (0.002 in.). The overall system falls within a true-position hole tolerance of ±0.4 mm (±0.016 in.). A typical cycle to drill one hole is as follows:

Operation	Time, sec
Advance drill to structure	2.0
Drill and countersink hole	5.0
Retract drill assembly	2.0
Advance to next hole	2.5
Total	11.5

The drilling operation is followed by manual riveting of the wing covers to the substructure. The accuracy and time-saving aspects of the scanning operation and the rapidity of the drilling cycle enabled us to transfer the Five-Axis Automated Assembly Fixture from the laboratory to production. An additional bonus, which should not be overlooked in any comprehensive economic analysis, is the repeatable reliability of the holes produced by this system that precludes or effectively minimizes extensive inspection and repairs.

Production Implementation

The Five-Axis Automated Assembly Fixture was implemented in production on the Grumman A-6E program. The A-6E (Fig. 11) is a two-seat, all-weather attack aircraft being manufactured for the U.S. Navy and the

Fig. 11 A-6E Intruder

Marine Corps at a rate of one and one half aircraft per month; the peak production rate was eight aircraft per month. Over 40 outer-wing assemblies (Fig. 12) have been produced by the Five-Axis Automated Assembly Fixture. This structure consists of five beams, eight ribs, and two covers ranging in thickness from 3.2 mm (0.125 in.) to 11.1 mm (0.438

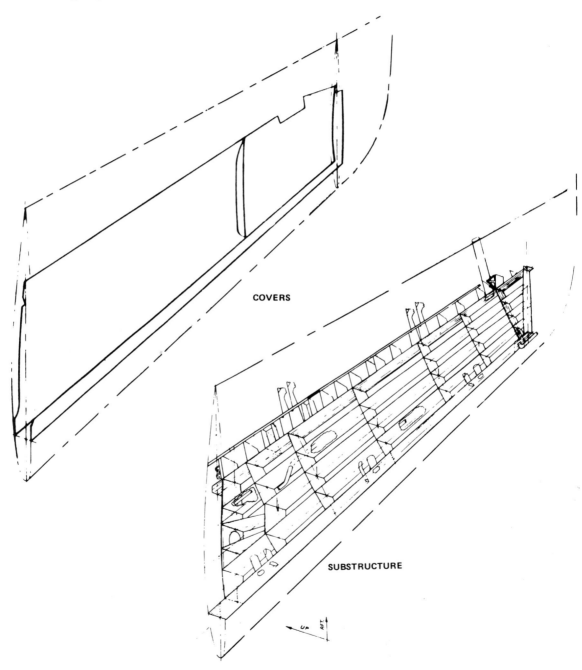

COVERS

SUBSTRUCTURE

Fig. 12 Exploded View of A-6E Wing

in.). About 2000 holes, ranging in diameter from 3.2 mm (0.125 in.) to

6.4 (0.250 in.), are drilled in this structure. Prior to introduction of

the automated equipment, the average production time to set up the hole

templates manually, locate the hole positions, and drill an A-6E wing

cover was 65 hours at Aircraft No. 600. The Five-Axis Automated Assembly

Fixture performed the same task in about 25 hours during the first

demonstration run. As shown in Fig. 13, about half of the time was

required to scan and drill; the remainder was used in set-up operations

such as camera re-use, drill unit installation and drill/countersink

changes. This results in a reduction of up to 60% in production labor

for the drilling operation at aircraft production unit No. 600.

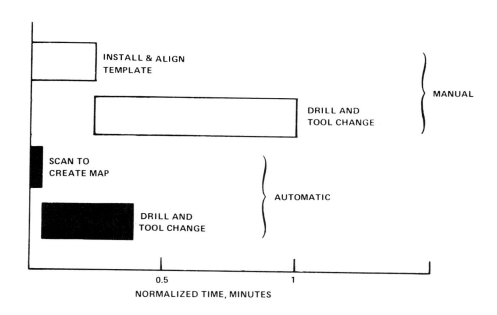

Fig. 13 Comparison of Scanning and Drilling Times

The Five-Axis Automated Assembly Fixture was also evaluated on the

Fairchild A-10 close-support aircraft (Fig. 14) in a joint Grumman-

Fairchild development program sponsored by the Air Force Materials

Laboratory, aimed at demonstrating additional production cost savings

Fig. 14 Fairchild A-10 Close-Support Aircraft

via innovative manufacturing technology. This 20-ft-long aluminum
structure, consisting of upper and lower covers and a substructure with
3 spars and 17 ribs (Fig. 15), requires the drilling of 3750 holes
having diameters ranging from 4.4 mm (0.172 in.) to 6.4 mm (0.250 in.).

The manual procedure for drilling the A-10 horizontal stabilizer
involves drilling pilot holes in the rib flanges prior to assembly of
the flanges in the box substructure. After the box substructure has been
installed in the assembly fixture, one cover is placed against the substructure so that pilot holes can be drilled in the cover from the rear
through the previously drilled pilot holes in the rib flanges. Full-
size holes and countersinks are then drilled from the front of the cover
using the pilot holes as guides. When one cover has been drilled, it

Fig. 15 A-10 Horizontal Stabilizer

is removed from the assembly fixture so that the other cover can be similarly drilled. The first unit drilled with the Five-Axis Automated Assembly Fixture was completed in 18.6 hr versus 33.6 hr for the manually drilled structure at Unit No. 118.

Impact on High-Material-Cost Structures

Another projected application for the Five-Axis Automated Assembly Fixture was the production program for the hybrid boron-graphite/epoxy B-1 horizontal stabilizer (Fig. 16). This swing-wing, intercontinental bomber (Fig. 17) is capable of flying at speeds up to Mach 1.6 at

Source: Proceedings of Autofact West, Vol. 2, 1980, 253-277

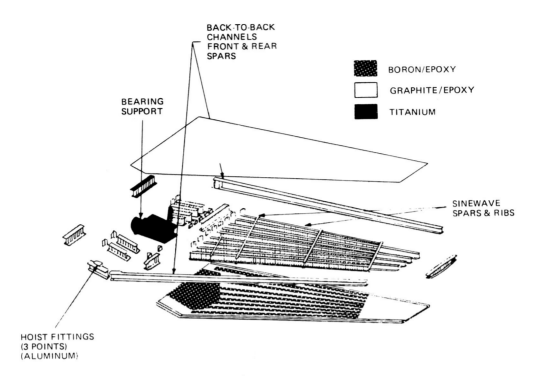

Fig. 16 B-1 Horizontal Stabilizer

Fig. 17 B-1 Bomber

high altitudes and at near-supersonic speed at low altitudes. The implementation of advanced composites in this structure resulted in weight savings of 243 kg (535 lb) per aircraft (16.2%) and the elimination of 10 000 fasteners (41%) compared to the baseline metal structure.

The drilling parameters for the cover-to-substructure attachments must be closely controlled to prevent delamination. Insuring proper hole location is also difficult because of the relatively large thickness tolerance of advanced composite parts. Since the substructure parts cannot be precisely located in the assembly fixture, the drilling operation becomes highly labor-intensive. About 2850 holes per cover (80% of the total number of holes) are drilled and countersunk through graphite/epoxy in one operation using carbide cutters mounted in a Winslow Spacematic Model M-62 drilling unit equipped with a vacuum foot (Fig. 18). Hole location in the interior spar-to-cover joints is controlled by strip templates pinned to the cover through the pilot holes which are drilled from the inside using a contoured bushing that picks up the corrugated web. After the interior has been drilled, additional strip templates are used to locate the attachments to the ribs relative to the flat webs.

The Five-Axis Automated Assembly Fixture is directly applicable for drilling and countersinking all of the holes in the graphite/epoxy regions of the B-1 horizontal stabilizer covers. After assembly of the substructure, the exact location of selected holes would be marked and the substructure scanned automatically to correct the computer program for variations in part thickness and alignment. The cover is then placed in position and the holes are drilled and counter-

Fig. 18 Winslow Spacematic Drill Unit

sunk in one operation, as described previously. Again, use of the Five-Axis Automated Assembly Fixture would improve the accuracy and perpendicularity of fastener placement, and would also improve hole quality.
The automated fixture would also be adaptable for installation of the tip cap and leading/trailing edges to the completely assembled torque box. These operations, including fastener installation, represent 18% of the manufacturing labor cost for the entire stabilizer (Fig. 19).
Thus, the anticipated savings in drilling costs, as verified on metal structures, are about 8 to 10% of the total stabilizer cost, which makes the advanced composite B-1 horizontal stabilizer even more cost-effective.

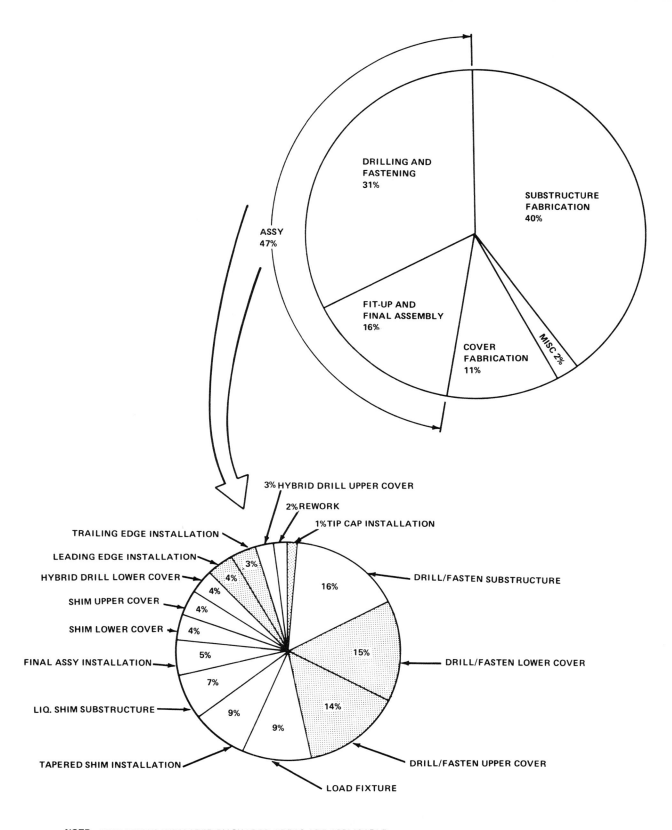

DRILLING AND
FASTENING
31%

SUBSTRUCTURE
FABRICATION
40%

ASSY
47%

FIT-UP AND
FINAL ASSEMBLY
16%

COVER
FABRICATION
11%

MISC 2%

3% HYBRID DRILL UPPER COVER

2% REWORK

1% TIP CAP INSTALLATION

TRAILING EDGE INSTALLATION

LEADING EDGE INSTALLATION

HYBRID DRILL LOWER COVER

SHIM UPPER COVER

SHIM LOWER COVER

FINAL ASSY INSTALLATION

LIQ. SHIM SUBSTRUCTURE

TAPERED SHIM INSTALLATION

DRILL/FASTEN SUBSTRUCTURE

DRILL/FASTEN LOWER COVER

DRILL/FASTEN UPPER COVER

LOAD FIXTURE

3%
4%
4%
4%
4%
5%
7%
9%
9%
16%
15%
14%

NOTE: OPERATIONS INDICATED BY SHADED AREAS ARE APPLICABLE
TO THE FIVE – AXIS AUTOMATED ASSEMBLY FIXTURE

Fig. 19 Manufacturing Breakdown for Advanced Composite B-1
Horizontal Stabilizer

Applicability and Benefits

Until the advent of the Five-Axis Automated Assembly Fixture, automation of the manufacturing operations for such aircraft structures as the A-6E wing and A-10 and B-1 horizontal stabilizers was not possible. The system, as demonstrated under U.S. Air Force auspices, can be applied to most assembly procedures that require multiple manipulations of drilling equipment and templates to locate fasteners precisely at the intersections of underlying substructure components that have been pre-assembled with their own network of locating tolerances.

Automated scanning and drilling operations would have the following advantages:

- 50 to 60% reduction in production labor
- Improved hole quality due to better tool platform
- Elimination of costly templates
- Predictable and consistent flow rates
- No cost penalty due to inexperience (learning curve effect)
- Less dependence on operator skill

Future Systems

The unit described is an initial step in utilizing computer technology and sensor information to provide a system with the necessary flexibility to be applicable to the relatively low aerospace production rates. Five years of production experience with the A-6E aircraft coupled with rapidly growing computer technology will help to provide a basis for future modifications utilizing microprocessors, direct links to CAM systems and assembly fixtures initially designed to use modules of such automated systems.

By J. Cusumano*, J. Huber*, and K.T. Marshall*

Ultrasonic drilling of boron fiber composites

Ultrasonic drilling of boron-epoxy composites is an excellent cost-effective method for more rapidly producing holes of better quality with a stationary drilling setup. It is anticipated that comparable advantages can be developed into portable ultrasonic equipment for drilling advanced composite laminates.

A technical feature
Gordon M. Kline, technical editor

Boron-epoxy composite aerospace hardware has progressed from the prototype stage to full-scale production. Such items as horizontal stabilizers, wing box girders, and fuselage components have proved the structural efficiency and demonstrated the weight savings possible. With the advent of better manufacturing techniques, the high material cost will be reduced sufficiently to allow more general use of the composite. A simultaneous cost reduction of the fabrication techniques is necessary to increase the cost effectiveness of the material.

One major area that exhibits great opportunity for reduction in fabrication costs is the method of drilling holes. A typical wing may require as many as 5000 holes in the boron-epoxy composite and the production rates and costs required cannot be obtained by using multiple high-speed steel drills or by the use of expensive diamond-impregnated core drills in conventional drilling fixtures. The Air Force Materials Laboratory[1] instituted a program of ultrasonic machining to reduce this production obstacle. This program established the optimum parameters for the greatest rate of material removal in reproducible methods that result in the lowest total cost per hole.

The drilling power equipment consists of a Branson Model UMT-3 rotary machine tool[2] with an ultrasonic machining head that includes an automatic and manual feed. This is mounted on a standard milling machine base with a compound work table. Rotary power is supplied by a $3/4$-hp. motor regulated by an autotransformer that allows speeds from 0 to 6100 r.p.m. The spindle end terminates in the drill chuck and coolant is supplied to the hollow core of the drill from the spindle coolant jacket that surrounds the lower end of the spindle and is fed through a side fitting. A small separate cabinet houses the ultrasonic power supply that is rated at a maximum output of

250 w. at a frequency of 20 kHz. A control knob allows an adjustment in the amplitude of the up-and-down motion of the drill with the maximum being just under 1 mil. A second knob allows the power supply to be "tuned" to the particular drill mass being used. This knob is rotated until the needle on the adjacent meter reads a minimum. This meter also shows the demand of power from the supply as the load on the drill becomes greater.

To measure the drill pressure and drill torque, a Lebow transducer, model 6421, was mounted on the compound work table. This instrument transmitted the loads electronically to two digital readout units. Thus visual monitoring was possible at all times during the drilling cycle. For a permanent record a multichannel strip recorder was connected to the two digital readout units.

Principles of operation

The ultrasonic "horn" (Fig. 1) that is built into the spindle of the drilling machine consists of a disk of piezoelectrical material, lead-zirconate titanate, which has a very high electromechanical conversion rating and is sandwiched between two thick disks of metal. When the regulated frequency current is fed to it, the whole system vibrates at some resonant frequency along its longitudinal axis and acoustically the motion is equal to a half-wave length. The metal disks are each represented by the spring and its mass. As each spring and mass has a natural period of oscillation, the two combined determine the resonant vibration frequency of the system or transducer. As the reciprocating motion of the cranks approaches the resonant frequency of the springs and their masses, the amplitude of the two masses at their midpoints is the largest. Only friction keeps the amplitude of oscillation of the two masses from continuously increasing. Electrical energy is fed to the rotating sandwich by means of sliding brushes. Drill masses or lengths have to be calculated so that the end of the drill falls at the end of the half-wave length for maximum amplitude. Bearings on the spindle conversely have to be at nodes on the axis to minimize wear and vibration.

Frequency. The frequency of operation is all important to the drilling efficiency. Three major frequencies were tried by using three spindles built to resonate at 16, 20, and 24 kHz., respectively. The lowest, 16, is about the threshold of most people's hearing and high frequency is limited by the size of the tool allowed. Also, the gripping of the drill bit becomes more difficult at the higher frequencies. The holding force for a 1-mil amplitude drill is about 750 lb. at 20 kHz.; the force increases to 1750 lb. at 40 kHz. If the frequency were doubled to 40 kHz., the output force would only be one-quarter of that at 20 kHz. and the energy delivered would be reduced eight times. Velocity of the tool remains the same for any frequency. Fig. 2 shows the relative positioning of the bearings on the spindle of the ultrasonic transducer born in as compared to the wave form. Testing has shown that the 20-kHz. sys-

*Advanced Materials & Processes Development, Grumman Aerospace Corp.

Based on a paper presented at the 28th SPI RP/Composites Institute Conference.

1—Fabrication Branch, Manufacturing Technology Div., Air Force Materials Laboratory, Air Force Systems Command, Wright-Patterson Air Force Base, Ohio.
2—Branson Sonic Power Co.

tem is the most efficient of the three frequencies.

Amplitude. Experience has shown that the hard brittle materials are very sensitive to the amplitude of the ultrasonic machining tool. Since ultrasonic amplitudes in excess of 1 mil shorten the tool life, only those of less than 1 mil were utilized. The actual amplitude of the drill is measured by means of a photoelectric sensor receiving reflected light from the blunt shiny surface of a cutoff drill as it vibrates. This signal is fed from the control box to an oscilloscope, where it is displayed and measured.

Cutting tools. Three categories of drill bits were used.

1) The Engis diamond core drill is manufactured in England and consists of a random grit (200 average) core built up on a steel tool electroplating. The diamond part is short, about 0.6 in. in length. Good drilling results were obtained with boron-epoxy-boron, boron-glass-boron, and boron-aluminum. Drilling boron-titanium materials caused the drill end to balloon out and groove the work. The drill wore off 0.0006 in./hole at 2.4 in./min. spindle feed when drilling boron-epoxy-boron.

2) Branson hot press core drill with a grit size of 60-80 has the hot pressed diamond core brazed concentrically to a steel tube. This drill also was successful in drilling the same materials as above and wore down 0.00065 in./hole at 2.4 in./min. spindle speed on boron-epoxy-boron. For boron-titanium-boron the wear rate was faster, 0.015 to 0.023 in./hole. Quality of the holes was excellent.

3) The Branson/habit diamond core drill, grit size of 60-80, also is constructed on a steel shank similar to the Engis drill, only the drilling surface is about 0.7 in. long. The results of drilling were similar to the Engis drill.

A reaming tool also was designed and fabricated. This drill bit consisted of a straight fluted shank impregnated with diamonds and with holes in the base of the flutes to allow the cutting fluid to carry away the residue and lubricate the cutting surface. Reamers in composites are used mainly to improve surface finish and tolerances, although they sometimes are used for slight mismatches.

The drill sizes used were $5/32$, $3/16$, $1/4$, and $3/8$ in. diameters with grit size from 20 to 100. All concentrations of grit were 100, except for the reamer which was random.

Chucks. If the manufacturing process step of brazing the core shank to the ultrasonic adapter could be eliminated, the cost of the assembled drill unit would be reduced by about 25%. The experimental ultrasonic tool chuck consists of a split male threaded sleeve that grips the drill shank and tightens on it as it is screwed into the threaded female holder. The shank is gripped tight enough so that the coolant flow is only through the center of the drill. Because of the thin wall of the shank, it is difficult to prevent collapse of the shank tube or to maintain concentricity.

Coolants. Plain water, water-soluble oils such as Crystal Cut 555 (5 to 10% by volume), and Freon TB-1 were used as coolants. During the coolant tests, feeds of gravity, 0.6 in./min., and 1.2 in./min. were tried. The addition of the Crystal Cut 555 did not reduce the torque and actually increased the amount of thrust required. Freon also did not improve the torque and thrust conditions over plain water and is much more expensive. No difference in hole quality was observed between the use of the three coolants. In drilling rates, the Crystal Cut 555 reduced the drilling time in boron-titanium by 60%, but did not improve drilling operations on any of the other material combinations. The recommended rate of flow of the coolant is 15 gal./hr. with a pressure of 40 p.s.i.g.

Where it would be detrimental to the composite laminate combination or the surrounding parts to use water, the dry drilling operation can be improved by using a "peck" drill cycle with air flow through the center of the drill. Optimum conditions are a "peck" drill rate of 120 strokes/min. with an air flow rate of 5 p.s.i. through the core of the drill and a gravity feed rate. Spindle speed should be about 4000 r.p.m. Because of boron dust, the operator should wear a face mask and utilize a vacuum cleaner attachment to the drilling setup.

Parameters. For any combination of advanced composites and other materials, it is necessary to optimize the following drilling parameters when using ultrasonic drilling techniques: 1) cutting feed rate; 2) cutting speed; 3) torque; 4) thrust; 5) drill amplitude; 6) frequency range; and 7) ultrasonic power.

The spindle speeds, in general, for the various diameter drills are: $1/4$ in., 4000 r.p.m.; $3/16$ in., 5300 r.p.m.; $5/32$ in., 6100 r.p.m. Items 5, 6, and 7 generally can be firmed up at

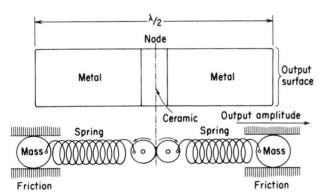

Fig. 1: Schematic of piezoelectric-type transducer.

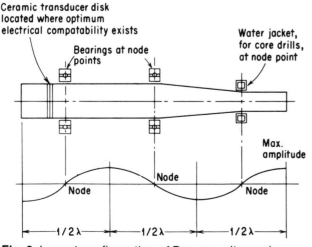

Fig. 2: Layout configuration of Branson ultrasonic transducer-horn UMT-3.

7 mils, 20 kHz., and 250 w., respectively. Thrust and torque loads are reduced over 50% by the use of ultrasonics over standard drilling with diamond core drills. Variables affecting the above parameters as well as cost and hole quality are: tool tolerance; tool wear; post processing, cleaning and deburring; alignment and spindle trueness; and tool chucking.

Hole inspection. Using existing nondestructive inspection techniques, the dimensional accuracy of the drilled holes was inspected for perpendicularity, hole taper, hole diameter and concentricity, out-of-roundness, and surface finish. Slower speeds gave better surface finishes. Perpendicularity depends on the work setup, and hole diameter and concentricity are almost entirely dependent on the quality of the drill and drilling machine. Quite a few drills were received in the off-center condition (0.002 in. T.I.R. maximum allowable). The out-of-roundness was always below the limits set (0.0015 in. maximum). Some hole taper was obtained if the drill was worn and did not penetrate through the hole far enough after breakout (0.25 in. suggested).

Materials. All advanced composite laminates should have a suitable backup material such as a paper-based laminate or other hard brittle material to prevent breakout. For formed sections an epoxy material should be cast or troweled against the back surface and then stripped off after drilling. Laminates that contain glass fiber or titanium layers are more easily drilled by the "peck" method through these materials. If the titanium splice plate in the laminate exceeds $1/8$ in., a step drilling technique with a multiple spindle machine is required. Thus, one would use an offset core drill to drill through the boron-epoxy; a slightly smaller cobalt rail drill to penetrate most of the way through the titanium, leaving 0.03 in.; an offset core drill to finish the titanium and go through the next layer of boron-epoxy; another cobalt rail drill for the next layer of titanium; and a core drill to finish the last layer of boron-epoxy and size the complete hole.

Feed rates. For gravity feed the spindle is allowed to settle by means of its own weight without any applied load. The hydraulic flow will have to be restricted to 3.0 in./min. The 0.6 in./min. rate is obtained by setting 80 p.s.i. on the top of the spindle feed piston and 40 p.s.i. on the pressure below the piston. The flow rate of the hydraulic pressure is set by timing the travel of the spindle over a measured distance.

Countersinking. Since the edge of the hole in the boron laminate is in contact with the countersink at always the same place for the same diameter hole, the wear is considerable. For example, without ultrasonics, the wear for countersinking thirteen $1/4$-in.-diameter holes is 0.006 in., whereas with ultrasonics the wear is only 0.004 in. after drilling twenty-four $1/4$-in.-diameter holes. Negligible wear occurs on the combination drill-countersink tools after 20 holes and better alignment results than when a drill and countersink are used.

Reaming. Excellent surface finishes were obtained by using the diamond-impregnated fluted reamer in the $1/4$- and $3/8$-in. diameter sizes. Wear was negligible on all laminate combinations except the boron-titanium laminates where it was noticeable.

Costs. The initial cost of the ultrasonic equipment is considerably higher than that for standard drilling machines, but the savings in tool life and greater production rates offset the investment in a short time.—END

DRILLING COMPOSITES WITH GUN DRILLS

R. T. Beall
Engineering Structural and Materials Laboratory
Lockheed-Georgia Company
Marietta, Georgia

Abstract

Gun drills have the ability to drill deep precision holes due to the provision for introducing a coolant to the drill tip which both cools and aids in removing chips and shavings. Lockheed has adapted such drills to portable equipment for drilling close tolerance holes in aluminum, graphite/epoxy composites, fiberglass composites and combinations of these materials. Hole quality is equal to or superior to all competitive drilling procedures in diameter control and surface finish. Finished holes are completed in one step, eliminating any requirement for piloting, step drilling and reaming.

1. INTRODUCTION

Graphite/epoxy composites have presented numerous problems in drilling precision load bearing holes. The material is abrasive. Carbide or diamond tools are usually required. Tool life is short and hole quality is seldom up to the standards desired based on typical aluminum drilling technology. These problems are compounded when deep holes are needed. The abrasive graphite/epoxy dust and drill chips can damage the hole wall.

Gun drills are, as the name implies, used for drilling gun barrels and have generally been limited to heavy fixed equipment. Lockheed adapted gun drills to portable equipment and has been using them to drill wing chordwise joint holes in aluminum laminations over three inches thick. Outstanding hole quality in diameter control and surface finish are produced. Holes are produced in one step with surface finishes of better than 20 RMS. No piloting or reaming is required. These drills have a single cutting flute and coolant is introduced directly to the cutting surface by a duct through the drill shank. Drill chips and coolant are constantly removed by a vacuum pick-up. This technology has now been adapted to drilling graphite/epoxy composites.

Reprinted with permission from National SAMPE Technical Conference, Vol. 11, 850-856, © 1979 Society for the Advancement of Materials Process Engineering

2. DRILLING EQUIPMENT DEVELOPMENT

A Quackenbush QDA27 drill motor was adapted for gun drilling as follows: The standard chuck was replaced with a Dresser fluid chuck No. G21258 so coolant could be introduced to the gun drill. Figure 1 shows the modified

ble when a mist of air and a light lubricant is used. The cannister to the left in Figure 1 is the mist dispenser. Excellent results have been obtained with a mist of air and LPS #1.* A drill speed of 2800 RPM and a feed of .0005 inches per revolution has been found to be satisfactory for both composite and aluminum.

FIGURE 1. VIEW OF MODIFIED QUACKENBUSH QDA27 GUN DRILLING A COMPOSITE JOINT SPECIMEN

QDA27 drill in use. The coolant is introduced to the fluid chuck through the fitting on the left side of the drill. A rubber flap seal allows the drill to feed while maintaining a vacuum seal for residue removal. Early trials showed it was possible to drill graphite/epoxy composite with only compressed air as a coolant. More extensive studies, however, have shown that superior drill life, hole finish and diameter control is possi-

Mist pressures of 40 to 70 psi are used.

Solid carbide gun drills** with the point configurations shown in Figures 2 and 3 have been found to be satisfactory for both graphite/epoxy composite and aluminum/composite combinations. Drills have been repointed up to 15 times and were still producing holes of excellent quality. Repointing is relatively easy but does require a special holding fixture.

*LPS Research Laboratories, Inc., Los Angeles, CA.
**Crafters Gundrills, Inc., San Fernando, CA.

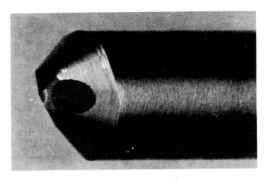

FIGURE 2. PHOTOMICROGRAPH OF 1/4 INCH DIAMETER GUN DRILL CARBIDE TIP

The results from a recent test are summarized in Table I and demonstrate some of the potentials of gun drilling graphite/epoxy composites. Strap joint specimens were fabricated of 24-ply laminates of Hercules 3501-6/AS for fatigue testing. Each specimen contained 12 fasteners. Total thickness of each hole through the adherend and strap was .260 inches. A total of 18 specimens were drilled before the drill was repointed. The first, ninth and eighteenth specimens were measured and examined in detail to determine hole quality deterioration due to drill wear.

TABLE I. DRILL LIFE STUDY

HOLE NO.	AVERAGE DIAMETER	RANGE
1 - 12	.2496	+ .0003
97 - 108	.2497	+ .0003
205 - 216	.2494	+ .0002
Drill Diameter - .2495		

All holes were examined by DIB enhanced X-ray. No delamination or cracking was detected. Wall finish was good with no apparent chipping or pullouts. Figure 4 shows the point after 108 holes, while Figure 5 shows the same point after 216 holes. While the wear land is quite pronounced, the drill was still producing

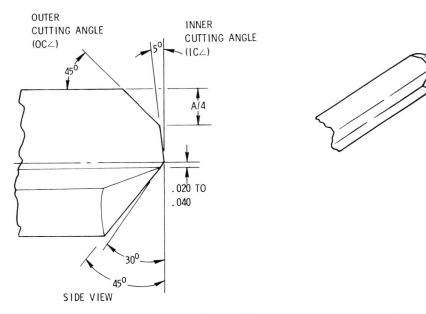

FIGURE 3. GUN DRILL POINT CONFIGURATION FOR CUTTING GRAPHITE/EPOXY AND G/E ALUMINUM COMBINATIONS

good holes. It had drilled 56 inches of composite.

FIGURE 4. GUN DRILL POINT SHOW-
ING WEAR LAND DEVELOP-
MENT AFTER 108 HOLES
(28 INCHES OF G/E COM-
POSITE)

FIGURE 5. GUN DRILL POINT SHOW-
ING WEAR LAND DEVELOP-
MENT AFTER 216 HOLES
(56 INCHES OF G/E COM-
POSITE)

Gun drilling has been used to produce the extremely high quality holes required for the studies reported by Freeman, et al[1]. The fibers in each ply had to be cleanly cut with no fracturing, tearing or pullouts.

3. DRILLING ALUMINUM-COMPOSITE COMBINATIONS

Lockheed has just finished fatigue testing a graphite/epoxy wing surface panel incorporating integral slots, ducts and metering holes as might be required for laminar flow control (NASA Contract NAS1-14631, "Evaluation of Laminar Flow Control System Concepts for Subsonic Commercial Transport Aircraft"). This surface panel is a hat-stiffened skin design with a design ultimate of 38 KIPs per inch. Testing consisted of two lifetimes of flight-by-flight spectrum fatigue cycling for a wing lower surface. End fitting design was a particular challenge. The concept chosen, while not considered to be a chordwise joint, does address many of the problems that would be encountered in a chordwise joint in a graphite composite wing box. Aluminum fittings were used in this test to reduce cost and process development requirements.

Figure 6 shows the inner surface of the surface panel with end fittings attached. Holes up to 1.75 inches in depth were required to bolt the end fittings to the

FIGURE 6. PHOTOGRAPH OF INNER SURFACE OF COMPOSITE WING SURFACE PANEL SHOWING HAT STIFFENERS AND FITTINGS. TOTAL PANEL LENGTH IS OVER SEVEN FEET

hat crowns and skin. Gun drilling was used to produce these holes. Figure 7 shows a section through a hat crown and the aluminum end fittings. Fiberglass pads were used on each side of the graphite/epoxy hat crown for isolation and gap accommodation. The total section is approximately 1.75 inches thick with the graphite/epoxy composite being 96 plies thick under the fitting. This hole was drilled in approximately two minutes. No pilot was used, nor was any finish reaming required. Surface finish of the as-drilled aluminum was 65 RMS, while the composite was 125 RMS. Table II summarizes the diameters of eight holes drilled for one fitting through a hat crown.

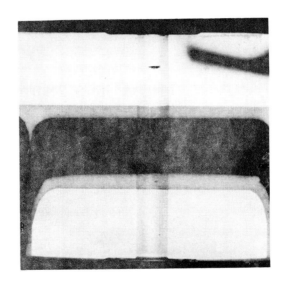

FIGURE 7. PHOTOGRAPH OF SECTION THROUGH HAT CROWN WITH END FITTINGS SHOWING GUN DRILLED HOLE

TABLE II. HOLE DIAMETERS OF ONE END FITTING

	HOLE NO. DIAMTER IN INCHES							
	1	2	3	4	5	6	7	8
Outer Aluminum	.2506	.2506	.2515	.2515	.2508	.2505	.2530	.2520
G/E Composite	.2507	.2508	.2530	.2510	.2510	.2510	.2512	.2510
Inner Aluminum	.2502	.2506	.2505	.2505	.2505	.2508	.2507	.2514
	DRILL DIAMETER - .2510							

It is representative of the results obtained on the panel.

Hole tolerance requirement was .250 to .253. It was found that results improved as operator skill and knowledge increased with experience. Results shown in Table II were from an early single element test specimen. Normally, ten to twelve holes could be drilled before repointing was required. A sharper drill is required to cut aluminum than graphite epoxy. Drills were changed as soon as any slowing of the drill speed was noted. Holes were drilled as small as 3/16 inch diameter but 1/4 inch was generally considered the minimum practical size for aluminum/composite combinations. Drill shanks of the smaller diameters can be over-torqued quite easily. This generally ceases to be a problem at diameters of 5/16 and greater.

A view of the end fittings being drilled on the surface panel skin is shown in Figure 8. Tooling consisted of a tapered drill block which was "C" clamped to the work piece until the end holes were drilled, then pinned through the holes. Over 500 holes were required to install end fittings on the panel. Three were damaged due to drill breakage and required oversize fasteners. The panel was exposed to humidity (160°F and 95% RH) to an average moisture level of 1% before starting cycling loading. No failures were experienced during fatigue testing. Narmco 5208/T300 graphite/epoxy was used to produce the surface panel.

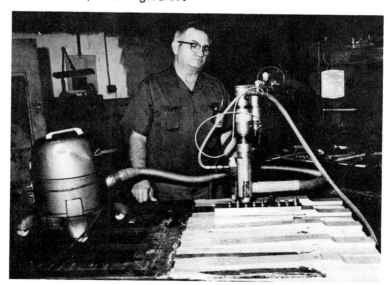

FIGURE 8. VIEW SHOWING SURFACE PANEL SKIN BEING DRILLED WITH GUN DRILL

4. BREAKOUT CONTROL

Backup is required to prevent splintering of graphite/epoxy composite when drilling with gun drills. Aluminum and wood were both satisfactory if tightly clamped to the exit surface. It was found that one ply of style 120 glass cloth integrally molded to the exit surface would generally prevent splintering, but one ply of style 181 glass cloth was found to be superior.

5. CONCLUSIONS

Gun drilling is a very effective procedure for drilling fastener holes in graphite/epoxy composite and aluminum composite combinations. The deeper the hole, the more likely that gun drilling will be the preferred method. The hard tooling required by a Quackenbush is the major disadvantage to using gun drilling. Initial drill cost is comparable to other special carbide drills that might be used on graphite/epoxy composites such as a Winslow Spacematic, but none of the competitive systems can be repointed as many times as a gun drill.

Some preliminary studies have been conducted on drilling titanium composite combinations. The present point configuration and feed rate is not satisfactory for titanium and will require additional study.

Reference:

(1) S. M. Freeman, C. D. Bailey, R. T. Beall, and J. M. Hamilton, "Detection and Verification of Internal Fiber Fracture Around Loaded Holes in Graphite/Epoxy Composites," 1979 SAMPE Technical Conference, Boston, Massachusetts.

Biography
R. T. Beall is a Development Engineer, Specialist in the Engineering Structural and Materials Laboratory of Lockheed-Georgia Company, Marietta, Georgia. He is a graduate of the Georgia Institute of Technology and has been employed by Lockheed for 28 years.

196 holes per shot in boron-epoxy

When drilling a multitude of holes in this composite material seemed inefficient, Sikorsky Aircraft turned to punching. It's glad it did

OF THE VARIOUS fiber-reinforced-resin composites used in the aerospace industry, none is more troublesome to drill than boron-epoxy. Even diamond-impregnated core drills, which are quite costly, wear out quickly when penetrating the abrasive fibers.

When design requirements for four 2.10-in.-wide x 90.5-in.-long, boron- and glass-reinforced-epoxy stabilator straps for the UH60A military helicopter called for 190-196 holes per strap, ultrasonic core drilling with the diamond tools was our initial approach. As production requirements for the aircraft increased, however, this method was abandoned in favor of building two dies at a total cost of $75,000 to punch all of the holes in each strap simultaneously.

No one regrets the changeover. In the two years or so that the dies have been in production, we've punched the holes in about 1000 straps and have accrued annual savings in direct labor and drill costs well in excess of the $226,000 initially estimated (Fig 1), providing a return on investment in just about four months.

The straps comprise as many as 14 layers of boron-epoxy and eight layers of E-glass-epoxy in laminate fashion: one layer of glass-reinforced material, followed by two boron-reinforced layers, then one glass-reinforced, two boron-reinforced, etc. In some of the glass-reinforced layers, fiber orientation is $+45°$ to the length of the strap; in others, $-45°$. Most of the boron fibers run lengthwise ($0°$), although some are oriented at $90°$—that is, transverse to length.

Besides this random array of continuous fibers, the straps vary in thickness—from 0.037 in. at the ends, tapering to a 0.164-in. flat area at the center. The holes, largely 0.176-0.180 in. dia except for a few 0.2045-0.2085 in. dia, are located throughout the strap at section thicknesses ranging from 0.099 in. to 0.164 in.

Two dies were needed because each of the two strap configurations has its own hole pattern. Two of each are required per aircraft. The one shown on the die after punching (see photo) has a very slight bow in the horizontal plane. The other is straight in the center portion and angles out somewhat at each end. A cross-section of the die appears in Fig 2.

Worn drills necessitated reaming

When we planned to drill the holes, we realized that drill life would be short—say, ten true holes per drill. Rather than having to replace the $60 drills this soon, we decided to run the drills longer and ream undersize holes to size. This, we believed, would markedly extend drill life. It would also, of course, necessitate the purchase of expensive, diamond-plated expansion reamers.

To meet production schedules while awaiting delivery and installation of ultrasonic-drilling equipment and the manufac-

By M.M. Schwartz, manager, Manufacturing Technology, and **S. Kosturak,** manufacturing engineer, Sikorsky Aircraft, United Technologies, Stratford, Conn

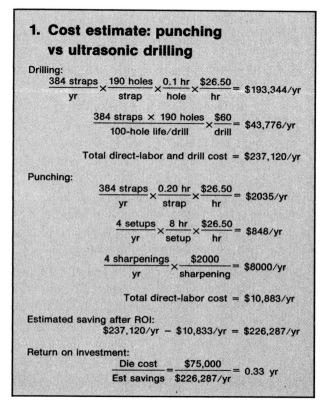

1. Cost estimate: punching vs ultrasonic drilling

Drilling:

$$\frac{384\ \text{straps}}{\text{yr}} \times \frac{190\ \text{holes}}{\text{strap}} \times \frac{0.1\ \text{hr}}{\text{hole}} \times \frac{\$26.50}{\text{hr}} = \$193{,}344/\text{yr}$$

$$\frac{384\ \text{straps} \times 190\ \text{holes}}{100\text{-hole life/drill}} \times \frac{\$60}{\text{drill}} = \$43{,}776/\text{yr}$$

Total direct-labor and drill cost = $237,120/yr

Punching:

$$\frac{384\ \text{straps}}{\text{yr}} \times \frac{0.20\ \text{hr}}{\text{strap}} \times \frac{\$26.50}{\text{hr}} = \$2035/\text{yr}$$

$$\frac{4\ \text{setups}}{\text{yr}} \times \frac{8\ \text{hr}}{\text{setup}} \times \frac{\$26.50}{\text{hr}} = \$848/\text{yr}$$

$$\frac{4\ \text{sharpenings}}{\text{yr}} \times \frac{\$2000}{\text{sharpening}} = \$8000/\text{yr}$$

Total direct-labor cost = $10,883/yr

Estimated saving after ROI:
$237,120/yr − $10,833/yr = $226,287/yr

Return on investment:

$$\frac{\text{Die cost}}{\text{Est savings}} = \frac{\$75{,}000}{\$226{,}287/\text{yr}} = 0.33\ \text{yr}$$

ture of related tools, we turned to a subcontractor that had the necessary equipment and had established speeds, feeds, and settings for boron-epoxy in an earlier development program.

After a number of straps had been so produced, it was determined that drill life was about 100 holes and that it took about 0.1 hr, specifically 6.5 min, to produce each hole. This involved an 11-step procedure, including not only drilling (2 min) but inspecting each hole and, when required, reaming (0.75 min) to size. Thus, with production rates of 384 straps per year at the time, total direct labor and drill costs alone would amount to roughly $237,000 annually.

As production rates for the UH60A increased, it became apparent that neither Sikorsky nor the subcontractor could economically meet schedule requirements by this approach. This, combined with the cost of the operation, prompted the company to seek alternatives. When the design-engineering section determined that critical weight distribution in the stabilator precluded switching to aluminum, titanium, or steel for these particular straps (others were converted to metal), the manufacturing-technology department investigated the feasibility of punching the holes.

All 190-196 holes in the composite straps are now die-punched simultaneously in this 1000-ton-rated brake press

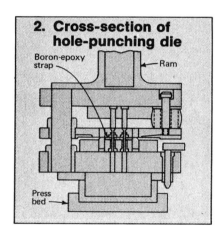

2. Cross-section of hole-punching die

Boron-epoxy strap

Ram

Press bed

3. How CPM 10V compares with other tool steels

Steel	Hardness, Rc	Wear resistance,[a] 10^10 psi	Impact strength,[b] ft-lb
CPM 10V	63	90	17
CPM 10V	60	75	26
CPM 10V	56	64	22
CPM 10V	52	51	25
D2	62	4	17
M2	64	8	17
M4	63	20	12
CPM M4	64	22	31

[a]Reciprocal of wear rate in unlubricated crossed-cylinder wear test involving contact with rotating tungsten-carbide cylinder.
[b]Charpy C-notch based on standard-size specimen and notch depth with ½-in. notch radius

Punching all of the holes in a single operation and performing related setup operations shouldn't take more than 0.20 hr/strap, we figured. Scheduling requirements were such that we probably would need four setups annually and four resharpenings of the dies. As noted in the cost estimate, this would involve a direct labor cost of approximately $11,000/yr. Thus, the resulting savings would more than justify die cost.

Because of its wear resistance and toughness (Fig 3), Crucible Steel's CPM 10V, a powder-metal tool steel, was chosen as the material for both dies and punches. A hardness of Rc 60, we believed, would optimize wear resistance and toughness.

Three punch-face configurations—dished, chisel, and flat—were evaluated; the flat proved best. Various punch-and-die clearances were also tested; 0.004 in. was the most suitable.

Tests also indicated that a lubricant on punches and dies would prolong tool life and ease punch withdrawal from the holes. Of the several tried, Du Pont's Microlon, a dry-film lubricant containing Teflon, provided the best results.

Punches were designed with a 0.75-in. usable length and would be capable of producing approximately 75 holes before requiring resharpening. Each regrind would remove 0.025 in. Therefore, estimated punch life was 2250 holes (0.750 in./0.025 in. x 75).

Punched samples were sectioned, inspected for delamination, and found to be essentially equivalent to the drilled and reamed straps. Consequently, the punched straps were accepted by the structural-analysis section for production use.

This does not mean, however, that hole punching would necessarily be allowed in all applications for boron-epoxy or, for that matter, for other composites. Because drilling is still generally preferred for structural reasons, each application must be considered individually.

The dies are mounted on a Cincinnati Inc brake press rated at 1000 tons. Set-up and punching of the straps is much the same as for punching metal except that blank-holding pressure is greater. Lubricant is spray-applied to both punches and die after each punching operation. Once the strap has been placed on the die, punching all 190-196 holes takes about 10 sec. ■

ULTRASONICALLY ASSISTED METAL REMOVAL

Janet Devine

ABSTRACT

Practical application of vibratory energy to various metalworking processes has been an objective of intensive study and development over a period of years in Government and industrial laboratories. This work has addressed the technical and operational problems relating to all metal removal processes, including lathe turning, boring, twist drilling, trepanning, and tapping.

1. INTRODUCTION

The application of ultrasonic energy to lathe turning, boring, twist drilling, and other metal removal processes has resulted in substantial improvements — in the process and in the product. As a result of investigations into these cutting processes, ultrasonic devices have been produced which can be readily and economically adapted to shop machines. The wide range of ultrasonically machinable materials we have processed, all with observed ultrasonic effects, is summarized in Table 1.

TABLE 1
ULTRASONICALLY MACHINED MATERIALS

Material	Process		
	Drilling	Turning	Boring and Trepanning
Ferrous Alloys			
Carbon Steel	x	x	x
Tool Steel		x	
Cast Iron	x		
Nonferrous Alloys			
Aluminum	x	x	x
Copper	x		
Titanium	x	x	
Nickel	x		
Ceramics			
Alumina		x	x
Magnesium			
Silicate		x	x
Graphite			x

The ability to apply ultrasonic energy to industrial metal removal first required the evolution of ultrasonic equipment capable of withstanding high tool forces — in the range of several thousand pounds — but that capability alone was not sufficient. There were three accomplishments which were critical to the practical, effective use of ultrasonics in the machine shop:

(1) Force-insensitive mounting systems
(2) Rigid tool holders
(3) Adjustability of tool cutting angles, as commonly used in shop work.

It was discovered that the ultrasonic power requirements for dramatic improvements in the removal of metals were low — in the range of 100 to 300 acoustical watts. Improvements in ceramics were achieved at even lower power levels.

This development effort has progressed, over the past 10 years, to our present level of partial standardization of ultrasonic prototype systems for heavy-duty shop equipment.

2. DISCUSSION

Ultrasonic Twist Drilling

Apart from drill material and lubricant refinements, the basic twist drilling process in use nearly a century ago remains unchanged, despite the fact that the normal twist drill is a very poor load carrying structure. Ultrasonic energy offers a new mode of delivering metal removal energy to the cutting edge, which is less demanding of the drill structure.

Thrust reductions encountered in ultrasonic twist

drilling have varied from 30 percent in copper and cast iron to 54 percent in titanium (6-4-2).

Torque reductions have varied from 25 percent in drilling mild steel to 50 percent in titanium (6-4) and 65 percent in 6061-T6 aluminum.

Rate increases have varied from a factor of 2 in drilling cold-rolled steel to a factor of 4 in titanium (6-6-2) and a factor of 5 in aluminum (1100-F).

The chips obtained with ultrasonic drilling, as illustrated in Figure 1, are characterized by greater curl radius and smoother chip surface and edges than chips obtained with conventional machining.

Figure 1. Chips Obtained During Machining With (Above) And Without Ultrasonic Assistance.

In twist drilling, it has been repeatedly observed that chip expulsion from the hole is facilitated with ultrasonics, alleviating the need for periodic retraction of the drill. Drill depth in titanium, for example, is normally limited to 2 to 4 times the drill diameter. With ultrasonic activation, depths of 8 times the diameter have been successfully drilled in titanium (6-4) without a drill retraction. In 6061 aluminum, no difficulty has been encountered in achieving depths of 20 times the diameter (i.e., 1/8-inch holes which were 2½ inches deep).

A concomitant effect is evidenced in the breakout patterns of the 6061 aluminum specimens shown in Figure 2. Note the pre-breakout bulging of the control specimen. With ultrasonics, breakout is continuous and sharp.

The data given in Figure 3 illustrate the tool loads encountered in twist drilling a titanium alloy. After 6 inches of non-ultrasonic drilling, torque and thrust loads had increased to an unacceptably high level. When the ultrasonic system was activated and using the same drill, these loads decreased markedly and remained within an

Figure 2. Breakout Patterns During Drilling Of Aluminum With (Above) And Without Ultrasonic Assistance.

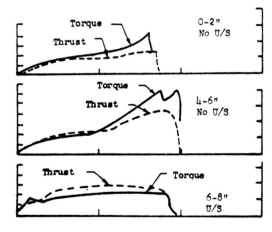

Figure 3. Tool Loads Encountered In Twist Drilling A Titanium Alloy With And Without Ultrasonic Assistance.

acceptable range for the remaining 44 inches of experimental drilling.

Although cutting tool wear is difficult to evaluate quantitatively, its qualitative effects include:
(1) Increased surface roughness on the machined part
(2) Increased dimensional variations
(3) Increased torque and thrust loads.

Ultrasonics has been demonstrated to alleviate these problems and extend tool life by reducing tool wear. Examination of the cutting edges of drills after equivalent periods of ultrasonic and non-ultrasonic cutting showed different wear patterns. As evident in Figure 4, the non-ultrasonically operated drill was worn at the outer periphery, while the ultrasonically operated drill wore evenly along the full cutting edge.

Figure 5 shows the tool load buildup experienced in drilling titanium (6-4) at an accelerated material removal rate. In non-ultrasonic drilling, torque and thrust had increased significantly by the end of the first 2-inch specimen. With ultrasonics, the tool loads remained low even after 50 inches of cutting without a tool change, and there was no indication that the terminal life of the tool had been approached.

Figure 4. Cutting Edges Of Drill Bits After Ultrasonic (Above) And Non-Ultrasonic Drilling.

Figure 5. Tool Load Buildup During Drilling Of Titanium.

Figure 6. Small-Diameter Ultrasonic R&D Drilling System.

Figure 6 is a small-diameter drill used to obtain measurements in drilling 15-mil holes.

Figure 7 shows a Universal-Automatic drill head modified for ultrasonic activation and capable of drilling carbon steel hole sizes up to 3/8-inch diameter. An advanced-model drill, Figure 8, is capable of drilling holes up to 3/4-inch diameter in mild steel.

Figure 7. Universal-Automatic Drill Head Modified For Ultrasonic Drilling.

Figure 8. Advanced Model Ultrasonic Drill.

Ultrasonic Lathe Turning

Ultrasonic machining systems for effectively turning outside diameters have been under development in our laboratories for a number of years. Our experience indicates that ultrasonic lathe turning results in quantitative and qualitative improvements in the process.

Quantitatively, this process has reduced the force required to turn Vasco steel by 30 percent and carbon steels C1010 and C1045 by 50 percent. Similar benefits have been noted in the turning of alumina, where force reductions of 30 percent have been noted.

This process has also produced marked increases in the cutting rate — by factors of 4 in aluminum (2024-T6), 3 in Vasco steel and 5 in cutting 1045 carbon steel. The turning of non-metallic materials has also shown increases, when assisted ultrasonically, by a factor of more than 2 in the turning of alumina and more than 4 in magnesium silicate.

In some instances, work could be accomplished only with ultrasonic assistance. For instance, a good cut was produced in low-porosity mullite with an ultrasonic tool post and a carbide-tipped tool, but when the ultrasonic power was turned off, the workpiece immediately shattered.

The qualitative improvements obtained with ultrasonically assisted turning have been no less impressive. Figure 9 shows the surface profile of machined 1018 carbon steel with and without ultrasonic assistance. The bottom photograph exhibits subsurface working effects evidenced in the plastic flow pattern and surface tearing. These effects are reduced to the point of elimination with ultrasonically assisted turning.

Conventional machining usually produces a glossy surface resulting from tearing, material enfoldment, and burnishing. Ultrasonically machined surfaces are generally characterized by a mat-like finish, with evidence of a more complete shearing of the chips from the bulk material (see Figure 10).

Normal machining of materials such as graphite, zirconia, and certain metals, which have a microstructure of strong particles held together in weak matrix boundaries, results in grain pullout. With ultrasonic activation of the tool, material is removed as fine grains.

With these improved surface characteristics, it seems likely that ultrasonically machined items will have an improved fatigue life. There is experimental evidence to this effect.

The ultrasonic tool post in Figure 11 is a fourth or fifth generation configuration. Readily mounted on any lathe, this model uses standard, interchangeable tools and incorporates electronic circuitry capable of providing automatic operation in production. Axial vibration at 28 kilohertz produces oscillations of the tool tip tangential to the surface of the workpiece.

Power levels are in the range of 375 acoustic watts. Several of these units are in industrial plants on pilot-production machining applications.

Figure 10. Steel Billet Machined Alternately With And Without Ultrasonic Assistance.

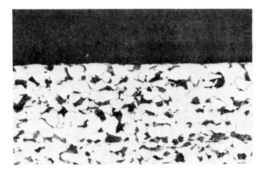

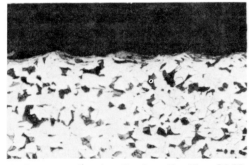

Figure 9. Sections showing Surface Profiles Of Machined 1018 Carbon Steel With (Above) And Without Ultrasonic Assistance (100X).

Figure 11. Modern Equipment Configuration.

Ultrasonic Boring and Trepanning

The benefits accrued in ultrasonic turning have also been evidenced in ultrasonic boring and trepanning. Experimentation has demonstrated a 56 percent reduction in the torque required for the machining of aluminum (2024-T6) and a 58 percent reduction in 1018 carbon steel. Material removal rates in both cases have been in the order of 2½ times that of conventional machining.

Similar improvements have been noted in the machining of non-metallic materials, notably an increased removal rate by a factor of 6.3 in the case of porcelain, 10 in zirconia, and 12 in graphite (particles R_C 60 in friable matrix).

The initial ultrasonic boring work involved artificial graphite, which was machined in a trepanning configuration with carbide-tipped cutters (Figure 12).

Figure 13. Commercial Model Boring Tool.

Figure 12. Ultrasonic Boring Of Graphite.

Figure 13 shows an ultrasonic boring tool employed in a lathe tool mode to facilitate torque and thrust measurements. The unit provides tool activation tangent to the cylinder via a torsionally activated ultrasonic system. The high stiffness requirements of the long support structure necessitated the use of a dual force-insensitive mount system. This versatile equipment, capable of use for both ID and OD machining, is currently being used in an industrial capacity.

Other Ultrasonic Machining Experience

One important advantage of ultrasonic machining which has not yet been discussed is its ability to reduce or completely eliminate machine tool chatter. This phenomenon has been noted in ultrasonic drilling, turning, boring, and other removal techniques.

A graphic illustration of the elimination of chatter was obtained by the single-point machining of hard (Knoop 2500), high-density, alumina. This was accomplished with single-point diamond tools, by attaching the specimen to the X-Y table of a precision surface grinder, as shown in Figure 14.

Figure 14. Array For Machining Alumina.

Using this experimental apparatus, the sample shown in Figure 15 was machined in a series of surface cuts. The chatter encountered with non-ultrasonic machining (shown by the dotted points of contact) immediately ceased when the tool was ultrasonically activated.

The small plane and saw frame shown in Figure 16 was instrumented to measure ultrasonic vs. non-ultrasonic tool forces, rates, etc. Primarily used for non-metallic cutting, this equipment has yielded data which has been extrapolated to hand-held tools. Most recently the apparatus has been used for the ultrasonic knife-edge cutting of Viton, a fluoro-carbon plastic filled with inorganic powder.

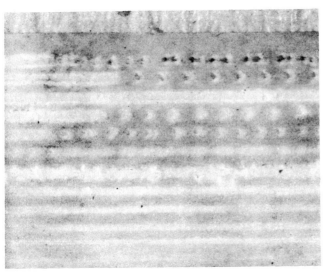

Figure 15. Single-Point Diamond Cutting Of Alumina.

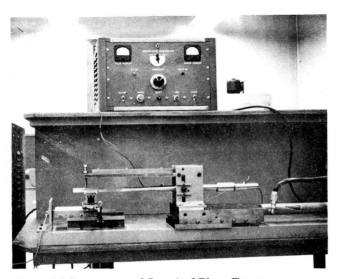

Figure 16. Instrumented Saw And Plane Frame.

3. CONCLUSIONS

Ultrasonically augmented point-machining offers substantial potential for improving material removal processes:

(1) Material removal rates are increased.
(2) Tool loads are reduced.
(3) Surface finish is improved.
(4) Subsurface deformation is reduced.
(5) The altered chip characteristics indicate that cutting is facilitated.
(6) Tool edge buildup is eliminated.
(7) Tool life is extended.

This potential is borne out by the recent comments of two of our customers who purchased ultrasonic point machining equipment. One, whose high-precision specialty machining work requires extraordinary surface integrity,

said that more than satisfactory results were obtainable with ultrasonics and that this had proved especially helpful with his most complicated items and difficult materials. Chips were noted to curl away from the newly machined surface, thus avoiding marring, work-piece breakage was reduced, and certain operational limitations were broadened. His production staff competed for the use of his several ultrasonic units. Another customer, who used ultrasonic drilling in an extensive investigation of process improvement, found that since ultrasonics facilitates chip clearance in deep drilling, far fewer tool retractions were required, and significant time savings were obtained.

The benefits of ultrasonic assistance to point cutting, then, lie in increased productivity for ordinary machining operations, solution of recalcitrant material removal problems, and production of superior machined items.

Janet Devine received her B.S. in Physics, Pure and Applied Mathematics in London, England. Her early experience included several years in the aerospace industry at British Aircraft Co., Ltd. Since joining Sonobond Corporation, she has been responsible for developing acoustical theory and practical application of new concepts utilizing ultrasonic technology. As Vice President of Product Development, she has conducted and directed programs relating to a fundamental understanding of ultrasonics and its application to metal processing such as machining, drilling, welding, and forming. Mrs. Devine has published numerous technical papers on ultrasonic developments. She is a member of the Acoustical Society of America.

SXA ALLOY DATA SHEET: **MACHINING OF SXA ALLOYS**

Silicon Carbide (S) isotropically reinforced (X) aluminum (A) alloys have been introduced by ARCO Metals Company. These alloys combine the high stiffness of silicon carbide and the excellent strength, corrosion resistance and fabricating characteristics of aluminum base alloys. Matrix alloys prepared by ARCO include 1100, 2024, 2219, 5083, 5456, 6061, 7075, 7050, X7090 and X7091. They can all be machined on standard equipment (lathes, drill presses, shapers, milling machines, and grinders). Figure 1 shows the results of various

Figure 1: Illustration of drilling, tapping and tapering operations.

drilling, tapping and tapering operations while Figure 2 shows a nut and bolt fabricated utilizing standard turning, milling, threading, tapping and knurling operations.

Lathe turning is one of the most common machining operations. Diamond composition or brazed carbide tools, as well as carbide insert tools are suitable for rough turning. Diamond tools have the best life and give the best surface finish. However, carbide tools show better resistance to interrupted-cut turning than diamond tools. Due to the silicon carbide in SXA alloys, the chip produced by the three types of tools is usually short and discontinuous, rather than the long and semi-continuous chip normally produced with wrought aluminum alloys.

Suggested turning and grinding practices for SXA Alloys are listed in the Table on the reverse side. Further information about aluminum — silicon carbide alloys and their application may be obtained by contacting Mr. Jack Cook at 803-877-0123.

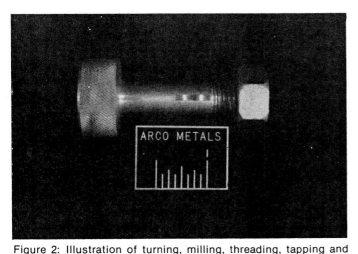

Figure 2: Illustration of turning, milling, threading, tapping and knurling operations.

Reprinted with permission from ARCO Metals Company, 1981

337

TABLE — SUGGESTED MACHINING PRACTICE FOR SXA ALLOYS

ROUGH TURNING

Tool: Diamond Composition, Carbide Inserts, or Brazed Carbide Tools. Types A-6 to A-8 or C-6 to C-8.

Rake Angle: Negative 5-7° *Headstock Speed:* 900-1500 RPM

Feed: 3-6 IPM *Cut:* 0.025-0.060 Inches

Cutting Fluid: None (See Note 1)

FINISH TURNING

Tool: Diamond Composition Tool, Types A-6 to A-8 or C-6 to C-8.

Rake Angle: Negative 5-7° *Headstock Speed:* 900-1500 RPM

Feed: 3-6 IPM *Cut:* 0.025-0.050 Inches

Cutting Fluid: None (See Note 1)

Resultant Surface Finish: 15-25 Micro-Inches

CYLINDRICAL GRINDING

Wheel: A-100-K-5-V or Similar Type

Headstock Speed: 400-500 RPM *Table Speed:* 3-4 IPM

Grind Depth: 0.002-0.006 Inches, Single Pass Only Required.

Grinding Fluid: None

Resultant Surface Finish: 10 Micro-Inches

Note 1: Aluminum/silicon carbide is commonly turned using no lubricants. When built-up-edge (BUE) may be a problem, such as at low silicon carbide concentrations, or when it is desired to cool the materials being turned, oil or water soluble lubricants may be used.

The suggested machining practices are to be considered of a developmental nature, and are intended as guidelines only. ARCO Metals Company makes no warranty, expressed or implied, as to their performance results.

DEVELOPMENT OF EFFECTIVE MACHINING AND TOOLING TECHNIQUES FOR KEVLAR COMPOSITE LAMINATES

Rod Doerr, Ed Greene, B. Lyon, S. Taha
HUGHES HELICOPTERS, INC.
Centinela and Teale Streets
Culver City, Calif.   90230

INTRODUCTION AND SUMMARY

This program was initiated to investigate/develop tooling and techniques needed to perform clean, efficient machining operations on Kevlar laminates. Kevlar is the trade name for an aromatic polyamide fiber developed by E. I. DuPont. This high strength, low weight reinforcement is presently being used in many structural and nonstructural aircraft applications. The investigation included two different Kevlar laminates, one with an Epoxy matrix and one with a polysulfone matrix.

Kevlar reinforced laminates have presented the aerospace industry with machining difficulties due to their tendency to fibrillate or "fuzz" when machined with conventional tools. (See Figure 1 below.) Considerable data was generated in attempting to solve this problem; much of it was conflicting and applied only to specific applications. The cost of production has suffered due to ill-defined parameters such as feeds, speeds, and tool configuration.

Figure 1.   Fibrillation or "Fuzzing"
Caused by Router Cut

Reprinted courtesy of Applied Technology Laboratories, U. S. Army Research and Technology Laboratories (AVRADCOM), June 1982, 45 pages (Loaned by Hughes Helicopters, Inc.)

In this program a cross section of conventional tools, specialty tools, and modified tools was systematically evaluated. Advanced machining techniques were also evaluated. Recommendations based on these studies were made and incorporated into a separate document, The Design Guide Handbook (Reference 1).

The overall objectives of this program are:

- Determine and evaluate state-of-the-art cutting tools and machining techniques

- Identify deficiencies of existing tools and techniques

- Identify and verify tools and machining techniques to eliminate/reduce deficiencies

The program was organized into the following major tasks:

Phase I - Technology Survey

Survey of literature and selected vendors/developers to determine the state of the art relative to tools, processes, and techniques.

Phase II - Development of Evaluation

Evaluation of tooling and equipment identified in Phase I; modification of selected tools; investigation of state-of-the-art cutting equipment; investigation of Polysulfone/Kevlar laminate machining.

Phase III - Manufacturing Methods Development

Evaluation of production scale-up problems; investigation and integration of automation into the production line.

---

[1] Design Guide Handbook Efficient Machining Methods and Cutting Tools for Kevlar Composite Laminates, November 1981, Hughes Helicopters, Inc.

PHASE I - TECHNOLOGY SURVEY

A DOD/NASA and HHI library literature search initiated the technology survey. This survey covered manufacturers' data, technical conferences, periodicals, and Government-sponsored studies. In addition, eight selected vendors were visted in order to determine the latest state of the art in Kevlar laminate machining. The information collected covers tools, processes, and techniques. Appendix A, Technical Data, and Appendix B, Vendors/Developers, present detailed listings of documents reviewed and vendors visited during this program.

In general, considerable composite laminate machining data was available. However, only a small portion of this information was specific for Kevlar. Kevlar specific tools were offered by a limited number of vendors. Of the 51 technical reports reviewed, only 6 reports contained data concerning conventional tools specifically designed or modified for Kevlar machining. Some specific information on Kevlar was available from machining literature covering nonconventional cutting equipment such as the water jet and the laser.

Reports on state-of-the-art cutting methods present some data specific to Kevlar, with three sources presenting material specific to Kevlar laminates or Kevlar and other composites. The Grumman Aerospace Corp's study, "Manufacturing Methods for Cutting, Machining, and Drilling Composites", includes some limited specific Kevlar laminate machining information within the considerable data covering composite machining techniques. The balance of the surveyed literature covers associated areas such as production scale-up and production automation information, and metal cutting tools that have possible use in Kevlar laminate machining.

The following reports were found to be valuable during this program. They contain information which applies specifically to machining cured Kevlar composite laminates:

- "Cutting and Machining of Kevlar Aramid and Its Composites"
  Louis H. Miner, Frank T. Penoza
  21st National Symposium SAMPE, Apr. 1976

- "Industrial Tools for Cutting Kevlar"
  Airtech International Inc.
  P.O. Box 2930, Court M-1
  Torrance, CA 90590

- "Industrial Tools for Cutting Kevlar Fabric and Laminates"
  Pen Associates, Inc.
  2639 W. Robino Drive
  Wilmington, Del. 19808

- "Speciality Tools for Machining Fiber Reinforced Composite
  Structures"
  Technology Assoc. Inc.
  P.O. Box 7163
  Wilmington, Del. 19803

- "A Guide to Cutting and Machining Kevlar"
  DuPont Inc. Textile Fibers Dept.
  Centre Rd. Building
  Wilmington, Del. 19898

- "Cutting Fabric, Prepreg and Kevlar Aramid"
  Sept. 19, 1975 DuPont Textile Fibers Dept.
  (Same as above)

- "Characteristics of Laser Cutting Kevlar Laminate"
  R.A. Van Cleave, Jan. 1979
  Prep for Dept. of Energy Contract DE-AC04-76-DP00613
  Bendix, Kansas City Div.

- "Laser Cutting of Kevlar Laminates"
  R.A. Van Cleave, Bendix Corp., Kansas City
  September 1977, BDX-613-1877

- "Flow Technology Report No. 7
  Waterjet Cutting of Advanced Composite Materials"
  Flow Industries, Inc.
  21414 68th Ave. South
  Kent, Washington 98031

- "Manufacturing Methods for Cutting, Machining and Drilling
  Composites"
  Vol. 1 Composites Machining Handbook, Vol. 11 Test and Results
  AFML-TR-78-103 Vol. 1 and Vol. 11, August 1978
  Grumman Aerospace Corp.
  Bethpage, N.Y. 11714

PHASE II - DEVELOPMENT OF EVALUATION

During Phase II, the performance of tools identified in Phase I was verified and their capabilities and limitations were further defined. Selected tools were modified and evaluated for increased efficiency. In addition, state-of-the-art machining procedures were examined. Tools and equipment were evaluated for machining rate and quality on both epoxy/Kevlar and polysulfone/Kevlar laminates.

TEST SPECIMEN PANEL DESCRIPTION

The epoxy/Kevlar panels were fabricated using vacuum pressure. Materials used were APCO 2434/2347, and style 281 Kevlar fabric. Thicknesses were 0.06, 0.12, 0.18 and 0.38 inch.

The polysulfone/Kevlar panels were autoclave cured using P1700 polysulfone and cross-plied unidirectional Kevlar for one panel and Kevlar 285 fabric for the other. Thicknesses were 0.06 and 0.12 inch respectively. The target resin content for all test panels was 40 ±5 percent.

MODIFIED TOOLS

Modified Drill Bits — A special tip grinding procedure was developed for modifying standard drills into drills capable of long run, quality Kevlar laminate drilling. Conventional drills are designed to start cutting the holes in the center. This forces chips against the walls of the hole, creating heat and swelling in a Kevlar laminate. The modified drills cut material toward the center of the drill and minimize packing. In addition, the modified drills cut with a shearing action at the hole diameter, eliminating fibrils and making a clean hole. The modified tip configuration is shown in Figure 2.

Three different standard drills were modified: one high speed steel (HSS) twist drill, one carbide twist drill, and one parabolic flute drill. (See Figure 3.)

The results of the testing of these three modified drills are presented in the following sections. The best results were obtained using the modified parabolic flute drill bit (Figure 3, bottom bit), probably because it offers more relief for the removal of chips. (See Figure 4.) These drill bits were effective both off-hand and bushing guided.

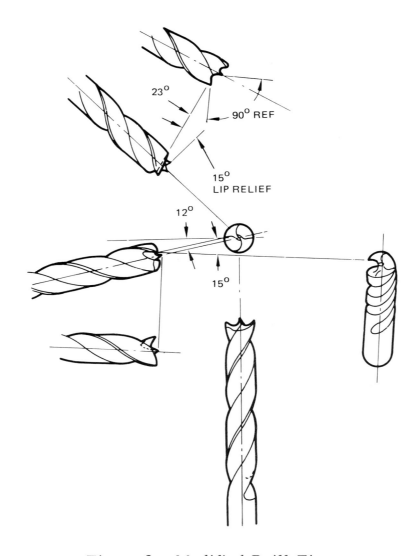

23°

90° REF

15°
LIP RELIEF

12°

15°

Figure 2. Modified Drill Tip

Figure 3. Modified Drill Bits

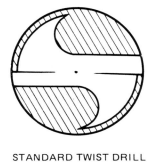

STANDARD TWIST DRILL

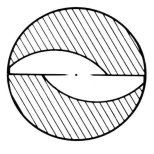

PARABOLIC FLUTE DRILL

Figure 4. Cross-Section View of Standard Twist and
Parabolic Flute Drills (Shows the Difference
in Chip Relief Area)

The modification of the standard drills for this program was done freehand
on a conventional bench grinder. However, in order to maintain the center
on the drill point and hence an accurate hole diameter, a grinding fixture is
recommended. Figure 5 illustrates the procedure for modifying a standard
drill.

The modified drill resembles the old "thin sheet" drill; however, there are
important differences that, while they are not easily noticed, greatly affect
performance. The steps illustrated in Figure 5 show the procedure for
modifying drill bits freehand.

The setup for Steps 1 and 2 shows the tool rest at a 15-degree angle. This
becomes the lip relief angle of the drill. The front view of the grinding wheel
shows a 35-degree angle line which is the angle at which the drill is fed into
the grinding wheel. The 35-degree angle allows the center of the drill to
extend beyond the forward cutting points of the land after the drill has been
ground.

In grinding Steps 1 and 2, the rotational position of the drill bit is very
important. The chisel edge must be in a vertical position so that the grinding cut is parallel to the chisel edge angle. After grinding one lip, the
drill is rotated 180 degrees to grind the opposite lip.

The final grinding steps, Steps 3 and 4, are done without the tool rest. The
drill bit is ground on each side to reduce the helix angle, forming a cutting
edge at the center of the web and bringing the chisel edge to a point.

Source: AVRADCOM, June 1982, 45 pages

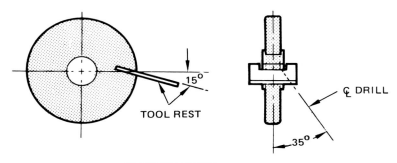

SETUP FOR STEPS NO. 1 & NO. 2

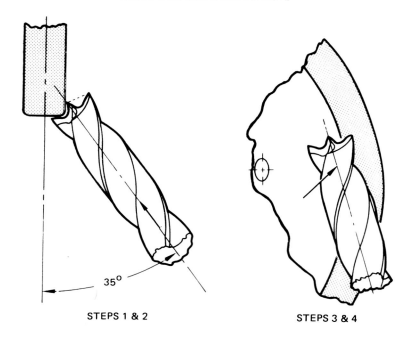

STEPS 1 & 2   STEPS 3 & 4

Figure 5. Grinding Procedure for Modifying Standard Drill

Modified Circular Saw Blade — A 60-tooth carbide tipped blade was specially ground to cleanly and efficiently cut Kevlar laminates. After several attempts, a grinding procedure was developed that resulted in a noticeable improvement in cutting over standard carbide tipped blades. Figure 6 shows the tooth angles of the specially ground blade.

Finishing Mill Cutter — A mill cutter was designed especially for edge finishing Kevlar laminates. (See Figure 7.) This mill cutting tool is limited in depth of cut, but is effective for cleaning up rough edges.

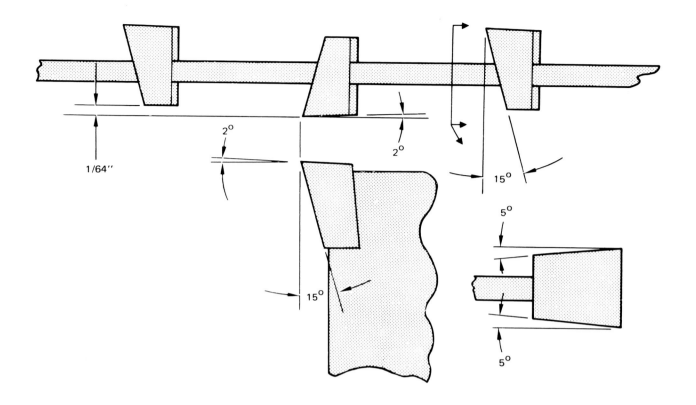

Figure 6.   Tooth Angles of Specially Ground Carbide
Tipped Circular Saw Blade

EVALUATION OF CONVENTIONAL TOOLS

Contrary to popular belief, Kevlar is not appreciably more difficult to
machine than graphite or fiberglass. Although the Kevlar fibers are tough,
they are easily cut by sharp tools with the proper cutting angles. It should
be noted, however, that since heat over 350° F tends to permanently swell the
edges of the cut, tools must be kept cool during the machining operations.

In the study of the conventional tools, a brief preliminary evaluation was
conducted on a wide variety of tools, and the tools that appeared less
efficient were dropped from evaluation. Although this type of evaluation was
necessary because of time constraints, some tools may have been prematurely
eliminated. (See Table 1, Conventional Tool Evaluation.)

Drill Bits — Among the drills evaluated were the Standard HSS twist, solid
carbide twist, single flute (spade), and standard parabolic flute; in addition,
the Standard HSS twist, solid carbide twist, and parabolic flute were each
evaluated with the modified tip.

Source: AVRADCOM, June 1982, 45 pages   347

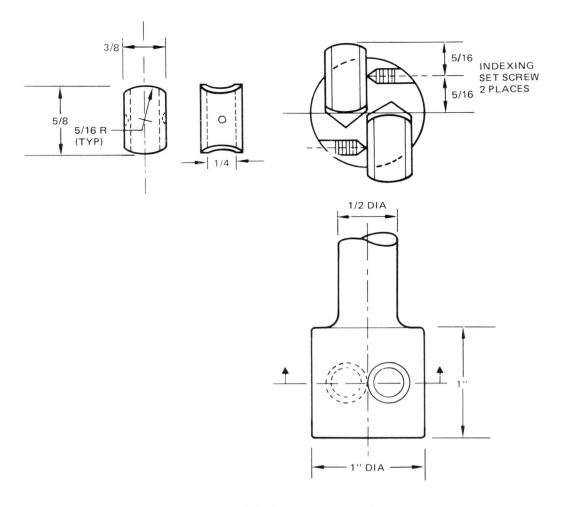

Figure 7. Mill Cutter, Special Design

Figure 8. Hole Quality Using Standard HSS Drill

TABLE 1. CONVENTIONAL TOOL EVALUATION MATRIX

TOOL DESCRIPTION	SIZE	SPEED	CUTTING RATE	EVALUATION QUALITY	EVALUATION COMMENTS
Circular Saw Carbide Tipped	7½" 40 Tooth	3500 SFPM	4 Mat'l 5 FPM	Good	Light fuzz downside only.
Remington Grit Edge Circular	10" Dia.	4500 SFPM	6 FPM	Good	Not recommended due to grit loading.
Cutoff Blade-Composition	10" Dia.	4500 SFPM		Poor	Generates heat and swells edges.
Band Saw Wavy Set	¼" 32 Tooth	650 SFPM	4 Mat'l 2 FPM	Good	
Band Saw Knife Edge	½"	1000 SFPM	10 FPM	Good	32 plies 281 style dry woven Kevlar.
Band Saw Scalloped Knife Edge	½"	1000 SFPM	10 FPM	Poor	32 plies 281 style dry woven fabric.
Saber Saw - Bosch T118A	1⅝" 20 TPI	191 SFPM	2 FPM	Good	
Saber Saw - Bosch T101P		191 SFPM	4 FPM	Good	
Saber Saw - Airtech AR3925		191 SFPM	2 FPM	Good	Special blade high cost.
Saber Saw - Bosch T130 Course		191 SFPM	1 FPM	Good	Grit coated blade for special applications
Saber Saw - Knife Edge Bosch 118A	2¼"	191 SFPM	10 FPM	Good	2 plies 281 style dry woven fabric.
Scroll Saw					Not available.
Drill HSS Standard Twist	⅜" 135°	600 RPM		Poor	Rejected.
Drill Solid Carbide Twist	⅜" 135°	600 RPM		Poor	Rejected.
Carbro Single Flute Drill	⅜"	600 RPM		Poor	Rejected.
Turboflute Drill	⅜"	600 RPM		Poor	Rejected.
Countersink 2 Flute Standard		600 RPM		Poor	Rejected.
Weldon Countersink		600 RPM		Good	Overheats fast.
Router - Opposed Helical	¼"	24,000 RPM	6"/min.	Good	Carbide bit broke-diameter too small.
Router - Diamond Cut Carbide	¼"	24,000 RPM		Poor	Rejected.
Hog Mill - Omcut	1"	4,000 RPM	6"/min.	Good	Suitable for heavy sections.
Bosch Nibbler Model 7501	14 GA	1460 Strokes per min.	2 FPM	Excellent	Clean and economical.
Circular Saw Carbide Teeth (Modified) Mod #1	10" 60 Tooth	4500 SFPM	6 FPM	Good	No advantage over standard blade.
Countersink Serrated		600 RPM	N/A	Poor	Unsatisfactory finish.
HSS Twist Drill Modified	¼", ⁵⁄₁₆", ⅜", ½"	600 RPM	24 sec/in.	Good	Excellent with backup.
Solid Carbide Twist Drill (Modified)	⅜"	600 RPM	20 sec/in.	Good	Excellent for graphite composite.
Turboflute or Paraflute Drill (Modified)	⅜"	750 RPM	24 sec/in.	Good	Excellent with backup.
Router - Special Design					
Circular Saw Carbide Teeth (Modified) Mod #2	10" 60 Tooth	4500 SFPM	6 FPM	Good	More effective than standard.

All of the unmodified drill bits except the spade bit produced unsatisfactory holes that were rough, fuzzy, and had backside delaminations. The spade drills produced satisfactory holes in the thinner laminates, but being centerless they required guides or bushings. In addition, the spade drill was delicate and subject to a high breakage rate.

The modified twist drill and the modified parabolic flute made satisfactory holes in laminate ranging in thickness from 0.06 to 0.38 inch. The drills performed satisfactorily off-hand with a drill press and using drill bushings. During Phase III tests, 100 holes were drilled in a 0.18-inch-thick laminate without a decrease in hole quality.

The modified parabolic flute is the most effective of all the evaluated drills, probably because of the greater chip relief area discussed previously. (See Figure 9.)

Countersinks — Three countersinks were evaluated: a standard severance two-fluted bit with micro stop, a four-fluted severance countersink with serrated lips (made especially for Kevlar), and the Weldon deburr and countersink tool. (See Figure 10.)

Of the three countersinks evaluated, the Weldon countersink tool made the cleanest countersink surface, as is shown in Figure 11. The standard two-fluted countersink left a rough countersink surface and dulled quickly. The serrated four-fluted countersink had better tool life but again made a rough surface.

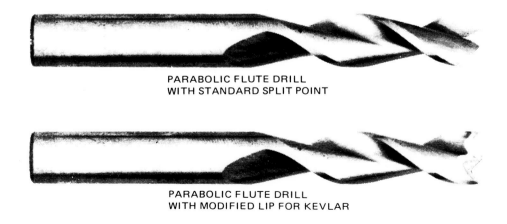

PARABOLIC FLUTE DRILL
WITH STANDARD SPLIT POINT

PARABOLIC FLUTE DRILL
WITH MODIFIED LIP FOR KEVLAR

Figure 9. Parabolic Flute Drill Modification

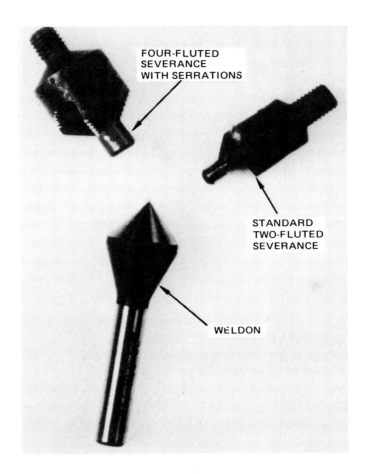

Figure 10. Countersinks

Figure 11. Countersink Evaluation

The Weldon countersink does not require modification. It is designed with the proper cutting configuration for Kevlar composites. However, care must be taken to prevent overheating, and a water coolant is recommended.

Router Bits and Mill Cutters — The result of an edge trimmed with a diamond cut carbide router (Figure 12) is shown in Figure 1. Most conventional routers produce the same results.

The opposed helical router (Figure 13), which is designed especially for routing Kevlar composites, produces a very satisfactory cut. However, care must be taken to keep the laminate on the cutting center line. In addition, the configuration makes the bit somewhat subject to breakage, and cutting speeds less than the usual 20,000 rpm are recommended for the smaller diameter bits.

Figure 12. Diamond Cut Router Bit

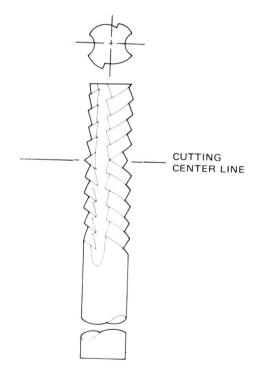

CUTTING
CENTER LINE

Figure 13. Opposed Helical
Router Bit

A 1-inch-diameter Hogmill (Figure 14) was used to route the edge of a 0.38-inch-thick Kevlar panel. The cutter appears to be a satisfactory method of removing thick sections of material, although the cut is left with slight lateral grooves and requires finishing.

The specially designed mill cutter discussed in the Modified Tools section and shown in Figure 7 is effective for finishing Kevlar laminate edges.

Circular Saws — Since circular saws make only straight cuts, they are somewhat limited in their application to aircraft composite production. These blades do have a wide thickness capability and easily cut the full range of thicknesses investigated in this program (0.06 to 0.38 inch).

For most applications a standard 60 to 64 tooth carbide tipped blade is sufficient. For more demanding jobs a carbide tipped blade ground to the configuration shown in Figure 6 will give superior results.

Best results were obtained at a cutting speed of 4,500 surface feet per minute (SFPM). Also, a water coolant is required for heavier cuts. A small amount of fuzz clings to the bottom edge of the cut, but it can be easily removed with a finishing wheel.

Grit-edged circular saw blades make a clean cut but become clogged with Kevlar fibers after a few inches.

The composition blade used for metal cutoff purposes is useful for cutting cross sections of investigative samples of Kevlar with metal inserts, but here again, the blade becomes clogged within a few inches.

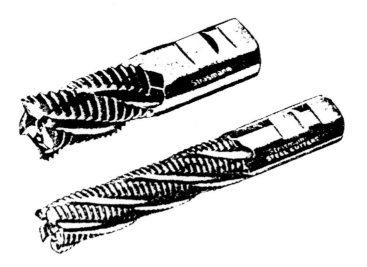

Figure 14. Hogmill

Source: AVRADCOM, June 1982, 45 pages

Figure 15 shows the carbide tipped and the grit-edged saw blades.

Sabre Saw Blades — Five sabre saw blades were evaluated using an air-driven sabre saw with a 1-inch stroke and a 2,000- to 2,500-stroke-per-minute capability. (See Figure 16.) This saw was also equipped with a blade orbiter; however, cuts made with the blades using straight reciprocation were cleaner and faster.

CARBIDE TIPPED   GRIT EDGED

Figure 15. Circular Saw Blades

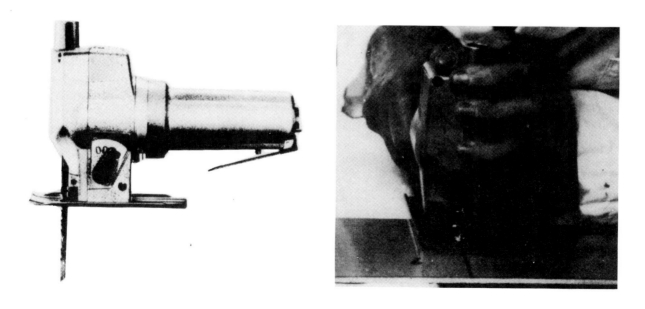

Figure 16. Sabre Saw

Of the five sabre saw blades investigated (see Figure 17), the 22-tooth-per-inch wavy set blade made one of the best cuts. A special Kevlar cutting blade with a repeated pattern of reversed teeth made an equivalent cut but is a much more expensive tool. A 10-tooth-per-inch wavy set blade made a faster but somewhat rougher cut. A coarse, carbide grit blade also made a satisfactory cut and did not have the severe clogging problem exhibited by the grit-edged circular saw blades.

An unsuccessful attempt was made to use a knife-edged blade for cutting thin laminates. However, this blade proved outstanding for cutting dry Kevlar fabrics.

Band Saw Blades — Most band saw operations are done as a rough trim close to a scribed line. A sander and finishing wheel are then used to complete the trimming to the line. With the exception of this inability to precision cut, band sawing is effective for Kevlar laminate trimming.

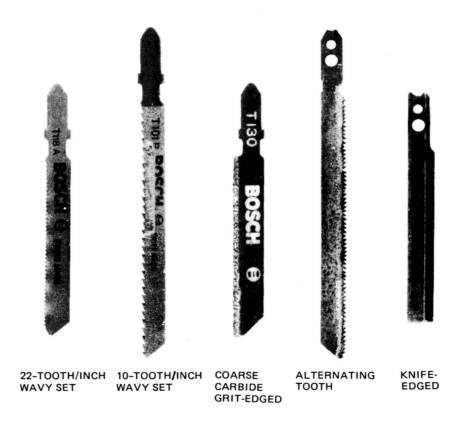

| 22-TOOTH/INCH WAVY SET | 10-TOOTH/INCH WAVY SET | COARSE CARBIDE GRIT-EDGED | ALTERNATING TOOTH | KNIFE-EDGED |

Figure 17. Sabre Saw Blades

Source: AVRADCOM, June 1982, 45 pages

Of the many band saw blades available, the standard blades perform as well as any. A 1/4-inch-wide, 32-tooth-per-inch wavy set blade is recommended for cutting Kevlar laminates up to 0.19 inch thick. A 1/2-inch-wide, 22-tooth-per-inch wavy set blade is recommended for thicker laminates (see Figure 18). Specially ground band-saw blades do not appear to have an appreciable advantage over the much less expensive standard blades discussed above.

As with the sabre saw blades, a knife-edged band saw blade is very effective for cutting stacks of dry Kevlar fabric.

Table 2 gives recommended cutting speed for various Kevlar laminate thicknesses.

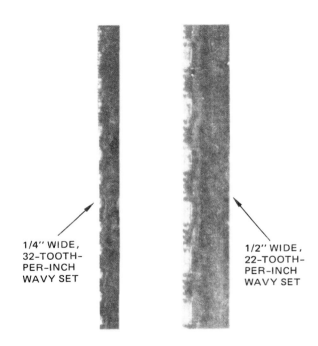

1/4" WIDE, 32-TOOTH-PER-INCH WAVY SET

1/2" WIDE, 22-TOOTH-PER-INCH WAVY SET

Figure 18. Band Saw Blades

TABLE 2. RECOMMENDED SPEEDS FOR
BANDSAWING VARIOUS LAMINATE
THICKNESSES

THICKNESS (IN.)	0.060	0.125	0.180	0.250	0.50	1.0
BAND SAW TYPE	SURFACE FEET PER MINUTE					
1/4" 32-TOOTH WAVY SET	1500	1200	900	600	–	–
1/2" 22-TOOTH WAVY SET	1500	1200	1000	1000	800	600

Nibbler — The nibbler (see Figure 19), which cuts by punching a series of interconnecting holes, is also an effective Kevlar laminate trimming tool. The nibbler operates at 1,400 to 1,600 strokes per minute and produces a kerf width of 0.2 inch. The nibbler is effective up to a thickness of 0.09 inch. The cut edge requires light cleanup.

Finishing Disks and Wheels — As noted previously, many cutting operations leave an irregular surface or some fuzz on the bottom side of the cut. The edges can be cleaned up using sanding disks and sanding wheels (see Figure 20).

Where very light cleanup is needed, a sanding wheel (synthetic fiber wheel shown in Figure 20) will do an adequate job by itself. For rougher cuts, sand off rough edges with a 16-grit sanding disk, smooth out with an 80-grit disk, and finish off with the sanding wheel.

EVALUATION OF STATE-OF-THE-ART MACHINING EQUIPMENT

Two advanced technology cutting methods are finding acceptance throughout industry: the laser and the water-jet cutter. The principal work of this program was directed to water-jet cutting technology because of both the availability of the water-jet to HHI and the belief that the water-jet is more adaptable to the aircraft production environment.

In addition to these rapid cutting systems, the slow cutting reciprocating diamond coated wire saw was also investigated.

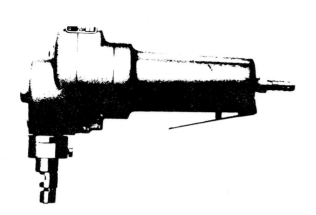 

Figure 19. Nibbler

Source: AVRADCOM, June 1982, 45 pages

Figure 20. Abrasive Disks and Finishing Wheel

Laser — Laser cutting of Kevlar/epoxy laminates produces a sharp,
clean edge with little discoloration at speeds unrivalled by other methods.
However, the depth of cut is restricted (approximately 3/8 inch for a
1,000 watt laser), and expertise is required by the operator because of the
dangers of high voltage, radiation exposure, and hazardous fumes.

The laser lends itself to automated cutting of flat or slightly contoured
parts, but has not to date been used for complex contour part machining.

Two references are presented below that detail laser cutting techniques
and methodology.

1. For general laser information:

   Lasers in Modern Industry
   By John F. Ready

Published by:
Society of Manufacturing Engineers
One SME Drive, P.O. Box 930
Dearborn, Michigan 48128

2.   For specific Kevlar laser cutting:

Laser Cutting of Kevlar Laminates
By R.A. Van Cleave
Ref: BDX-613-1877
         U.S. Dept. of Commerce
         National Technical Information Service

Water-Jet — Water-jet cutting is accomplished by severing material with a highly columnated stream of water, traveling at high speed and under extreme pressure. The schematic below (Figure 21) illustrates the basic components of the water-jet cutter. The major components are a low pressure pump, intensifier, accumulator, on-off valve, nozzle, and catcher (drain).

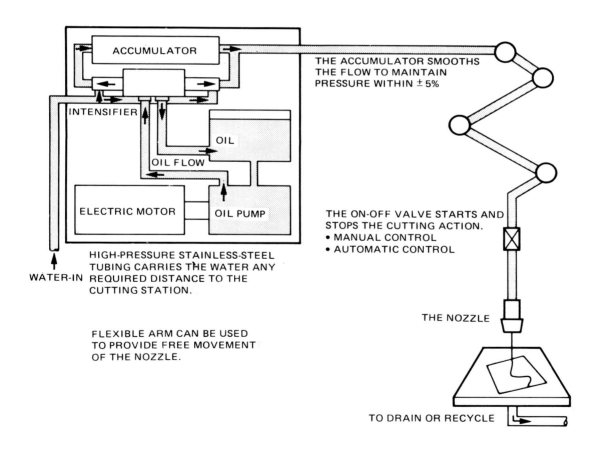

Figure 21.   Water-Jet Schematic

HHI evaluated two different water-jet setups. The major portion of the work was done with a water router. This is a hand-held device with an overhead counterbalance support (see Figure 22).

This system was effective to about a 1/8 inch thickness. Thicker specimens began to show roughness and backside delaminations. In order to eliminate nozzle maintenance and orifice wear, the pressure setting is limited to 50,000 psi, resulting in a working pressure of about 45,000 psi.

The water router was evaluated for cutting speed and quality of cut through the full range of Kevlar laminate thicknesses under investigation. The results of this evaluation are presented in Table 3, Water-Jet Router Evaluation. The test results show the quality of the cut beginning to degrade at 0.18 inch thickness with unsatisfactory cuts at 0.31 inch thickness.

The second water-jet evaluated was a fixed position nozzle with a variable-speed feed table. The pressure on this equipment was set at 55,000 psi. This fixed position water-jet was easily able to cut the maximum thickness laminate investigated on this program (0.38 inch) and appeared to have capabilities up to 1.0 inch thickness. See Table 4, Fixed Heat Water-Jet Evaluation.

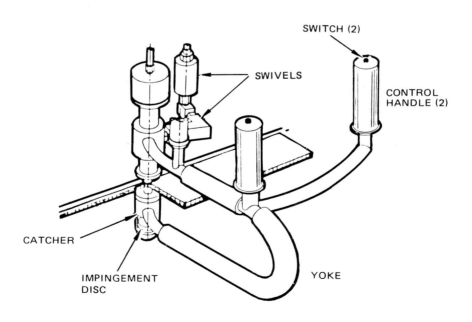

Figure 22. Water-Jet Router

TABLE 3. WATER-JET ROUTER EVALUATION

THICKNESS (IN.)	JEWEL HOLE DIA. (IN.)	PRESSURE (PSI)	CUTTING TIME (SEC)	AREA OF CUT (IN²)	CUTTING TIME FOR 100 IN² (MIN)	QUALITY
0.125	0.012	45,000	41	6.	11.4	FAIR
0.125	0.012	45,000	28	6.	7.77	FAIR
0.125	0.012	45,000	57	6.	15.83	FAIR
0.180	0.012	45,000	100	8.64	19.29	POOR
0.180	0.012	45,000	90	8.64	17.36	POOR
0.180	0.012	45,000	105	8.64	20.25	POOR
0.060	0.012	45,000	14	2.88	8.1	GOOD
0.060	0.012	45,000	11	2.88	6.3	GOOD
0.060	0.012	45,000	8.5	2.88	4.9	GOOD
0.060	0.016	45,000	14	2.88	8.1	GOOD
0.060	0.016	45,000	9	2.88	5.2	GOOD
0.060	0.016	45,000	12	2.88	6.9	GOOD
0.180	0.016	45,000	134	8.64	25.84	FAIR TO GOOD
0.180	0.016	45,000	131	8.64	25.27	GOOD
0.180	0.016	45,000	66	8.64	12.73	GOOD
0.125	0.016	45,000	45	6.	12.49	GOOD
0.125	0.016	45,000	47	6.	13.05	GOOD
0.125	0.016	45,000	44	6.	12.22	GOOD
0.310	0.012	45,000	140	14.88	15.68	POOR
0.310	0.016	45,000	186	14.88	21.04	POOR

TABLE 4. FIXED HEAD WATER-JET EVALUATION

Thickness (in.)	Working Pressure (psi)	Orifice Size (in.)	Cutting Speed (FPM)	Results
0.06	50,000	0.007	10	Good
0.38	50,000	0.012	1	Good

A new development in water-jet cutting is the oscillating head. The water-jet head is oscillated in such a manner that a tapered kerf is cut in the laminate; in this way the jet is not dissipated against the walls as rapidly and thicker cuts are possible (see Figure 23).

Although the oscillating head is not needed to cut the maximum thickness investigated in this program (0. 38 inch), it may be advantageous for much thicker cuts.

Reciprocating Wire Saw — Kevlar laminates of any thickness can be cut with a reciprocating wire saw. (See Figure 24.) This reciprocating mechanical wire cutter uses diamond impregnated wire, sized 0. 003 through 0. 016 inch diameter, that reciprocates on a grooved drum. The saw cuts with a type of lapping action and does not generate appreciable heat. It does not load, and produces a cut of very high quality.

Unfortunately, the very slow cutting speed of this equipment relegates it to specialized cutting applications such as sample cutting. The ability of the wire saw to cut Kevlar laminates in combination with any hard material makes it a good method for cutting investigative samples.

Specifics of the panel cutting operation are shown in Table 5.

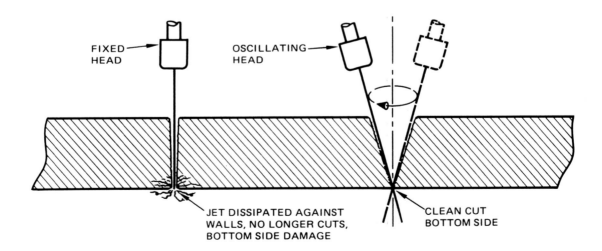

Figure 23. Theory of Oscillating Head

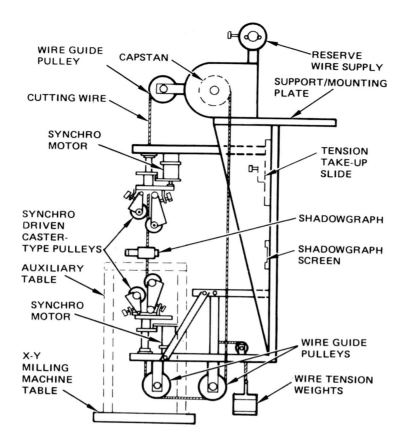

WIRE GUIDE PULLEY

CAPSTAN

CUTTING WIRE

SYNCHRO MOTOR

SYNCHRO DRIVEN CASTER-TYPE PULLEYS

AUXILIARY TABLE

SYNCHRO MOTOR

X-Y MILLING MACHINE TABLE

RESERVE WIRE SUPPLY

SUPPORT/MOUNTING PLATE

TENSION TAKE-UP SLIDE

SHADOWGRAPH

SHADOWGRAPH SCREEN

WIRE GUIDE PULLEYS

WIRE TENSION WEIGHTS

Figure 24. Reciprocating Wire Saw

MACHINING POLYSULFONE/KEVLAR PANELS

Fabrication of the two polysulfone/Kevlar test panels was described on page 13. Each panel was made with a different reinforcing system - one having woven fabric and the other a cross-plied unidirectional reinforcement.

The woven Kevlar fabric/polysulfone panel machined easily using the same tools that were acceptable for machining the woven Kevlar/epoxy panels; however, a little care must be exercised to prevent generation of excess heat. The unidirectional Kevlar/polysulfone test panel, on the other hand, does not machine well using the tools in a normal manner. It was very difficult to prevent delamination of the outer unidirectional plies on the surfaces adjacent to the cut or hole. The only acceptable cuts were made by using the nibbler and the water-jet. It is expected that the reciprocating wire saw and the laser would work as well.

Acceptable holes were made only by casting plaster on each side of the panel and drilling with the modified tip parabolic flute drill.

Source: AVRADCOM, June 1982, 45 pages

TABLE 5. RECIPROCATING SAW CUTTING EVALUATION

SAMPLE CUT NO.	THICKNESS AND LENGTH (IN.)	TIME OF CUT (MIN)	TIME PER 100 IN^2 (HRS)	AREA OF CUT (SQ IN.)	WIRE DIA (IN.)	TENSION ON WIRE (GR)	WEIGHT ON WIRE (GR)	WEIGHT ON YOKE	SPEED SETTING	SLURRY USED	COOLANT USED
1	0.38 x 6	50	36.549	2.28	0.010	3500	40	1 MAIN WT + 100 G	80	SILICONE	WATER
2	0.38 x 6	27	19.7	2.28	0.010	3500	60	1 MAIN WT + 100 G	80	SILICONE	WATER
3	0.195 x 6	16	22.79	1.17	0.010	3500	60	1 MAIN WT + 100 G	80	SILICONE	WATER
4	0.121 x 6	8	18.36	0.726	0.010	3500	60	1 MAIN WT + 100 G	80	SIICONE	WATER
5	0.38 x 6	21	15.35	2.28	0.010	3500	160	1 MAIN WT + 100 G	80	SILICONE	WATER

PHASE III - MANUFACTURING METHODS DEVELOPMENT

During Phase III production scale-up, characteristics of hole drilling in relation to the quality, cost, and tool durability were evaluated. Tool life under normal operating conditions, tool life using special tooling materials, and tool life improvement when using coolants and special coating were evaluated.

Production application problems were also addressed and concepts were developed for production machining of complex aircraft components. Automation methods were studied and integration of automated machining into labor efficient production lines was conceptualized.

DRILL LIFE STUDY

During Phase II testing it was determined that the modified tip drill bits out-performed all other drill bits evaluated on this program. In order to determine relative production life expectancy of modified tip drill bits with different materials in different configurations, three distinct bits were evaluated: high speed steel (HSS) standard twist, cobalt standard twist, and HSS parabolic flute. One hundred holes were drilled with each bit into an 0.18-inch-thick laminate. (See Figure 25.) In all three cases the one-hundredth hole had the same quality and took the same time to drill as the first hole. The average time for drilling each hole was 6 seconds. The drills were cooled with water as frequently as was necessary.

From these drilling exercises, it was concluded that the working life of the modified HSS drill was adequate for most Kevlar laminate drilling operations. The more expensive drill bit materials such as cobalt or solid carbide need only be used when drilling Kevlar hybrids or combinations with fiberglass or graphite.

The modified parabolic flute drill was the least expensive and produced the cleanest holes. This is probably due to its greater chip relief area.

MACHINING BACKUP, COOLANTS, AND COATINGS

Machining Backup — Even with the best of cutting tools some of the top and bottom Kevlar fibers adjacent to the cut or hole may be broken loose when machining without a backup. Ideally, the laminate should be clamped between two flat plates of hardwood, or acrylic sheets. Laminates that

Source: AVRADCOM, June 1982, 45 pages

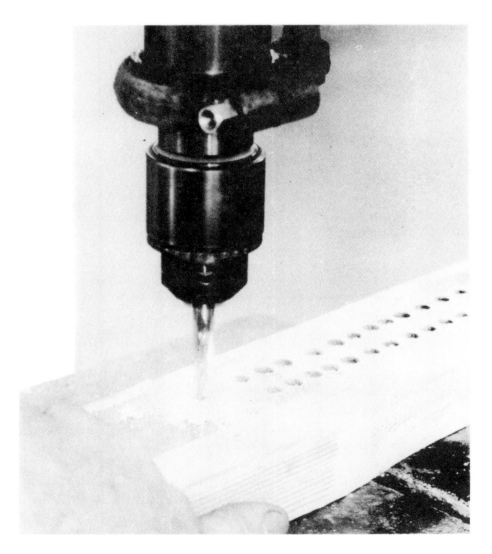

Figure 25. Drill Exercise, 100 Holes

are machined with a down acting shear component, such as the circular saw, band saw, or drill, perform with less bottom side fiber dislocation or delaminating if a backup board is used.

Coolants — All Kevlar laminate cutting tools should be held below 350°F during machining operations to prevent swelling of the cut edge. This can generally be accomplished by keeping the tool sharp and tool speed at a low operating level. However, when maximum production is required a coolant will allow machining at a faster rate.

Water has been found to be an inexpensive and entirely satisfactory coolant for Kevlar laminate machining. It can be applied as either a spray or periodic dip.

<u>Coating and Release Agents</u> — Dry lubricants are useful on circular saws, band saws, drills, and countersinks. Dry lubricants prevent galling and allow chips to move more freely without sticking to the tool.

Tools may also be coated with liquid release agents such as Release-all. Best results are obtained by applying three coats of the release agent and then baking for 1 hour at 300°F.

Drill bushings are very damaging to drills, and drill guides (adapters which locate the drill without touching the lands) should be used whenever possible. Bushings, when used, should always be kept well lubricated by using dry lubricants or fluorocarbon sprays. New bushings should be broken in with old drill bits.

PRODUCTION SCALE-UP AND AUTOMATION

<u>Production Cutting System</u> — The tedious tasks of trimming Kevlar composites with saws and hand grinding are being eliminated by the use of optically or numerically controlled machining equipment. The water-jet setup with either of these control systems could handle the thickness range investigated in this program (0.06 to 0.38 inch) and eliminate the dust control problem associated with cutting and grinding. The Phase II investigation indicated a controlled feed water-jet has Kevlar laminate thickness cutting capabilities up to approximately 1 inch. Figure 26 shows an optically controlled system for water-jet cutting flat Kevlar laminates, and Figure 27 shows a biaxial, tape controlled water-jet trimming system.

<u>Robotics</u> — The employment of lightweight robotics for the trimming and drilling of complex Kevlar parts is in the early stages of development. Present off-the-shelf equipment lacks the necessary precision needed for jig-free machining. However, considerable effort is being expended by the robotics industry to develop the necessary precision and repeatability.

It is noted that the milling, routing, drilling, and countersinking tools identified in Phase II would be suitable for robotic machining, as would the laser and the water-jet.

Figure 28 shows a typical robotic station employing a rotary cutter.

Figure 26. X-Y Optical Tracing System, Water-Jet
Cutting of Flat Sheet

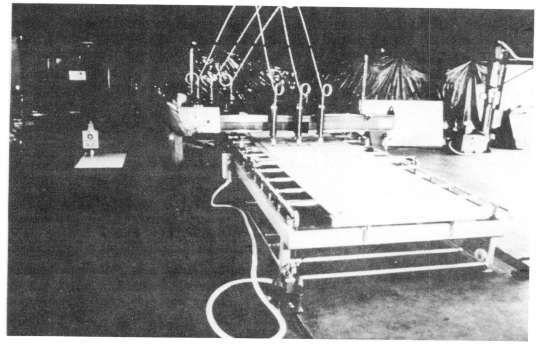

Figure 27. Tape Controlled Water-Jet Trimming System

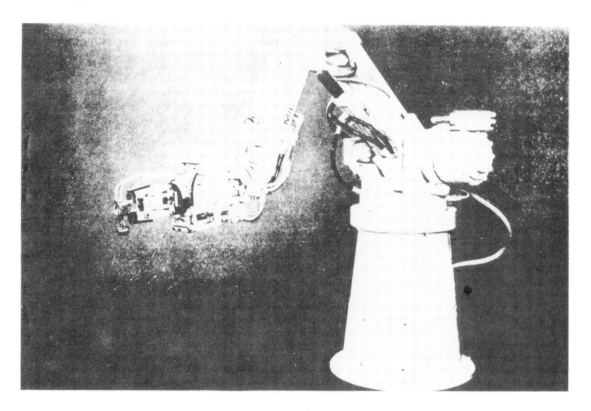

Figure 28. Robotic Trim Station

INTEGRATION OF AUTOMATION INTO ADVANCED COMPOSITE PRODUCTION LINES

The performance evaluation in the first segment of this program has shown that tools and equipment presently available have suitable quality and durability for incorporation into automated production lines. The use of a spray mist of water will be adequate in most cases to keep the tools cool and achieve long tool life. Both the water-jet and the laser appear to be adaptable to automated machining.

In this section we will discuss the integration of automated Kevlar composite machining equipment into typical automated composite fabrication lines. The discussion can best be presented by breaking it into three paragraphs:

- Present Day - Equipment available for production today.

- Near Term - Equipment that will be available in 3-5 years.

- Future - Equipment that will be production ready in 10-15 years.

Source: AVRADCOM, June 1982, 45 pages

Present Day — Presently available are the numerical control (N/C) (tape controlled) and optical tracing trimming equipment. Figure 29 shows a five-axis N/C water-jet trimming station integrated into a present-day production line along with an N/C controlled pattern cutter.

Near Term — In the near term, the automated composite production line is expected to incorporate computer storage and self-contained tooling. The contribution in the automated machining area in addition to N/C machining will be robotic drilling. As near-term robotics will not yet have the accuracy and repeatability for precision aircraft hole patterns, this equipment will be used in conjunction with drill templates for accurate location. Figure 30 shows the robotic drilling station integrated into the composite production line as well as the N/C trimming station.

Future — Future composite production lines will further reduce costly labor by introducing robotic handling, automated curing, and automated non-destructive testing. The automated machining contribution will be in the form of robotic trimming and drilling of sufficient accuracy such that no trimming or drilling jigs will be required. Figure 31 shows the future composite production line with the automated production equipment integrated into the system.

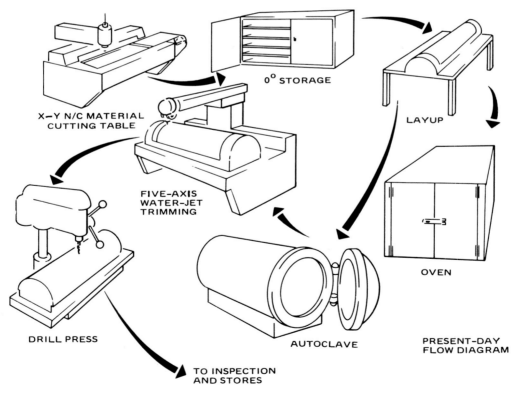

Figure 29. Present-Day Automated Composite Production Line

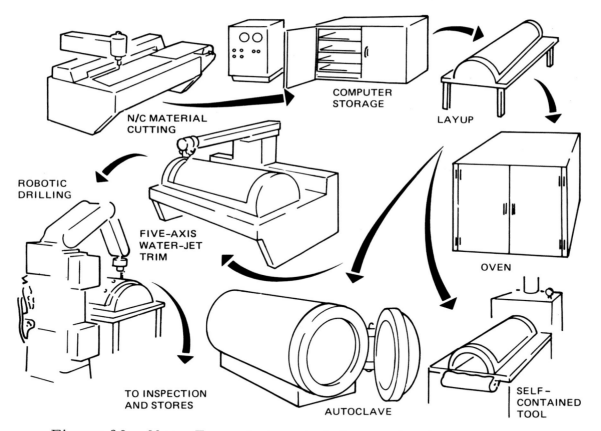

Figure 30.   Near-Term Automated Composite Production Line

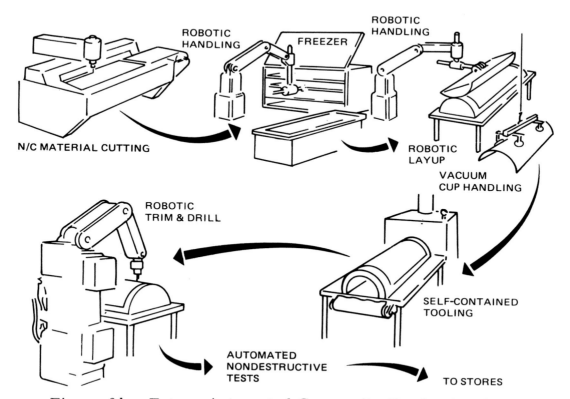

Figure 31.   Future Automated Composite Production Line

Source: AVRADCOM, June 1982, 45 pages

CONCLUSIONS AND RECOMMENDATIONS

The technology survey of current tools and practices performed during this program shows that for many applications existing tools and technology are adequate. In the area of cutting, recommended tools and practices are adequately developed and verified for cutting of raw materials (yarns, rovings, and pre-impregnated fabrics). However, during the evaluation phase of this program, many of the recommended tools for cutting laminates proved ineffective for laminates reinforced with Kevlar 49. Notable exceptions were the laser, water-jet, and reciprocating wire saw.

In the area of machining of cured Kevlar/epoxy and Kevlar/polysulfone laminates (drilling, countersinking, counterboring, and milling operations) special tools are required in many cases and only a limited range of feeds and speeds may be successfully used. Particular attention must be given to cutting the outermost layers of the composite laminate. Many of the tools recommended for these operations were found to have limited cost effectiveness in a production environment.

Evaluation of cutting tools and systems was followed by a study of advanced machining methods that are presently being introduced and those envisioned for the future. Near-term applications such as multi-axis and automated trimming and robotic drilling were found to be in the initial stages of development. Both the water-jet cutter and the laser that were found to be efficient cutting systems for both Kevlar/epoxy and Kevlar/polysulfone laminates may be readily adapted for automated trimming systems.

Outstanding problems on which resources could be allocated and approaches recommended include the following:

a. Complex Shapes

   Kevlar machining activity to date has concentrated on the machining of simple shapes which are either flat or of simple contour. Techniques for machining the complex shapes found commonly among aircraft components are needed.

b. Thick Laminates

   The present program performed extensive studies on laminates up to 0.38 inch thick and limited study on cutting laminates up to 1.0 inch thick. Additional study is needed to develop techniques to cut and machine laminates of 1/2- to 2-inch thickness, for both straight and complex cuts.

c.   Automated Trimming

In order to push forward the concepts already under initial study, two areas need investigation and development.

1.   Universal, automated trim stations with capabilities for trimming different configurations of complex parts

2.   Specialized trim stations with accurate, low-cost tools designed for automated trimming of a specific part (applicable to inline production tooling)

d.   Routing/Milling Techniques

The routing/milling tools developed under this program have characteristics (rough cut, required precise cutter positioning) that limit their use. Further work is needed to locate/develop more versatile cutters.

APPENDIX A

LITERATURE SURVEY

PAMPHLETS AND CONFERENCES

"A Solution to the Problem of Drilling Reinforced Plastics"
35 Annual Tech Con. 1980
Reinforced Plastics/Composites Inst.
Society of the Plastics Ind. Inc.

"Fabrication of Low-Cost Primary Aircraft Structures with Thermoplastic
Composites"
Bicentennial of Materials
Processings, Oct 12-14, 1976

"Problems in Testing Aramid/Epoxy Composites"
Lawrence Livermore Lab, L.I. Clements

"Cutting and Machining of Kevlar Aramid and Its Composites"
Louis H. Miner, Frank T. Penoza
21st National Symposium SAMPE
Apr. 1976

"Producibility Aspects of Advanced Composites for an L-1011 Aileron"
Society of Manufacturing Engineers
20501 Ford Rd.
Dearborn, Michigan 48128

"The History and Present Status of Synthetic Diamonds"
K. Nassau, Bell Labs
Murry Hill, NJ 07974

"Polycryshalline Diamond: The Uptime Tooling"
MAN Technical Report
Feb. 1981

"Nontraditional Machining Guide"
Machinability Data Center MDC76-101
Dept. of Defense Information Analysis Center
4-7-81

"Advanced Laser Cutting for Metal and Other Materials"
Society of Manufacturing Engineers AD 80-258
S.A. Ali, Hughes Aircraft

PAMPHLETS AND CONFERENCES (CONT)

"High Performance Aramid Fabric with Kevlar 29 and Kevlar 49"
Ind. Textiles Div., Fothergill and Harvey
Lancashire, England (1976)

"Cutting by Water Jet"
Society of Manufacturing Engineers MR 80-902
Dr. J. Olsen, Flow Ind. Inc.

"Thermal Machining Processes"
Society of Manufacturing Engineers
L. of C. Cat Card No. 79-62917

"Kevlar Composites"
Kevlar Composites Symposium
December 2, 1980, El Segundo, CA

"Lasers in Modern Industry"
Society of Manufacturing Engineers
L. of C. Cat Card No. 79-66705

"International Industrial Diamond Conference"
Industrial Diamond Assoc. of America
October 1969, Chicago, IL

PERIODICALS

"New NDE Techniques Finds Subtle Defects"
Materials Engineering
Sept. 1980, Vol. 59

"Using Water as a Cutting Tool"
American Machinist
April 1980 Pg. 123-126

"Industrial Tools for Cutting Kevlar"
Airtech International Inc.
P. O. Box 2930, Court M-1
Torrance, CA 90509

"Automation Holds the Key to Future of Composites"
Product Engineering
Sept. 22, 1969, Pg. 52 and 53

"Outline of Machining Tips for Graphite Composite"
Product Engineering — P. West
40:52, Sept. 22, 1969

"Diamond Tools Solve New Machining Problems"
by Robert N. Stauffer
Manufacturing Eng. Jan. 1979

GOVERNMENT-SPONSORED STUDIES

"Producibility and Serviceability of Kevlar - 49 Structures Made on Hot Layup Tools"
Hughes Helicopters, Inc. DAAG46-74-C-0100, Apr. 75
Head, Leach, Goodall, Sitterly

"Laser Welding and Cutting of Metals"
Proposed Mfg. Methods and Tech. Programs for 1979
Hughes Helicopters, Inc.

"Manufacturing Methods for Cutting, Machining and Drilling Composites"
Vol. I, Composites Machining Handbook,
Vol. II, Test and Results
AFML-TR-78-103 Vol. I and Vol. II, August 1978
Grumman Aerospace Corp.
Bethpage, N.Y. 11714

"Characteristics of Laser Cutting Kevlar Laminate"
R.A. Van Cleave, Jan. 1979
Prep for Dept. of Energy Contract DE-AC04-76-DP00613
Bendix, Kansas City Div.

"Manufacturing Methods for Machining Processes for High Modulus Composite Materials"
Vol. I, Composite Machining Handbook, Born
General Dynamic AD-766-332
AFML-TR-73-124, Vol. I, May 1973

Same as Above
Vol. II Material Removal Tests and Results, F. Hanley
Convair Aerospace Div. Fort Worth
AFML-TR-73-124, Vol. II, May 1973

"Ultrasonic Machining"
Technical Report AFML-TR-73-86
April 1973, W. Grauer
Grumman Aerospace Corp.

"Laser Cutting of Kevlar Laminates"
R.A. Van Cleave, Bendix Corp., Kansas City
September 1977, BDX-613-1877

GOVERNMENT-SPONSORED STUDIES (CONT)

"Ultrasonically Assisted Machining of Aircraft Parts"
J. DeVine, P.C. Krause - Sonobond Corp.
AVRADCOM Report No. TR80-F-18, October 1980

"Study of the Influence of Hole Quality on Composite Materials"
J.J. Pengra, Lockheed, CA
Contract NAS1-15599, February 1980

"Investigation of Hole Preparation and Fastener Installation for Graphite/ Epoxy Laminates"
E.F. Condon, McDonnell Aircraft
Tech Report N00019-79-C-0293

Design Guide Handbook
Efficient Machining Methods and Cutting Tools for Kevlar Composite Laminates
Hughes Helicopters, Inc., Nov 1981
Contract No. DAAK51-81-C-0008

VENDOR LITERATURE

"Industrial Tools for Cutting Kevlar Fabric and Laminates"
Pen Associates, Inc.
2639 W. Robino Drive
Wilmington, Del. 19808

"Speciality Tools for Machining Fiber Reinforced Composite Structures"
Technology Assoc. Inc.
P.O. Box 7163
Wilmington, Del. 19803

"A Guide to Cutting and Machining Kevlar"
DuPont, Inc. Textile Fibers Dept.
Centre Rd. Building
Wilmington, Del. 19898

"Environmental Effects on Aramid Composites"
DuPont, Inc., Textile Fibers Dept.
Centre Rd. Building
Wilmington, Del. 19898

"Cutting Fabric, Prepreg and Kevlar Aramid"
Sept. 19, 1975
DuPont Textile Fibers Dept.
Centre Rd. Building
Wilmington, Del. 19898

"Laser Shop Cutting"
MG Cutting Systems
Dw of CLR-0 Inc.
W1441 N 9427 Fountain Blvd.
Menomonee Falls, Wis. 53051

"Cutting With Wire"
Laser Technology, Inc.
10624 Ventura Blvd.
No. Hollywood, CA 91604

"Flow Technology Report No. 7 Waterjet Cutting of Advanced Composite Materials"
Flow Industries, Inc.
21414 68th Ave. South
Kent, Washington 98031

VENDOR LITERATURE (CONT)

Laser Inc. Sub of Flow Industries, Inc.
"Lasers for Industry"
"Everpulse and Everlase Machine Tools PR 980"
"Everlase, Laser Machine Tools"
Coherent Inc.
3210-C Porter Dr.
Palo Alto, CA 94304

"Weldon Tools Cat No. 12"
"Weldon C.S. and Deburring Tools"
Weldon Tool Co.
3000 Woodhill Rd.
Cleveland, Ohio 44104

"Hendricks Panel Saws"
Hendrick Mfg. Corp.
32-36 Commercial St.
Salem, Mass 01970

"Grit-Edge Tungsten Carbide Saw Blades"
Abrasive Product Sale
Remington Arms Co., Inc.
939 Barnum Ave.
Bridgeport, Conn. 06602

"Special Carbide Tipped Tooling"
Guhring, Inc.
1445 Commerce Ave.
Brookfield, WI 53005

"Industrial Diamond Tool Blanks"
Megadiamond Ind. Inc.
3418 N. Knox Ave.
Chicago, Ill. 60641

"Bendix Drill"
Bendix Ind. Tools Div.

"High Speed Steel End Mills"
Crest-Kret, Multi-Flute
Weldon Tool Co.
3000 Woodhill Rd.
Cleveland, Ohio 44104

VENDOR LITERATURE (CONT)

"Acra-Bore"
Roase-Acia Tools
2212 N. Wayne Ave.
Chicago, Ill. 60614

"High Pressure Water Jet Technology - Principles and Applications"
T.J. Labus, IIT Research Inst.

"Suggestions for Machining Scotchply" B78-2
Industrial Specialties Div., 3-M
Saint Paul, MN

APPENDIX B

VENDORS/DEVELOPERS VISITED DURING
PHASE I LITERATURE SEARCH

- Pen Associated, Inc., Wilmington, Delaware
  (F. Penoza)

- Technology Assoc., Inc., Wilmington, Delaware
  (J. Madden)

- Kevlar Special Products, E.I. DuPont, Wilmington, Delaware
  (K. Keith)

- Laser Technology, Inc., North Hollywood, California
  (H. McLaughlin)

- Airtech International, Inc., Carson, California
  (R. Moore)

- Grumman Aerospace Corp., Bethpage, N.Y.
  (S. Trink)

- J.K. Smith Diamond Industrial Products, Philadelphia, PA
  (R. Sgro)

- Flow Industries, Kent, Washington
  (H. Parks)

APPENDIX C
MACHINE AND TOOL SOURCES

Nibbler

Bosch Model 7501

    Robert Bosch Corporation
    2800 South 25th Avenue
    Broadview, Illinois 60153

    Triumph America Inc.
    Farmington Industrial Park
    Farmington, Conn. 06032
    (203) 677-9741

Wire Cutter

    Laser Technology, Inc.
    10624 Ventura Blvd.
    North Hollywood, CA 91064
    (213) 763-7091

Water Jet Cutter

    Flow Systems, Inc.
    21414 - 68th Ave. South
    Kent, Washington 98031
    (206) 938-FLOW   Telex: 152983

Mill Cutter

    Airtech International, Inc.
    2542 East Del Amo Blvd.
    P.O. Box 6207
    Carson, CA 90749
    (213) 603-9683 Telex: 194757

Circular Saw

Simonds Ref. No. 08882

    Simond Cutting Tools
    15619 Blackburn Ave.
    P.O. Box 505
    Norwalk, CA 90650
    (213) 802-2689

Drills

    Airflute HSS and Airtip Solid Carbide

        Airtech International, Inc.
        2542 East Del Amo Blvd.
        P.O. Box 6207
        Carson, CA 90749
        (213) 603-9683 Telex: 194757

Band Saw

        The L.S. Starrett Company
        Athol, Maine 01331

        Simonds International
        Fitchburg, Massachusetts 01420
        (617) 343-3731 Telex: 928414

Finishing Wheel

    Scotch-Brite 4" General Purpose

        Building Service and Cleansing
        Productions Division/3M
        3M Center
        St. Paul, Minnesota 55101

Special Tools - Sabre Saw Blade and Opposed Helical Router

        Pen Associates, Inc.
        3639 W. Robino Drive
        Wilmington, DE 19808
        (302) 995-6868

        Airtech International, Inc.
        2542 East Del Amo Blvd.
        P.O. Box 6207
        Carson, CA 90749
        (213) 603-9683 Telex: 194751

        Carbro Corporation (Router Only)
        15724 Condon Ave.
        Lawndale, CA 90260
        (213) 675-0355

Machining Boron-Epoxy Composites

JAY H. DORAN and FISKE HANLEY

THE hard, abrasive nature of boron-epoxy composite and the dissimilar machining characteristics of the two materials composing boron-epoxy/titanium laminate require new cutting tools and machining methods. This paper discusses six commonly performed material removal operations: drilling, reaming, countersinking, routing, milling, and sawing. General machining problems and specific problems encountered during machining tests are covered. (Because drilling is the most commonly performed composite machining operation, the program as well as the paper emphasize drilling studies.) Based on project results, recommendations for cutting tool designs and machine tool operation are presented. Realistically achievable tool lives and typical operation costs for drilling are outlined.

Proposed research for establishing specifications for portable equipment for drilling boron-epoxy/titanium is briefly discussed. The effort involves cutting force analysis and design of a high pressure, through-the-tool coolant system.

BACKGROUND OF THE STUDY

The high strength-to-weight and stiffness-to-weight ratios exhibited by fibrous reinforced plastics are contributing to increasing usage of advanced composite materials for aircraft structures. Boron-epoxy, one of the earliest of these new engineering materials, has been of interest to aerospace engineers for more than five years.

Boron-epoxy composite is composed of 0.004 in. diam filaments of boron coated on a 0.0005 in. diam tungsten substrate. These filaments are placed parallel to one another in an epoxy resin system to form a tape, 3 in. wide by 0.005 in. thick, Fig. 1. Multilayer structures with various strength characteristics can be designed by varying the orientation of the fibers.

The extreme hardness (9.5 Mohs Scale) and abrasiveness of boron-epoxy make it difficult to machine.

JAY H. DORAN and FISKE HANLEY are Manufacturing Research Engineer and Manufacturing Research Supervisor, General Dynamics Corp., Convair Aerospace Div., Fort Worth Operation, Fort Worth, Texas. This paper was presented at the Third Air Force Metalworking Technology Conference, March 13-17, 1972, Los Angeles, Calif.

The problem is compounded when boron-epoxy is bonded in a multilayer structure with a metal, such as a titanium alloy. The dissimilar machining characteristics of the composite and the metal require that compromise cutting tools and techniques be used.

Because of increasing use of composites in aircraft structures and because of the need for new material removal technology, the Air Force Materials Laboratory awarded a boron-epoxy machining program to the Fort Worth operation of the Convair Aerospace Division. This program was directed at improving production tools and techniques for machining boron-epoxy and boron-epoxy/titanium laminate. Operations of principal interest were drilling, reaming, countersinking, routing, milling, and sawing. The original program was completed in July 1971, but additional research is anticipated.

Workpieces selected for material removal studies were 0.08 and 1.00 in. thick boron-epoxy and 1.00 in. thick boron-epoxy/titanium laminate — Types A, B, and C in Fig. 2. The original laminate consisted of two layers of boron-epoxy and two layers of Ti-6A1-4V, annealed. Because of new trends in design concepts, laminate design was later changed to multilayer, $\frac{1}{2}$ in. thick materials. The three new designs (types D, E, and F in Fig. 2) consisted of layer thicknesses of 0.04, 0.06, and 0.08 in.

Material removal parameters of principal interest

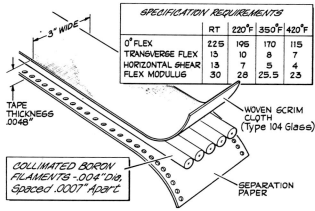

Fig. 1—Boron composite tape.

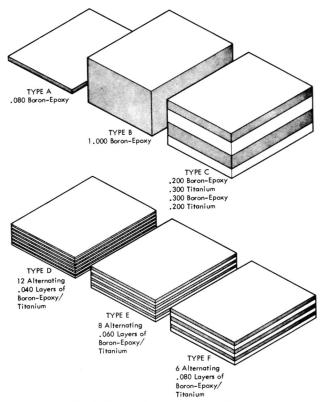

TYPE A
.080 Boron-Epoxy

TYPE B
1.000 Boron-Epoxy

TYPE C
.200 Boron-Epoxy
.300 Titanium
.300 Boron-Epoxy
.200 Titanium

TYPE D
12 Alternating
.040 Layers of
Boron-Epoxy/
Titanium

TYPE E
8 Alternating
.060 Layers of
Boron-Epoxy/
Titanium

TYPE F
6 Alternating
.080 Layers of
Boron-Epoxy/
Titanium

Fig. 2—Material removal workpieces.

Fig. 3—Buffalo drill press.

were cutting tool material, tool geometry, spindle speed, and feed rate.

In the initial program phase, Norton Co. was awarded a subcontract for evaluating and recommending cutting tool materials. Norton's research centered around diamond abrasives, but also included an investigation of Borazon and aluminum oxide materials. Company engineers recommended using blocky diamonds, either natural or synthetic, in a matrix of the hardest metal available. Diamond-plated or coated products were recommended for some operations.

Convair Aerospace used Norton's recommendations for procuring tools, purchasing them from Fish-Schurman Corp., Norton Co., O'Rourke Diamond Co., and Permattach Diamond Tool Corp.

The second phase of the contract involved material removal tests conducted at the Fort Worth operation. Objectives were to select cutting tool and machine tool parameters for cost-effective cutting of boron-epoxy and boron-epoxy/titanium.

DRILLING TESTS

A major portion of the program involved drilling tests, conducted on a No. 18 Buffalo drill press with through-the-spindle cooling and an air-hydraulic feed system, Fig. 3. Rotary ultrasonic drilling was also investigated extensively with a fixed-based unit manufactured by Branson Sonic Power Co., Fig. 4. Convair Aerospace worked with Branson and Aro Corp. in developing a portable rotary ultrasonic unit, Fig. 5.

Tools $\frac{1}{4}$ in. and $\frac{1}{2}$ in. in diameter were evaluated

for drilling boron-epoxy and boron-epoxy/titanium laminate, Fig. 6. Researchers evaluated core drills, with and without large diamonds set in the tool end, and solid diameter drills with large axial diamonds. Tools were purchased from several companies, and therefore included various bond formulations. Bond content was proprietary, but most were carbides of undetermined composition.

Each tool was used to drill a predetermined number of holes under a specific setting of operating conditions. Core drills were periodically measured to determine change in the tool length, establishing a wear rate. Hole diameter and tool diameter changes were used as indicators of solid tool wear. Holes drilled with all tools were measured for roundness and diameter, data being recorded on a standard form for later analysis.

Boron-Epoxy Workpieces

Tool testing revealed no problems in drilling boron-epoxy work pieces, except that high pressure coolant was required when drilling thick material. Tests of spindle speeds in the range of 200 to 800 sfm revealed little relationship between tool life and speed.

Portable drilling of boron-epoxy was hindered primarily by inability to get adequate cutting fluid to the tool-workpiece interface. This inadequacy will have to be solved before production-level drilling is feasible.

Rotary ultrasonic drilling of boron-epoxy using the Branson UMT-3 revealed two major problem areas. Cutting tool designs incorporated two braze joints,

Fig. 4—Branson UMT-3 rotary ultrasonic drill.

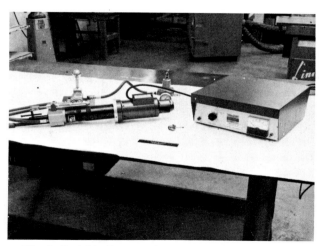

Fig. 5—Portable Branson/Aro rotary ultrasonic drill.

Fig. 7; the upper joint was particularly susceptible to failure due to stresses created by the 20 kHz ultrasonic vibrations. The design was revised to eliminate this joint. The other joint failed when extreme caution was not exercised by the diamond tool manufacturer when making the drill. The second problem, encountered when drilling $\frac{1}{2}$ in. diam holes in boron-epoxy/titanium, will be discussed later.

Three prototype portable rotary ultrasonic drills, representing an evolutionary process, were evaluated during the program. Though all units were adequate for drilling thin boron-epoxy, none permitted use of sufficient coolant pressure to cool the tool-workpiece interface when cutting thick material.

Boron-Epoxy/Titanium Workpieces

Drilling boron-epoxy/titanium laminated proved to be the most challenging operation investigated. All efforts prior to this program resulted in almost immediate catastrophic tool failure because the diamond tool could not penetrate titanium and because much heat was generated. Solution of these two problems provided the key to drilling composite/metal laminate.

Testing revealed that grinding through the titanium was not an insurmountable task if the drill thrust force (and thus the feed rate) could be controlled adequately. The hollow spindle of the Buffalo drill press permitted use of coolant at pressures of 75 to 100 psi, providing adequate cooling.

Because the ultrasonic drill could not develop sufficient torque to penetrate the metal layers in a timely manner without stalling, use of this machine was discontinued. Portable ultrasonic drilling was not successful because of inability to obtain adequate coolant at the cutting zone, an inadequacy of both conventional and portable equipment.

Drill life data was not as conclusive as anticipated because of data scatter due largely to inconsistent diamond tool performance and tool quality.

In summary, research indicates that a coolant

should be used when drilling composite materials thicker than $\frac{1}{8}$ in. Cutting fluid pressures greater than 75 psi are required to prevent thermal damage to the workpiece and tool when drilling thick boron-epoxy and boron-epoxy/titanium.

Single tool penetration of boron-epoxy/titanium is possible, providing that low feed rates and high coolant pressures are maintained. Feed rates of $\frac{1}{8}$ ipm or less are mandatory, necessitating use of air-hydraulic feed equipment.

Recommended cutting tools, speeds, and feeds for drilling boron-epoxy and boron-epoxy/titanium are shown in Table I. Also indicated are reasonable drilling times, tool life predictions, and drilling costs based on program data.

COUNTERSINKING

Countersinking, commonly performed in conjunction with drilling, can be classed as a hole preparation operation. Major objectives in the countersink evaluation were to determine the effect of operating parameters and tool design factors on tool life.

Diamond plated and diamond impregnated $\frac{5}{8}$ in. diam, 100 deg countersinks were evaluated for countersinking $\frac{1}{4}$ in. holes at various speeds and feeds using boron-epoxy and boron-epoxy/titanium workpieces. All work was accomplished with stationary equipment.

Information gained during this evaluation revealed that feed rates had a marked effect on tool life. Boron-epoxy/titanium countersinking proved extremely difficult, resulting in rapid tool failure and work of poor quality. No satisfactory countersinking method was established. Undoubtedly the secret lies in low feed rates and use of copious cutting fluid.

REAMING

Holes are commonly reamed to improve surface finish and achieve closer tolerances in diameters. Reaming of boron-epoxy and boron-epoxy/titanium was investigated to determine surface finish and diameter tolerance improvements that could be attained.

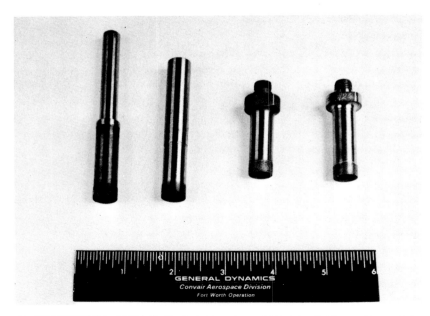

Fig. 6—Diamond impregnated drills.

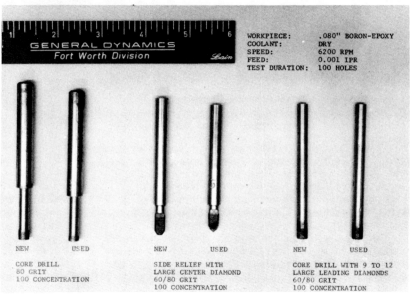

WORKPIECE:	.080" BORON-EPOXY
COOLANT:	DRY
SPEED:	6200 RPM
FEED:	0.001 IPR
TEST DURATION:	100 HOLES

NEW   USED   NEW   USED   NEW   USED

CORE DRILL   SIDE RELIEF WITH   CORE DRILL WITH 9 TO 12
80 GRIT   LARGE CENTER DIAMOND   LARGE LEADING DIAMONDS
100 CONCENTRATION   60/80 GRIT   60/80 GRIT
                      100 CONCENTRATION   100 CONCENTRATION

All tests were conducted with stationary equipment. Expansion reamers $\frac{1}{2}$ in. in diameter with large set diamonds, Fig. 8, were evaluated at various speeds and feeds.

Results indicated that reaming did not improve surface finishes significantly, nor did it provide closer dimensional control than a core drill, properly used. Based on these results, reaming is not recommended as a standard procedure for hole preparation.

MILLING

Type 11A1 cup wheels, diamond impregnated and diamond plated, were evaluated for trimming edges of boron-epoxy and boron-epoxy/titanium to close tolerances. (In effect, this operation was similar to grinding except that a cut 0.030 in. in depth was employed.)

A high-speed Bridgeport mill with power feed table was used. Because of the high cost of the composite material and the long tool life anticipated, tests were short, making it difficult to establish recommended feeds, speeds, and cutting tools. Though it was also difficult to determine relative merits of plated and impregnated tools, several performance characteristics were noted. As anticipated, plated tools cut more freely, therefore generating less heat. Impregnated tools, however, yielded better finishes on metal and composite. Impregnated tools would probably cost less per unit of material cut than plated tools, but test results do not provide conclusive evidence.

Cutting rates less than 2500 sfm and $5\frac{3}{4}$ ipm resulted in increased smearing of titanium. Cutting fluids were mandatory when milling boron-epoxy/titanium.

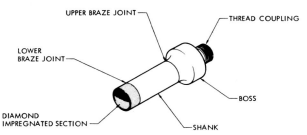

Fig. 7—Diamond impregnated core drill with adapter for Branson UMT-3 rotary ultrasonic drill.

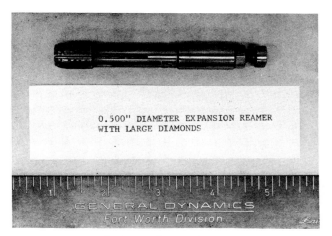

0.500" DIAMETER EXPANSION REAMER WITH LARGE DIAMONDS

GENERAL DYNAMICS
Fort Worth Division

Fig. 8—Diamond expansion reamer.

ROUTING

Routing was also evaluated as an edge trimming operation. For these tests, a reciprocating, high-speed machine was used. Equipment parameters of primary interest were spindle speed, feed thrust force, and reciprocating versus nonreciprocating action. Impregnated and plated tools of various grit sizes were tested.

Problems with cutting tool quality severely affected results. Excessive runout and poor finishes on shank surfaces were characteristic of these $\frac{1}{2}$ in. diam tools. Furthermore, diamond sections of impregnated tools were poorly brazed to the steel shank, tools failing catastrophically under no load conditions in one instance.

Test results indicate that reciprocation provides a definite advantage. This axial motion aids in cleaning the tool, and also distributes wear more evenly over the length of the tool. Tool speeds of 2000 to 3000 sfm produced satisfactory results. Boron-epoxy (1 in. thick) and boron-epoxy/titanium ($\frac{1}{2}$ in. thick) were fed at $\frac{3}{4}$ ipm and $\frac{1}{4}$ ipm, respectively, with acceptable results. Maximum feed is dictated primarily by set-up rigidity and spindle motor power.

CIRCLE SAWING

A common table saw was used to evaluate diamond impregnated and diamond plated cutoff wheels, 6 in. in diameter and $\frac{1}{16}$ in. thick, Fig. 9. Once again, the limited quantity of material available prevented extensive testing.

Table I. Factors for Drilling Composites

Material	Speed, sfm	Feed, ipr Equiv.	Unit Tool Cost	Tool Recommendation	Drilling Time, min	Tool Life, Holes	Cost per Hole
Boron-epoxy, 0.08 in.	300 to 600	0.0005	$35, $\frac{1}{4}$ in.	core with end set diamonds	0.1	300 to 400	$0.10 to $0.20
Boron-epoxy 1.00 in.	300 to 600	0.0005	$45, $\frac{1}{2}$ in.	core with end set diamonds	1.0	75 to 100	$0.75 to $1.00
Boron-epoxy/ titanium multi-layer, 0.50 in. total	200 to 500	0.00005	$35*, $\frac{1}{2}$ in.	core	4.0	30 to 50	$1.50 to $2.00

*Tool design differs from that recommended for boron-epoxy.

Fig. 9—Typical circle saw blades.

Machine parameters of major interest were blade speed and feed thrust force. Diamond plated and diamond impregnated blades with continuous and segmented edges were evaluated.

Heat generation proved to be the major problem. The machine had no provision for flood coolant application; however, a spray mist was used to reduce heat. Nevertheless, blades would warp when sawing thick material if the workpiece was fed too rapidly. The inability to get sufficient cooling at the cutting zone was the primary reason for inability to saw boron-epoxy/titanium laminate. Limitations of most sawing equipment makes it impractical to circle saw boron-epoxy/titanium. Wheel speeds of 4000 to 6000 sfm and feeds of 5 ipm were satisfactory for cutting thin boron-epoxy. This method of cutting is a relatively fast, low cost approach.

BAND SAWING

Band sawing tests were conducted with a DoAll saw specifically designed for a diamond blade and a flood coolant, Fig. 10. The machine had a constant blade speed (3000 sfm), but was equipped with a variable-rate power feed table.

Diamond-plated saw blades of several grit sizes,

Fig. 10—Diamond band sawing boron-epoxy/titanium.

and with continuous and segmented edges, were evaluated for cutting thick boron-epoxy and boron-epoxy/titanium.

Band sawing was judged the best method of any tested for trimming edges of boron-epoxy/titanium. Surface finishes of 20 μin. AA were readily achieved as were tolerances of ± 0.002 in. When proper feed rates were used no titanium smearing was evident. Feed rates up to 2 ipm yielded satisfactory surfaces when sawing the laminate.

Insufficient material was available to test each blade to failure. However, several trends were apparent, and subsequent use of these tools has revealed additional information. A relatively coarse 40 to 60 grit blade cut much faster than tools with finer diamonds. Surface finish, even when the coarse blade was used, was quite satisfactory. A continuous-edge blade is preferable because of fatigue breakage experienced with segmented blades

CONCLUSIONS

This program examined the cutting tool and machine tool parameters associated with the most commonly performed operations for machining composite materials. Major findings are summarized below:

1) Diamond abrasive tools are required for cutting boron-epoxy and multilayer boron-epoxy/titanium. Impregnated tools should last longer than plated tools in most applications, and (despite higher initial cost) should result in lower overall expense for material removal.

2) Use of cutting fluid is mandatory for boron-epoxy/titanium machining, and is recommended for cutting boron-epoxy. Flood cooling reduces heat buildup in the cutting tool and workpiece, thereby increasing tool life and preventing thermal damage to the workpiece. Water soluble oils are satisfactory, providing that they reach the cutting zone in sufficient quantity to dissipate heat.

3) Single tool drilling of boron-epoxy/titanium is feasible with extremely low feed rates and adequate flow of cutting fluid. Portable drilling equipment must be developed to meet these requirements.

4) Ultrasonic action applied to a rotary diamond drill provided no significant cost savings when used for drilling composites. Present equipment is inadequate for production drilling.

5) Band sawing with a diamond-plated band is characterized by excellent surface finishes and good dimensional control. Although relatively slow, this operation produced the most acceptable cuts achieved during the program.

6) Axial reciprocation of the diamond tool is required to produce good router cuts in boron-epoxy/titanium. Reciprocating action aids in cleaning the tool and distributing tool wear.

7) Equipment protection to prevent abrasive boron dust from wearing moving machine parts is recommended. Way covers, coolant filters, and so forth should be used to prevent abnormally high wear of equipment.

8) Tool quality must be improved if diamond abrasive products are to be used for precision production removal of material. Inconsistent tool quality contributed to data scatter, seriously hindering data analysis. Catastrophic failure occurred much too frequently.

9) A boron machining handbook was published as a part of the program documentation. This handbook describes recommended cutting tools, machine tools, and procedures for performing common material removal operations. Quality control, safety, and equipment protection are also discussed.

PLANS FOR FUTURE WORK

Additional research will begin shortly, the major objective being to analyze drilling torque and thrust force. The goal is to specify, assemble, and test a portable drill motor for drilling boron-epoxy/titanium. This unit must provide extremely low feed rates and permit through-the-tool flow of high pressure coolant.

Boron-epoxy, boron-epoxy/titanium, boron-aluminum, and boron-aluminum/titanium workpieces will be used for this effort.

ACKNOWLEDGMENT

This report is based on Air Force Technical Report AFML-TR-71-129, Vol. I and II, prepared for Manufacturing Technology Division, Air Force Materials Laboratory, Air Force Systems Command, United States Air Force, Wright-Patterson AFB, Ohio 45433.

A Guide to Cutting and Machining Materials Containing Kevlar®Aramid Fiber

Introduction

KEVLAR® is a DuPont Company registered trademark for its family of unique aramid (generic designation for aromatic polyamide) fibers. KEVLAR* was introduced commercially in 1972 and is now available in three types offering industry new levels of reinforcing fiber performance in applications for which KEVLAR is qualified.

The three types are:

- **KEVLAR 49:** has high tensile strength and high tensile modulus and is designed for use in plastic reinforcement.
- **KEVLAR 29:** has the same tensile strength, but only about two-thirds the tensile modulus of KEVLAR 49, and is used in a variety of industrial applications.
- **KEVLAR:** has properties similar to KEVLAR 29, but is designed for rubber reinforcement applications.

The guidance provided here applies to all types of KEVLAR.

*DuPont registered trademark.

Table I illustrates that KEVLAR 29 and KEVLAR 49 combine the properties exhibited by high modulus reinforcing fibers such as graphite with those commonly associated with organic industrial fibers such as polyester. The higher elongation of KEVLAR and its tendency to yield in compression without fracture result in a fiber that is tougher and less brittle than other commonly available reinforcing fibers. As will be discussed later, this behavior creates special difficulties for those involved in cutting and machine operations. KEVLAR aramid exhibits high thermochemical stability. At elevated temperatures, it neither shrinks nor melts but rather degrades to a char beginning at or above 800° F (425° C) when exposed to a normal atmosphere. KEVLAR resists attack by many common chemicals; in general only hot, concentrated mineral acids or bases significantly degrade the fiber. KEVLAR® exhibits a small, negative coefficient of thermal expansion along the longitudinal axis. KEVLAR has excellent electrical and thermal insulating properties. The latter contributes to increased difficulty in certain machining operations.

TABLE I
COMPARATIVE FIBER MECHANICAL PROPERTIES

	KEVLAR*[1] 29	KEVLAR*[1] 49	High Strength Graphite[1]	Glass Fiber[1] "E"	Glass Fiber[1] "S"	DACRON**** Polyester[2] Type 68
Fiber Density						
—lb/in^3	0.052	0.052	0.063	0.092	0.090	0.050
—g/cm^3	1.44	1.44	1.75	2.55	2.49	1.38
Break Elongation						
—%	4.4	2.9	1.25	3.5	3.5	14.5
Tensile Strength						
—psi × 10^3	525	525	450	350	575	162.5
—GPa	3.62	3.62	3.1	2.41	3.96	1.12
Specific Tensile Strength[3]						
—10^6 in	10.1	10.1	7.1	3.8	6.4	3.3
—10^7 cm	2.5	2.5	1.8	0.9	1.6	0.8
Tensile Modulus						
—psi × 10^6	12	18	32	10	12.4	2
—GPa	83	124	221	69	85	13.8
Specific Tensile Modulus[3]						
—10^8 in	2.3	3.5	5.1	1.1	1.4	0.40
—10^8 cm	5.7	9.0	12.6	2.7	3.5	1.0

[1]ASTM D2343 resin impregnated strand test
[2]ASTM D885 textile test
[3]Specific property = property divided by material density

*KEVLAR is DuPont's registered trademark for an aramid fiber
**DACRON is DuPont's registered trademark for its polyester fiber

Cutting Yarn, Roving, Fabric or Resin Preimpregnated Materials Containing KEVLAR®

KEVLAR is an inherently "tough" material. Tools to be used for cutting or trimming yarns, rovings and woven or felted fabrics must be sharp, clean and properly aligned. Shears must have minimum clearance between cutting surfaces. Resin preimpregnated materials (prepregs), i.e., materials to which a resin has been added and the coating partially cured as a preliminary step to laminating operations, require special precautions. Tools should be cleansed frequently to avoid a buildup of the tacky, partially cured resin which can cause a loss in cutting action. Many common industrial solvents are acceptable for this purpose. Among these are acetone and methyl ethyl ketone (MEK). The manufacturer is encouraged to practice good industrial hygiene by minimizing skin contact and providing adequate ventilation in the work area to assure compliance with the recognized exposure limit for the solvent being used.

Table II provides guidance concerning the applications and sources for a variety of manual and power shears and rotary action cutters for use with yarns, rovings, fabrics and tapes containing KEVLAR aramid. Shears with serrated edges which prevent the material being cut from slipping, such as those offered by Pen Associates or Technology Associates, cut materials containing KEVLAR very well. These shears are also available with a proprietary wear-resistant surface treatment which extends the useful life of the tools. All items listed in the table including those with the protective treatment can be resharpened and reconditioned.

Ultrasonic tools have been developed which can cleanly and easily cut materials containing or composed of KEVLAR fibers. In the spinning of KEVLAR yarns, wraps of yarn sometimes occur on forwarding rolls, and these are very difficult to remove with a knife. A hand-held ultrasonic tool facilitates removal because ultrasonic vibrations of the tip of the cutting tool cut KEVLAR easily. Excellent control of the cutting is possible because the user needs only to guide the tool effortlessly. Also, mounting the ultrasonic tool on the end of a sucker-gun enables cutting and reoving roll-wraps whether or not the roll is running. Both table-mounted and hand-held units successfully cut ropes of KEVLAR or cut and trim fabrics. Advantages are that ultrasonically cut edges of ropes and fabrics are compacted without unraveling and that they look like a fused solid.

The table-mounted unit used in these developments was a Branson Model 500. In it, an ultrasonic horn operating at 20kHz with high energy output was pressed against a stationary sharp edge. The hand-held unit was a modified Branson K-144D ultrasonic plastic welder in which the tip of the tool was either a blade or a hollow punch. This unit operated at 40kHz and at a lower energy output than for the table model.

Multiple plies of fabric with or without resin preimpregnation can be processed using a cutting die or with abrasive grit coated, toothless band or sabre saw blades fitted to conventional equipment. Refer to Table III for source information. These blades can cut up to 50 plies of a fabric containing KEVLAR that weighs 5-10 ounces per square yard (170-340 grams per square meter). Suggested set-up conditions for sawing operations are:

1. Carbide grit-coated band saw
   - 4500-6000 SFPM (1370-1980 SMPM)* blade speed
   - Bottom backing only of posterboard or heavy cardboard.
2. Carbide grit-coated sabre saw
   - 2500 strokes per minute blade speed
   - Support both surfaces with posterboard or heavy cardboard.

Information on die cutting of fabrics containing KEVLAR may be obtained from:

- USM Corporation
  9 Valley Street
  Hawthorne, NJ 07506

  201/423-1434

- Ontario Die Company
  119 Rogers Street
  Waterloo, Ontario
  Canada
  519/576-8950.

*SFPM = surface feet per minute
SMPM = surface meters per minute

Machining Composite Materials Containing KEVLAR Aramid

The machining of fiber reinforced plastics (FRP's), including those containing KEVLAR, involves requirements and conditions significantly different from those of machining metals:

- FRP's must be machined within a limited temperature range. Even though the reinforcing fibers will tolerate higher temperatures, the curing temperature of the resin matrix, typically 250° F (121° C) and occasionally 350° F (177° C), should not be exceeded to avoid deterioration of the matrix properties in the cutting zone.
- The generally low thermal conductivity of FRP's, and especially those containing KEVLAR, favors overheating in the cutting zone during machining operations. Since little heat dissipates into the material, most must be conducted away by the tool itself.

- The matrix resins commonly used in FRP's exhibit high thermal coefficients of expansion. Although this behavior may be modified by the presence of the reinforcing fiber, maintenance of dimensional accuracy is more difficult with these materials than with metals.
- The impact of fluid absorption on the material properties of FRP's has not been studied in detail. This factor has to be considered when using tool coolants.

Extensive testing has shown that machining of FRP materials containing KEVLAR® imposes tool requirements which are different from those of glass or carbon fibers. As discussed in Section I, KEVLAR aramid exhibits high tensile strength, but low resistance to compressive stress. For this reason, there is a tendency for the KEVLAR® fibers within the resin matrix to deflect when compressive forces develop during the machining process. The structure of the composite, including the orientation and types of the fiber and fiber layers, the volume content of fiber, and the properties of the matrix material, influences this behavior. Machine tool performance in terms of both quantity of work per tool and overall quality of result will be enhanced by the use of tools with very sharp edges whose working surfaces are kept clean of resin build-up as discussed in Section I. In circumstances demanding highest quality, consideration should be given to dedicating certain tools for use only with composites containing KEVLAR.

In many important respects, the machining characteristics of FRP's containing KEVLAR are similar to those of wood to which they bear some structural similarity in the following sense. The structures of both materials are characterized by the presence of highly oriented fibrous material, embedded or bound together in a matrix material of different properties. For structures of this sort, optimum cutting or machining results are facilitated when the process proceeds in such a way that the fiber component is pre-loaded in tension and then cut with a shearing action. For tools with a rotary action, this means the fibers must be drawn from the periphery of the opening being formed toward the center of rotation by the action of the tool.

In undertaking operations involving FRP's, the manufacturer must continually be aware of the diversity of material properties that result from the use of composite technology in commerce. Tools and techniques that work well with one composite material may not suffice for use with another. Some composites will contain a single reinforcing fiber; tools and techniques that have proved useful for those containing only KEVLAR® aramid will be discussed here. Increasingly, KEVLAR is being combined with other reinforcements such as glass or carbon fiber in so-called "hybrid" composites. In such hybrids, tools and techniques which have worked well with composites containing KEVLAR alone should receive first consideration, keeping in mind that the abrasive nature of glass and carbon fibers will impose special difficulty in the maintenance of sharp tool edges that facilitate cutting of the KEVLAR fibers.

High speed steel (HSS) tools typically will provide adequate, although limited, service life with composites reinforced solely with KEVLAR if care is exercised to avoid overheating in use. Life can be extended economically by coating the tools with titanium carbide (TiC) or titanium nitride (TiN). When KEVLAR is hybridized with glass or carbon fiber, the abrasive nature of these fibers requires the use of TiC or TiN coated tools at a minimum. Carbide or carbide insert tools will provide much better service life and machining quality when working with these hybrid composites.

When used in a hybrid composite, the KEVLAR fiber is likely to constitute the surface layers where its combination of strength and toughness enhance the impact resistance of the material. In such cases, as well as where KEVLAR is the sole reinforcement, particular attention must be given to cutting cleanly the outermost layers of KEVLAR in the composite. Where operations require the use of a tool that does not adequately restrain the fibers in this layer, a sacrificial external backup support such as a thin sheet of glass fiber-reinforced plastic or oil-free pressed board should be used to minimize fraying or fuzzing of the surface fibers. Where such backup support cannot be used, leaving the glass fiber-reinforced peel ply on the composite surface during the machining operation will promote better hole quality. In some cases the composite may contain a thin, glass fiber reinforcing layer in critical machining areas to facilitate good quality finish.

All the machining operations required for the fabrication of structures using composite materials reinforced with KEVLAR have been accomplished with varying degrees of success. In general, these operations can be divided into three categories:

Conventional Tools and Methods	Conventional Tools Special Methods	Specialty Tools
Counterboring	Band Sawing	Drilling
Milling	Radial Sawing	Countersinking
Grinding	Sanding	Routing
	Chamfering	Sabre Sawing
	Turning	Razor Blade Cutting

The following sections describe in some detail tooling, set-up, tool speeds and feed rates suggested for most general cutting and machining tasks involving composite materials reinforced with KEVLAR®. This guidance is summarized in Table III. Table IV provides information concerning sources from whom tools described here can be purchased. More specialized operations not covered herein should be discussed with:

E. I. du Pont de Nemours & Co., Inc.
Textile Fibers Department
Composites Marketing Group
Centre Road Building
Wilmington, DE 19898
302/999-2680.

A. DRILLING

Many different drill bit designs have produced acceptable holes in composite materials containing KEVLAR. With one exception to be discussed below, these tools are characterized by geometrically defined cutting edges. Typically, drills with geometrically undefined cutting surfaces such as those with diamond or carbide grit coatings are not recommended. Not all tool designs can be adapted to all composite materials or drilling set-ups encountered in actual practice. Techniques that provide optimum results with available tools as well as the potential limitations of these tools will be discussed here.

1. **Single Point Drills:** Single point designs with symmetric cutting edges can be used off-hand. Those with asymmetric cutting edges must be in a drill press or, in the case of hand-held motors, with a drill-bushing to insure proper tool alignment is maintained in operation. The conventional twist drill is the most common example of the symmetrical single point tool. These tools can be used for drilling solid laminate materials containing KEVLAR aramid that have structural resin content levels (typically 40-60 volume percent). Care should be taken to insure very sharp cutting edges. Even so, considerable feeding force will be required to penetrate the laminate. This can cause delamination of the composite, especially on the exit side. Firm backup of the laminate surfaces should be provided when using conventional twist drills to minimize development of reinforcing fiber fuzz on the entrance side and delamination of the composite at tool exit as discussed above. When drilling through a composite material into an underlying metal structure for alignment purposes, additional backup material on the exit side may not be required.

Conventional twist drills are not recommended for use with "cored" laminates which typically are thin solid laminate skins bonded to foam or honeycomb core material. In this construction it is impossible to provide backup on the interior skin surfaces. Therefore, fiber fuzz and composite delamination may occur, causing poor dimensional control in the drilled hole. Conventional twist drills are also not recommended for use with very low resin content (less than 25 volume percent) composites that contain KEVLAR (see Core Cutting Tools, below).

A variety of asymmetric single point drills, including "spade" drills, modified spade drills and single flute designs with off-center cutting edges, have been proposed for use with composites containing KEVLAR. The spade drill in its simplest form involves machining a flat section across a rod of tool material, of diameter equal to the hole to be drilled, at an included angle of approximately 15 degrees. Effective material cutting occurs over a comparatively small radius at the tip of this tool. New tools with sharp edges can provide good quality entrance holes, with minimum fiber fuzzing and upset due to delamination. However, hole quality deteriorates fairly quickly as the cutting edge wears. Feeding force tends to be high, requiring backup support to minimize exit side delamination and fiber fuzzing. A modified spade drill design in which the section across the rod is not flat but with a slight concave radius provides some increase in effective cutting radius, with some potential to improve tool cutting life at the expense of making a somewhat more fragile tool. The spade drills have poor chip or plug removal capability and should only be used on thin, solid laminate materials.

The single flute drill is effective with solid laminates of greater thickness than the spade drills because of better chip-removal ability. This tool has a more aggressive cutting action than conventional twist drills. It is recommended for drilling through laminate material into metal supports. Backup support is required on both entrance and exit sides of the laminate to avoid fuzz and delamination.

As with conventional twist drills, the spade and single flute designs are not recommended for use with very low resin content laminates.

2. **Multiple Point Drills:** Modification of conventional twist drill geometry from a conical to a brad point or "fish-tail" profile provides a means of realizing the tool geometry required to load the reinforcing fibers in tension while cutting them in shear as discussed in the introduction. Brad point drills with positive cutting angles cut aggresively and require controlled rate of feed to produce the best quality holes. With suitable feed rate control,

backup support may not be required to produce acceptable hole quality. Brad point designs with negative cutting angles require feed pressures approaching those of conventional twist drills so backup support should be used to avoid exit side delamination. As before, feed rate control will produce best quality holes.

Brad point drills work well on solid composites over a wide range of thicknesses. They are generally less useful with cored laminates since the pilot or center point in some designs is too prominent. In such cases, the tool cuts a disk of material which, trapped by the pilot point, fouls the cutting action, requiring the tool be removed for cleaning. As with twist drills discussed above, the multiple point tools are not recommended for use with low resin content laminates.

The effective cutting edge radius of brad point designs is somewhat larger than for the spade designs discussed above, but such designs are equally subject to rapid loss of sharpness. These tools are not recommended for drilling through composite materials into metal supports.

3. **Core Cutting Tools:** Openings larger than 0.5 inch (12.7 mm) in diameter in solid or cored laminates require the use of a core cutting tool with geometrically defined cutting edges. These tools should be used with a drill press, although piloted versions are available for off-hand work. Backup support is recommended to minimize fiber fuzz development or delamination. Core cutting tools with geometrically defined cutting edges, e.g., hole saw, are not recommended for use with very low resin content laminates.

Core cutting tools with geometrically undefined cutting edges are also useful in drilling holes up to 0.5 inch diameter in certain composites. These tools are simply a length of tubing, the wall of which has been sharpened to an edge, which in use is rotated against the material to be cut. This type of tool works best when used in a drill press. These tools cut almost totally in compression and for this reason are very useful in producing holes in very low resin content laminates, often used in structural ballistic applications. Depth of cut is limited because of plug clearing problems. High feed pressures are required and backup support is recommended to maintain hole quality on the exit side. Slow speeds and low feed rates are recommended to minimize overheating. Metal supports cannot be drilled with this tool.

Core cutting tools with undefined cutting edges can also be used with thin solid laminates provided adequate backup support is used on the exit side. They are not recommended for thin-skinned cored laminates because they cause delamination of the interior surfaces.

4. **Specialty Tools:** A single-point, self-centering serrated drill which combines up and down cutting flutes can be used to produce acceptable holes in laminates containing KEVLAR® aramid where it is difficult or impossible to access the exit side of the material to provide backup support, or where other considerations prevent use of tools already discussed. This highly specialized tool requires considerable skill to produce a clean, properly dimensioned hole and is therefore not recommended for general use. To obtain clean entrance and exit surfaces, this drill must rapidly penetrate the work piece until the center line of the tool is approximately mid-way between the entrance and exit surfaces of the laminate. After initial penetration, and while continuing to operate the drill motor, the tool is moved rapidly in an orbital manner about its center line for several seconds allowing the cutting edges to shear the uncut fibers toward the center of the laminate thickness. During this process, radial movement of the tool must be minimized to insure a properly dimensioned hole.

This tool is not recommended for use with low resin content composites nor for drilling into underlying metal support.

B. **Countersinking**

As a routine machining operation, countersinking of composites containing KEVLAR aramid imposes greater demands on tool selection and maintenance than does drilling because it is impractical to provide entrance side backup support to minimize fuzzing or delamination.

Many conventional radial cutting edge countersinks, whether of 1, 2, 3 or 4 flute designs executed in HSS or full carbide or with carbide or polycrystalline diamond inserts, have produced very good to excellent quality holes except that they often leave an unacceptable border of fuzz around the perimeter of the countersunk hole. These designs are recommended for use with composites containing KEVLAR only if: (1) such hole quality does not interfere with the intended use, and (2) secondary finishing can be committed to upgrade the result. These tools can give acceptable results where the composite contains a glass fiber-reinforced ply over the KEVLAR.

Where quality standards require fuzz-free holes, and design considerations preclude the use of a glass ply overlay, countersinks with curved or sickle-shaped cutting edges are recommended. These tools embody the principles discussed above in that they provide a compressive, shearing action as the tool rotates. Because they frequently have a positive cutting angle, these tools cut very aggressively and should be used only with a micro-stop limit device to control hole depth.

Low tool speeds are recommended for all counter-sinking in composites containing KEVLAR® aramid.

C. ROUTING

As a routine machining operation, routing of composites containing KEVLAR imposes the greatest demands on tool selection and maintenance. This is because it is much more difficult to design tools that restrain the surface fibers of the laminate in tension while cutting them with compressive shear, the principle which has proved effective in refining drill and countersink tools for use with composites containing KEVLAR.

As a general rule, conventional straight flute or diamond pattern routers often used with metals or glass or carbon fiber composites will not produce acceptable cut edges in composites containing KEVLAR becase these designs leave excessive fiber fuzz or cause delamination at the cut edge. However, where provision for backup support can be made or where a glass fiber surface ply is provided, cut edges that require only minimal secondary finishing in the form of light, wet or dry sanding can be routed.

Several specialty router bits have been developed which minimize fuzz generation in the absence of backup support or a glass fiber-reinforced surface ply. These tools are designed to direct the cutting force toward the center of the laminate thickness, thereby cutting the surface fibers in compression. Some designs feature complex cutting edge geometry that provides alternating up and down cutting action. In some cases, the cutting edge geometry of the router will feature an axis of symmetry which requires that the relationship between the tool and laminate thickness remain constant during the cutting operation. These tools are best used in pin router tables or numerically controlled routing machines. They are diffifult to control in off-hand, trim operations or for contour routing.

Other router designs which provide for alternating up and down cutting but with uniform geometry along the tool axis can cause severe vibration to develop in the work piece or the tool as it rotates. Thin, solid laminates as well as cored laminates with thin skins are sensitive to this condition which can result in inferior cut edge quality or tool damage. Rigid tool and work piece mounting, including the use of backup support, will optimize edge quality.

The router bits developed for use with composites containing KEVLAR aramid function best when used with rotational speeds of 20,000 revolutions per minute or higher. They are more effective when used in trim cutting either off-hand or in an appropriate fixture because the designs that efficiently cut the surface fibers tend to be deficient in chip removal ability. The more open trim cutting aids in chip removal. On the other hand, plunge cutting generally results in shorter tool life and poor cut edge quality for the same reason since chip flow is less free in the more confined environment.

D. SAWING

1. **Band sawing:** Band sawing can be accomplished by cutting with a fine-tooth blade, preferably using water as a coolant. Some edge fuzzing may remain, mostly on the blade exit side, but can be removed by light, wet or dry sanding. The band should be run in reverse, such that the heels of the teeth enter the composite first.

The band saw teeth should be honed. Honing can be done while the blade is turning. A standard shop mechanic's honing stone ¾ × 2 × 6 inches (19 × 51 × 152 mm) should be used. Apply the hone to the side of the band to form flats of 0.002-0.006″ (0.05-0.15 mm) on both sides of the teeth.

2. **Circular sawing:** Straight cuts on laminates up to ¼ inch (6.4 mm) thick can be made with some edge fuzz remaining on the exit side.

Blade design (suggested), 10-inch (254 mm) diameter.

(Also available in 8-inch (203 mm) diameter) 0.050-inch (1.3 mm) kerf.

250 teeth. Hone teeth as described for band sawing (above).

Alternate top and face bevel.

Feed rate—up to 36 inches/minute (915 mm/min.), (depending on thickness).

Blade speed—4500 to 6500 SFPM (1370-1980 m/min.).

Coolant preferable—water soluble; 150 to 200 parts water to one part concentrate (can be cut dry if necessary).

3. **Sabre sawing:** A sabre saw blade has been developed which cuts the outermost fibers on both sides of the laminate toward the interior, leaving an excellent cut edge.

4. **Hole sawing:** Tungsten carbide insert or grit-edged hole saws up to 3½ inches (89 mm) in diameter will cut laminates of KEVLAR aramid up to 1-inch (25 mm) thick. Fuzzing at entrance and exit surfaces can be reduced by use of backup support, but the inside diameter wall of the hole is smooth.

5. **Edge trimming:** Abrasive-grit, toothless band-saw blades have been developed for rough cutting the edges of composites up to ¾-inch (19 mm)

thick. Some fuzzing will remain at the exit surface. These blades have good wear life. While use of a coolant is preferable to keep blade free of debris (150 to 200 parts water to one part concentrate), dry cuts can be made with occasional wire brushing to keep blade clean.

E. CUTTING

1. **Metal cutting shears (electric):** Electric metal cutting shears designed to cut ferrous and non-ferrous sheet metal will cut laminates of KEVLAR® up to 0.1 inch (2.54 mm) thick. Pattern lines are easily followed. Cut edges are smooth with very little fuzzing. Lineal speed 30 to 40 inches/minute (760-1015 mm/min) can usually be maintained with slight forward pressure.

Power shears which have been modified to minimize working clearance between the blades can also be used to cut KEVLAR in all styles of fabric from 1 to 12 plies in thickness.

2. **Power nibblers:** Power nibblers designed to cut ferrous and non-ferrous sheet metal can be used to cut composites containing KEVLAR aramid up to 0.10 inch (2.54 mm) thick with a minimum of residual fiber fuzz on the cut edges. Because of design of the cutting die, cut edge quality is directional; one side always shows some delamination of the composite. Modification of the die set to further reduce clearance between die and anvil will minimize both edge fuzz and delamination. Pattern lines can be easily followed. Lineal speed of 5 ft. (1.5 m/minute) can usually be maintained with slight forward pressure. Cutting radius is limited by the size of the cutting die. Contact vendors to discuss tool modifications to facilitate use in connection with pattern jigs.

3. **Razor cutting:** Tungsten carbide razor blades honed to a sharp, durable edge are useful for cutting fabrics and prepregs of KEVLAR, and for trimming thin composite edges. The blades and holders can be supplied to satisfy individual requirements.

4. **Ultrasonic cutting:** Ultrasonic tools are also useful to trim, to punch holes or otherwise machine FRP's which contain KEVLAR. A table-mounted unit has been used to cleanly cut straight edges of composites, and both table-mounted and hand-held units have produced holes of varying sizes and geometries.

F. GRINDING

(Outside diameter, surface, chamfering)
Composites of KEVLAR can be ground or chamfered by strict adherence to specific conditions (TABLE III). A glazed work surface with a minimum of uncut fibers is obtained. Coolant is necessary to prevent wheel loading.

G. SANDING

Sanding laminates containing KEVLAR aramid can be done wet or dry. Wet sanding aids in cooling and prevention of waste buildup between abrasive particles. Positioning the composite so that the cutting action is *always* directed towards the interior of the material is critical to assure clean trimmed edges.

Power belt sanders with aluminum oxide or silicon carbide grit paper of 80-180 coarseness can be used at high surface speed, either dry or with light water mist, to minimize paper loading (Table III). Because of its small dimensions, the Dotco pneumatic sander has proven very effective for secondary finishing operations on composites containing KEVLAR.

H. TURNING

A cutting tool made of C-2 category cemented tungsten carbide brazed to a steel shank (Figure 1) has been developed for turning rods and tubes reinforced with KEVLAR. The tool is ground with a diamond wheel to give correct relief and sharpness of the cutting edge. Reference Table III for operating conditions.

Figure 1

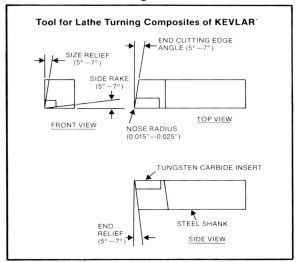

I. LASER AND WATER JET CUTTING

Lasers and water jets are proven tools for cutting multiple plies of fabric or prepreg of KEVLAR, or for finishing composites reinforced with KEVLAR. For further information, contact the equipment vendor (Table IV).

TABLE II
SHEARS AND ROTARY CUTTERS FOR CUTTING MATERIALS THAT CONTAIN KEVLAR® ARAMID FIBER

Manufacturer/Vendor and Model #	Manufacturer/Vendor Code	Yarn or Roving	Light & Medium Weight Fabric, Felt -7 oz.yd² (237 g/m²)	Heavy Weight Fabric & Woven Roving +7 oz/yd² (237 g/m²)	Felt	Prepreg Fabric	Prepreg Tape	Multiple Fabric Plies	Cured Laminates 0.050 in. (1.25 mm)
HAND SHEARS									
Technology Associates, Inc. MPO-152 1.5-inch (38 mm)	26			X	X				X
Technology Associates, Inc. CS1-161 (coated*) LDS 2-151 2-inch (51 mm) (uncoated)	26	X	X			X	X		
Pen Associates, Inc. WR-4S-2 (coated*) 2.5 inch (64 mm) PA1-4A-2 (uncoated)	21	X							
Pen Associates, Inc. WR-10M-2 (coated*) 2-inch (51 mm) PAI-10M-2 (uncoated)	21			X	X				X
Pen Associates, Inc. WR-8L-2 (coated*) 2-inch (51 mm) PAI-8L-2 (uncoated)	21	X	X			X	X		
Pen Associates, Inc. WR-10E-4 (coated*) 4-inch (102 mm) PAI-10E-4 (uncoated)	21	X	X			X	X	X 3-ply med. wt.	
Technology Associates, Inc. ECB4-153 4-inch (102 mm)		X	X			X	X	X 3-ply med. wt.	
Pen Associates, Inc. WR-12C-6 (coated*) 6-inch (152 mm) PAI-12C-6 (uncoated)	21		X	X	X	X		X 4-ply med. wt.	
Technology Associates, Inc. EHD 6-155 6-inch (152 mm)	26		X	X	X	X		X 4-ply med. wt.	
Wiss #4 I.S. (12-inch—305 mm) Upholstery, carpet & canvas shears	29		X	X	X				
POWER SHEARS									
United Cloth Cutting Machine Co. Electric shears Model No. LIL 25159	28		X					X 4-ply med. wt.	
Pen Associates, Inc. electric WR-180 or air WR-580, Power shears Pen Associates, Inc.	21		X	X	X	X	X	X 12-ply med. wt.	X .100 in (2.5 mm)
ROTARY CUTTERS									
Maiman "Mini-Sphere"	16, 21		X					X 4-ply med. wt.	
Maiman "Roto-Sphere"	16, 21			X	X			X 8-ply med. wt.	
Eastman Machinery Co. P3-Chickadee Cutter	9		X	X	X			X 8-ply med. wt.	
Black & Decker—Model #7975 Variable-speed reversing rotary power cutter as distributed by Pen Associates, Inc.	5, 21		X	X	X				

Manufacturers' addresses appear in Table IV of this brochure.
*Coated for increased durability.

TABLE III
SUGGESTED OPERATING CONDITIONS FOR TOOLS
USED IN MACHINING COMPOSITES CONTAINING KEVLAR® ARAMID FIBER

Tool Description	Vendor[1] Code	Suggested Speed	Feed Rate	Remarks
1. Single Flute Drill	24	200-600 SFPM (60-182 SMPM)[2]	0.002-0.005"/rev (0.5-0.13 mm/rev)	Drill locater bushing required. Controlled feed rate recommended. Backup support required.
2. Brad Point Drill	1, 21, 24	6000-25000 RPM (1800-7600 SMPM)	0.002-0.006"/rev (1.5-0.15 mm/rev)	Controlled feed rate required. Cannot be used to drill metals.
3. Spade Drill	21, 26	250-450 SFPM (60-135 SMPM)	0.005-0.010"/rev. (.13-0.25 mm/rev)	Primarily used on thin materials 0.020" (0.51 mm) thick. Backup may be required. Drill locater bushing required.
4. Self-Centering Drill	21	25-200 SFPM (7-60 SMPM)	0.005-0.010"/rev (0.13-.25 mm/rev)	Special operational techniques required. See text.
5. Core Drill	6	30-60 SFPM (10-20 SMPM)	0.006-0.010"/rev (0.15-0.25 mm/rev)	Backup required. Used on low resin content laminates only.
6. Hole Saw	14, 18, 21, 22	150-300 SFPM (50-100 SMPM)	.005-0.010"/rev (0.13-0.25 mm/rev)	Backup required.
7. Counter Sinks	1, 3, 12, 14, 31	50-250 RPM		Micro stop tooling required.
8. Band Saw Conventional	21	3000-6000 SFPM (900-1800 SMPM)	Up to 36"/min[3] (1 m/min)	Raker or straight set. 14-22 teeth/in (5-9 teeth/cm). Hone
Carbide Grit	22	3000-6000 SFPM (900-1800 SMPC)	Up to 36"/min[3] (1 m/min)	See text, use with hybrid laminates.
9. Circular Sawing	21, 22	3000-6000 SFPM (900-1800 SMPM)	Up to 36"/min[3] (1 m/min)	Coolant[4] may be required, particularly for heavy cuts.
10. Sabre Sawing	21, 26	2500 Strokes/min	Up to 36"/min[3] (1 m/min)	Specialty designs available.
11. Router Bits	20, 21, 23, 24, 26, 27, 30	20,000-27,000 RPM	Up to 60"/min[3] (1.5 m/min)	Some secondary edge finishing may be required, depending on end use application.
12. Edge Sanding (Power)	7, 8, 19, 21	4000-5000 SFPM (1200-1500 SMPM)	—	80-180 grit aluminum oxide or silicon carbide. Belt sanding preferred over disk sanding.
13. Lathe Tools	—	250-500 SFPM (75-150 SMPM)	0.0015-0.0024"/rev (0.04-0.06 mm/rev)	Depth of cut 0.01-0.02". (0.25-0.50 mm)
14. Grinding	—	4500-6500 SFPM (350-2000 SMPM)	75-175 SFPM (23-53 SMPM)	Use aluminum oxide or silicon carbide wheel. Depth of cut: rough 0.001-0.003" (0.025-0.075 mm). Depth of cut: finish 0.0005-0.001" (0.013-0.026 mm). Traverse: ¼-½ wheel width/pass.
15. Milling	21	Up to 4000 RPM	Up to 12"/min (0.3 m/min)	Secondary edge finishing required.
16. Knibbler	2	2500 strokes/min	Up to 60"/min[3] (1.5 m/min)	0.1" (2.54 mm) thick material maximum.
17. Water Jet	10, 17, 25	For specific recommendations contact manufacturer.		
18. Laser	4, 11, 13, 15	For specific recommendations contact manufacturer.		
19. Power Shears	TABLE II	—	30-40"/min[3] (760-1015 mm/min)	0.1" (2.54 mm) thick material maximum
20. Tungsten Carbide Razor Blades	21	—	—	—

The above cutting information should be considered as a general guide. Variations both within and/or beyond the suggestions listed should be done in order to optimize the tool life and/or cut edge quality with regards to each different composite.

NOTES: [1]Consult Table IV (following) for addresses
       [2]SFPM = Surface Feet Per Minute
        SMPM = Surface Meters Per Minute
[3]Depends on material thickness
[4]Suggest water soluable coolant, 150-200 parts water—1 part coolant concentrate.

Source: A Guide to Cutting and Machining Kevlar Aramid Composites, 1983, 12 pages

TABLE IV
MANUFACTURERS/VENDORS OF TOOLS FOR CUTTING KEVLAR® ARAMID AND MACHINING COMPOSITES REINFORCED WITH KEVLAR® ARAMID

Vendor Code	Manufacturer/Vendor	Vendor Code	Manufacturer/Vendor	Vendor Code	Manufacturer/Vendor
1	Airtech International 2542 East Del Amo Blvd. P.O. Box 6207 Carson, CA 90749 (213) 603-9683	11	Flying Machines Inc. (Contract processor) 2029 Research Dr. Livermore, CA 94550 (415) 447-0656	21	Pen Associates 2639 W. Robino Drive Wilmington, DE 19808 (302) 995-6868
2	Atlas Copco 24404 Indo Plex Circle Farmington Hills Michigan 48018 (619) 521-4522	12	Guehring 1455 Commerce Avenue Brookfield, WI 53005 (414) 784-6730	22	Remington Abrasive Pdts. 939 Barnum Avenue Bridgeport, CT 06601 (203) 333-1112
3	ATI Industries 2425 W. Vineyard Avenue Escondido, CA 92025-2591 (619) 746-8301	13	GTE-Sylvania Electric P.O. Box 188 Mountainview, CA 94040 (415) 966-2636	23	Spacematic Cutting Tools 138 Hulls Farm Rd. Southport, CT 06490 (203) 254-1863
4	Bendix Corp. Kansas City, MO 64141 (816) 997-2000	14	International Carbide 1111 N. Main St. Wauconda, IL 60084 (800) 323-7440	24	Starlite Industries 1111 Lancaster Avenue Rosemont, PA 19010 (215) 527-1300
5	Black & Decker Corp. 701 E. Joppa Road Towson, MD 21204 (301) 828-3900	15	Laser Applications (Contract processor) 2904 Stafford Street Baltimore, MD 21223 (310) 362-6100	25	Surgino USA, Inc. 3501 Woodhead Drive Northbrook, IL 60062 (312) 498-1370
6	Comp Tool 12139 North Linden Road Clio, MI 48420 (313) 687-5450	16	H. Maiman Co., Inc. 119 West 40th Street New York City, NY 10018 (212) 921-0765	26	Technology Associates P.O. Box 7163 Wilmington, DE 19803 (302) 475-6219
7	Dotco Ohio Route 18E Hicksville, OH 43526 (419) 542-7711	17	McCartney Mfg. Co. 635 W. 12th Street Baxter Springs, KS 66713 (316) 856-2151	27	Ultra Tool Intl. 5451 McFadden Ave. Huntington Beach, CA 92649 (800) 854-2431
8	Dynabrade Inc. 72 E. Niagara St. Tonawanda, NY 14150 (716) 694-4600	18	Morris Tool Ltd. G-4006 Corunna Rd. Flint, MI 48504-5891 (313) 732-3550	28	United Cloth Cutting Machine Co. 1123 Broadway New York City, NY 20010 (212) 242-2050
9	Eastman Machinery Co. 779 Washington Street Buffalo, NY 14240 (716) 856-2200	19	Nitto Kohki USA 111 Charlotte Place Englewood Cliffs New Jersey 07632 (201) 568-7980	29	Dayton-Price, Ltd. One Park Avenue New York City, NY 10018 (212) 532-8470
10	Flow Industries 21414 68th Ave., S. Kent, WA 98031 (206) 854-1370	20	Onsrud Cutter 800 Liberty Drive Libertyville, IL 60048 (312) 362-1560	30	Randall-Midwest Co. 9741 James Ave. S. Bloomington, MN 55431 (612) 881-7997
				31	Weldon Tool Co. 3002 Woodhill Road Cleveland, OH 44104 (216) 721-5454

This source list provides guidance to a representative group of suppliers. It does not necessarily include all companies manufacturing or selling the products listed nor is it intended as an endorsement of any specific supplier or manufacturer.

INDEX